Light Scattering Media Optics

Problems and Solutions (Third Edition)

Springer
Berlin
Heidelberg
New York
Hong Kong
London
Milan
Paris
Tokyo

Alexander A. Kokhanovsky

Light Scattering Media Optics

Problems and Solutions (Third Edition)

Published in association with
Praxis Publishing
Chichester, UK

Dr Alexander A. Kokhanovsky
B. I. Stepanov Institute of Physics
National Academy of Sciences of Belarus
Minsk
Belarus

and

Institute of Remote Sensing
University of Bremen
Otto Hahn Allee 1
D-2 8334 Bremen
Germany

SPRINGER–PRAXIS BOOKS IN ENVIRONMENTAL SCIENCES
SUBJECT *ADVISORY EDITOR*: John Mason M.Sc., B.Sc., Ph.D.

ISBN 3-540-21184-5 Springer-Verlag Berlin Heidelberg New York

Springer-Verlag is a part of Springer Science+Business Media (springeronline.com)

Bibliographic information published by Die Deutsche Bibliothek

Die Deutsche Bibliothek lists this publication in the Deutsche Nationalbibliografie; detailed bibliographic data are available from the Internet at http://dnb.ddb.de

Library of Congress Control Number: 2004105866

Third edition 2004
Second edition published 2001
First edition published 1999
Printed in Germany

Cover design: Jim Wilkie
Project Management: Originator Publishing Services, Gt Yarmouth, Norfolk, UK

Printed on acid-free paper

To Marina, Maria and Andrew

Contents

Preface to the third edition

The third improved edition of this book differs from the second edition in its expansion of the last chapter – devoted to applications. Also, the text was improved in many places to make the presentation more transparent. However, I have kept the chapter ordering unchanged. So, as in the previous edition, I consider single light scattering, radiative transfer, densely packed media, and applications following the same pattern.

I am indebted to John P. Burrows, Wolfgang von Hoyningen–Huene, Vladimir V. Rozanov, and Eleonora P. Zege for many important discussions on light scattering and radiative transfer.

Preface to the second edition

The first edition of this book was warmly welcomed by the light scattering community. The main idea behind preparing this second edition was to widen the range of solutions considered. More illustrations and physical discussions were added. However, the main idea of the book – the presentation of the principal approximate solutions of light scattering media optics in a short, instructive, and user-friendly form – was preserved.

I am grateful to Prof. Alan R. Jones (Imperial College, UK), Prof. Teruyuki Nakajima (University of Tokyo, Japan), and Prof. Reiner Weichert (Technical University Clausthal, Germany) for many helpful discussions about light scattering problems and continuous support. Special thanks are due to Prof. A. P. Ivanov and researchers of the Light Scattering Media Optics Laboratory in Minsk (B. I. Stepanov Institute of Physics, Belarus), where I spent 20 years studying various aspects of light interaction with inhomogeneous media. I also acknowledge the support given by INTAS grant 99/652.

Last, but not least, I would like to underline that this book would never have been prepared without the help and understanding of my family to whom I dedicate it.

Preface to the first edition

The main subject of this book is the interaction of light with disperse systems, defined as ensembles of small particles in a host medium, within the framework of physical optics. In principle, particles can have arbitrary size, chemical composition, internal structure, shape, and concentration. The light field is characterized by the Stokes vector or the density matrix.

Generally speaking, there are two methods for handling the problem. The first one is to solve Maxwell's equations for the system of N particles. This method can be used only for small numbers of particles N because of limits imposed by modern computers. Another approach calls for the replacement of a disperse system by a continuous medium with some phenomenological propagation constants. The last method is of frequent use in physical sciences. For instance, the electrodynamics of continuous media is based on essentially the same approach. Namely, the system of oscillating charges is replaced by a continuous medium with a priori defined values of the specific conductivity, dielectric permittivity, and magnetic permeability. These quantities are calculated within the framework of quantum mechanics.

In the case of disperse systems, we should calculate propagation constants (e.g. extinction and scattering coefficients) in the framework of classic electrodynamics of continuous media and consider light propagation in the framework of radiative transfer theory afterwards. Such a simplified approach is possible only if the wavelength of incident radiation is much smaller than extinction and scattering lengths, defined as the inverse values of extinction and scattering coefficients. This often holds for the optical band of the electromagnetic spectrum.

Both electrodynamics of continuous media and radiative transfer theory can be considered as special branches of mathematical physics with their own characteristic methods and techniques. For example, the main task of the famous book by S. Chandrasekhar (1950), devoted to radiative transfer problems, was to present the subject of radiative transfer as a branch of mathematical physics.

My objective here is completely different. Exact methods are beyond the scope of this book. The main subject is introduction to the approximate methods of light scattering media optics. This is important for the derivation of simple equations, which relate microphysical (size, shape, concentration of particles, their optical constants) and optical (the intensity or Stokes vector of reflected or transmitted light) properties of disperse systems. But the main key reason for my approach is that any field of theoretical physics is not completely developed if simple solutions of a general problem at some limiting cases are not presented. Accordingly, the book is divided into the following five chapters.

The main definitions are presented in the introductory Chapter 1.

Chapter 2 is devoted to scattering theory, based on Maxwell's equations. The main task of this part is to present approximate methods and results for the calculation of propagation constants. The special cases of small and large, optically soft and hard particles are considered.

The subject of Chapter 3 is approximate results of the radiative transfer theory. Approximate formulae are presented for small and large values of the optical thickness of a scattering layer, weak and strong absorption of light in a medium, and isotropic and highly anisotropic scattering of light by particles.

Complex problems of close-packed media, where the use of the radiative transfer equation is in question, are touched upon in Chapter 4.

Chapter 5 is the outcome of Chapters 2–4. Here the simple approximate relationships derived are used for the solution of a number of applied problems. For the most part, attention is given to geophysical applications. Indeed, approximate methods are most suitable to the solution of geophysical problems. The investigator in the laboratory can prepare a special light-scattering medium and describe its optical properties with exact theoretical methods. This is impossible in the case of natural media, which are characterized by an extremely complex structure. Exact knowledge of the microstructure parameters of such media (particle size distributions, chemical compositions of particles, their shapes and concentrations) are not available a priori. Moreover, these parameters rapidly change in space and time. Thus, we arrive at the important conclusion that, in the field of geo-optics, results obtained by exact and approximate methods can have about the same accuracy (when compared with measurements). This gives the book its special value for tackling atmospheric optics problems and provides a rationale for its inclusion in the Wiley-Praxis Series in Atmospheric Physics and Climatology. However, I do hope that this chapter and the book as a whole will be useful for engineers and scientists who deal with applied light-scattering problems in different branches of modern physics, geophysics, biophysics, and chemical physics.

Finally, I thank Eleonora Zege, who taught me light scattering.

List of figures

List of tables

List of abbreviations

AM	Analytical Method
APS	Amplitude-Phase Screen
CP	Circular Polarization
D	Dust
DDA	Discrete Dipole Approximation
DSCS	Differential Scattering Cross-Section
DWS	Diffusion Wave Spectroscopy
FM	Fitting Method
GOA	Geometrical Optics Approximation
IP	Inverse Problem
ISPs	Irregularly Shaped Particles
LDR	Lattice Dispersion Relation
LP	Linear Polarization
MCRTA	Monte Carlo Ray-Tracing Algorithm
MCRTC	Monte Carlo Ray-Tracing Code
MODIS	Moderate Resolution Imaging Spectroradiometer
MS	Multiple Scattering
MTF	Modulation Transfer Function
OC	Oceanic
OTF	Optical Transfer Function
POLDER	Polarization and Directionality of the Earth's Reflectance
PSD	Particle Size Distribution
PSF	Point Spread Function
RTE	Radiative Transfer Equation
S	Soot
SAA	Small-Angle Approximation
SCIAMACHY	Scanning Imaging Absorption Spectrometer for Atmospheric Chartography
SS	Single Scattering
VSC	Volume Scattering Coefficient
WS	Water Soluble

1

Introduction

1.1 LIGHT FIELD

The main subject of this book is the calculation of characteristics of a light field in disperse media. The intensity and polarization characteristics of a light field are usually characterized in terms of the Stokes vector-parameter $\mathbf{I} = (I, Q, U, V)$, which can be easily measured (Rozenberg, 1977; Bohren and Huffman, 1983). This vector-parameter is defined as follows.

Let us consider a plane electromagnetic wave, propagating in the direction, specified by the vector $\vec{e}_3$. It is well known that the electric vector $\vec{E}$, associated with this wave, is transverse to the propagation direction. Thus, we obtain:

$$\vec{E} = E_1\vec{e}_1 + E_2\vec{e}_2 \tag{1.1}$$

where complex oscillating functions E_1 and E_2 represent components of the electric vector in the plane perpendicular to the propagation direction along vectors $\vec{e}_1$, $\vec{e}_2$. It follows for unit vectors $\vec{e}_1$, $\vec{e}_2$, $\vec{e}_3$ that:

$$\vec{e}_2 \times \vec{e}_1 = \vec{e}_3 \tag{1.2}$$

It is common to define a plane of reference in electromagnetic scattering problems (van de Hulst, 1957). This plane contains the incident and scattered light beams. Let us assume that the vector $\vec{e}_1$ is parallel to this plane and $\vec{e}_2$ is perpendicular to the reference (or scattering) plane.

Components of the Stokes vector $\mathbf{I}$ (Perrin, 1942; Chandrasekhar, 1950; Born and Wolf, 1965; Ivanov, 1969) are defined in terms of the components E_1, E_2. Among different definitions we choose the following (the constant multiplier is omitted for the sake of simplicity):

$$I = E_1E_1^* + E_2E_2^* \tag{1.3}$$

$$Q = E_1E_1^* - E_2E_2^* \tag{1.4}$$

$$U = E_1 E_2^* + E_1^* E_2 \tag{1.5}$$

$$V = i(E_1 E_2^* - E_1^* E_2) \tag{1.6}$$

The values of I, Q, U, V completely characterize the arbitrarily polarized light beam in terms of the intensity, degree of polarization and characteristics of the polarization ellipse (the ellipticity, the azimuth, and the direction of rotation). Light beams with the same values of the Stokes vector cannot be distinguished by polarization measurements determining quadratic quantities (e.g. $\langle \vec{E}\vec{E}^* \rangle$). However, these radiation fluxes can differ, since they can have different field correlators.

The components of the Stokes vector can be written in terms of the amplitudes a_1, a_2 and phases σ_1, σ_2 of a simple electromagnetic wave as well:

$$I = a_1^2 + a_2^2 \tag{1.7}$$

$$Q = a_1^2 - a_2^2 \tag{1.8}$$

$$U = 2a_1 a_2 \cos(\sigma_1 - \sigma_2) \tag{1.9}$$

$$V = 2a_1 a_2 \sin(\sigma_1 - \sigma_2) \tag{1.10}$$

where we used the following representation of complex amplitudes E_1 and E_2 (see equations (1.3)–(1.6)):

$$E_1 = a_1 e^{-i(kz - \omega t + \sigma_1)}, \qquad E_2 = a_2 e^{-i(kz - \omega t + \sigma_2)} \tag{1.11}$$

Here $k = 2\pi/\lambda$, λ is the wavelength, z is the distance along the propagation direction $\vec{e}_3$, $\omega = kc$ is the circular frequency, c is the speed of light, t is time.

Note that amplitudes and phases in equations (1.11) are not constants for real light beams (Rozenberg, 1977). Thus, equations (1.3)–(1.10) should be averaged taking into account many vibrations (Chandrasekhar, 1950).

As mentioned before, the components of the Stokes vector **I** completely define the characteristics of the polarization ellipse. We can obtain (Chandrasekhar, 1950; Bohren and Huffman, 1983):

$$I = a^2, \qquad Q = a^2 \cos 2\beta \cos 2\Psi, \qquad U = a^2 \cos 2\beta \sin 2\Psi, \qquad V = a^2 \sin 2\beta \tag{1.12}$$

where $a^2 = a_1^2 + a_2^2$. Equations (1.12) provide the geometrical interpretation of the Stokes vector **I**. Thus, the angle of the polarization plane (or the azimuth) Ψ and the ellipticity angle β can be found from the following equations (see equations (1.12)):

$$\Psi = \frac{1}{2} \arctan \frac{U}{Q}, \qquad \beta = \frac{1}{2} \arctan \frac{V}{\sqrt{Q^2 + U^2}} \tag{1.13}$$

The angle β can be calculated also from the following formula (Chandrasekhar, 1950):

$$\beta = \frac{1}{2} \arcsin \frac{V}{\sqrt{Q^2 + U^2 + V^2}}$$

It should be pointed out that the ellipticity ε of the polarization ellipse is equal to $\operatorname{tg}\beta$. There are two values of β, Ψ, which satisfy equations (1.13). If we choose the smaller value of β, the value of $\cos(2\Psi)$ should have the same sign as Q (Chandrasekhar, 1950). Note that values of $\operatorname{tg}\beta = 1, -1$ correspond to right-hand and left-hand circularly polarized electromagnetic waves, respectively. Equations (1.12) hold only for completely polarized waves. It follows in this case that:

$$I^2 = Q^2 + U^2 + V^2 \tag{1.14}$$

Thus, only three parameters are independent in this particular case.

In most cases light is partially polarized and we obtain (Chandrasekhar, 1950):

$$I^2 \geq Q^2 + U^2 + V^2 \tag{1.15}$$

The degree of polarization P is defined as follows:

$$P = \frac{\sqrt{Q^2 + U^2 + V^2}}{I} \tag{1.16}$$

We can represent the partially polarized light beam with the Stokes vector $\mathbf{I}$ as a combination of the unpolarized light beam with the Stokes vector $\mathbf{I}_u$ and the completely polarized light beam with the Stokes vector $\mathbf{I}_p$:

$$\mathbf{I} = \mathbf{I}_u + \mathbf{I}_p \tag{1.17}$$

where

$$\mathbf{I}_u = (I(1-P), 0, 0, 0) \tag{1.18}$$

and

$$\mathbf{I}_p = (PI, Q, U, V) \tag{1.19}$$

Thus, the degree of polarization P determines the amount of the polarized radiation in the light beam. Values (1.13) describe the ellipsometric parameters of the polarized component (1.19). Note that the decomposition (1.17) is not unique. For instance, we can consider the partially polarized light beam as the sum of the unpolarized light flux with the Stokes vector $\mathbf{I}_u$, the flux of the linearly polarized radiation with the Stokes vector $\mathbf{I}_l$, and the circularly polarized light flux with the Stokes vector $\mathbf{I}_c$:

$$\mathbf{I} = \mathbf{I}_u + \mathbf{I}_l + \mathbf{I}_c \tag{1.20}$$

where

$$\mathbf{I}_u = \left(I - \sqrt{Q^2 + U^2} - |V|, 0, 0, 0\right) \tag{1.21}$$

$$\mathbf{I}_l = \left(\sqrt{Q^2 + U^2}, Q, U, 0\right) \tag{1.22}$$

$$\mathbf{I}_c = (|V|, 0, 0, V) \tag{1.23}$$

Parameters

$$P_l = \frac{\sqrt{Q^2 + U^2}}{I}, \qquad P_c = \frac{V}{I} \tag{1.24}$$

determine the degree of the linear and circular polarization, respectively. Both of them can be easily measured. It follows from equations (1.16), (1.24) that:

$$P = \sqrt{P_l^2 + P_c^2} \tag{1.25}$$

The angle $\Psi = \frac{1}{2}\arctan(U/Q)$ determines the direction of oscillations of the linearly polarized component. With $P_c > 0$, the rotation of the electric vector is clockwise looking in the direction from which the light is coming (a right-hand elliptical polarization). With $P_c = 0$, the light beam is a mixture of unpolarized and linearly polarized light. With $P_c < 0$, the light beam can be considered as a mixture of unpolarized, linearly polarized, and left-hand circularly polarized light. Note that in most cases (e.g. in the problems of geophysical optics), the value of P_c is small.

The radiation density matrix $\hat{\rho}$ can be defined in terms of the Stokes parameters as follows (Kuscer and Ribaric, 1959):

$$\hat{\rho} = \frac{1}{2}\begin{pmatrix} I + Q & U + iV \\ U - iV & I - Q \end{pmatrix} \tag{1.26}$$

Thus, the description of the light field in terms of the Stokes vector $\mathbf{I}$ is equivalent to the description of the light beam in terms of the density matrix (1.26). The advantage of the density matrix formalism is due to the fact that components of the matrix (1.26) transform as tensor components on the rotation of co-ordinate systems. The possibility of using familiar tensor techniques simplifies the analytical calculations both in scattering and radiative transfer problems (Dolginov *et al.*, 1995).

In the above representation, which is called the Linear Polarization (LP) representation, two linearly polarized states (with perpendicular planes of polarization) serve as a basis, namely $\vec{e}_1 = (1, 0)$ and $\vec{e}_2 = (0, 1)$ (see equation (1.1)). Sometimes it is more appropriate to take a different basis, namely two circularly polarized states, represented by the complex unit vectors

$$\vec{e}_+ = \left(\frac{1}{\sqrt{2}}, \frac{i}{\sqrt{2}}\right) \quad \text{and} \quad \vec{e}_- = \left(\frac{1}{\sqrt{2}}, \frac{-i}{\sqrt{2}}\right)$$

It follows in the framework of the so-called circularly polarized (CP) beams representation that:

$$\vec{E} = E_+\vec{e}_+ + E_-\vec{e}_- \tag{1.27}$$

The components of the Stokes vector in this representation are expressed as follows (Kuscer and Ribaric, 1959):

$$\frac{Q + iU}{2}, \quad \frac{V - I}{2}, \quad \frac{I - V}{2}, \quad \frac{Q - iU}{2}$$

They transform independently when the reference plane is rotated along the direction of propagation. This simplifies the solution of radiative transfer problems.

General laws of the transformation of the Stokes vector in an arbitrary isotropic symmetric and asymmetric media were considered in the valuable paper of Perrin (1942).

1.2 PARTICLES

Particles in a light – scattering medium are characterized by their concentrations, shapes, orientations, sizes, structures (uniform or nonuniform), and chemical compositions. The latter determines the complex refractive index m or correspondent tensors (for anisotropic, gyrotropic, or chiral particles) of scatterers. Particles of different sizes, shapes, structures, and chemical compositions can be present simultaneously in an elementary volume of a disperse medium. Thus, local optical characteristics of this volume are average values for distributions of all these parameters.

For uniform isotropic spheres, only the particle size distribution (PSD) $f(a)$, the refractive index $m = n - i\chi$ and the number concentration of particles N are important. The refractive index m of selected atmospheric aerosols, water droplets, and ice crystals is presented in Appendix 1. The characteristic values of the number concentration N for different particulate media are given in Table 1.1. The volumetric concentration of particles C_v is related to the value of N by the following equation:

$$C_v = N\langle V\rangle$$

where $\langle V\rangle$ is the average volume of particles in a unit volume of a medium. It follows for spherical particles that:

$$\langle V\rangle = \frac{4\pi}{3}\int_0^\infty a^3 f(a)\,da$$

The PSD is normalized by the following condition:

$$\int_0^\infty f(a)\,da = 1 \tag{1.28}$$

where a is the radius of a particle. The most frequently used PSDs of spherical particles are presented in Table 1.2 together with simple equations for the average

Table 1.1. The number concentration of particles N in different disperse media (Landolt-Bornstein, 1988).

Disperse medium	N (cm^{-3})
Hailstones	10^{-5}
Rain	10^{-3}
Water clouds	10^{-2}–10^{2}
Aerosols	10^{2}–10^{5}

Table 1.2. Particle size distributions.

$f(a)$	B	$\langle a \rangle$	a_{ef}	Δ	Δ_{ef}	$\langle a^n \rangle$
Gamma distribution $Ba^{\mu} e^{-\mu(a/a_0)}$	$\frac{\mu^{\mu+1}}{a_0^{\mu+1}\Gamma(\mu+1)}$	$a_0\left(1+\frac{1}{\mu}\right)$	$a_0\left(1+\frac{3}{\mu}\right)$	$\sqrt{\frac{1}{\mu+1}}$	$\frac{1}{\mu+3}$	$\left(\frac{a_0}{\mu}\right)^n \frac{\Gamma(1+\mu+n)}{\Gamma(1+\mu)}$
Lognormal distribution $\frac{B}{a}\exp\left(-\frac{\ln^2 \frac{a}{a_m}}{2\sigma^2}\right)$	$\frac{1}{\sqrt{2\pi}\,\sigma}$	$a_m\, e^{0.5\sigma^2}$	$a_m\, e^{2.5\sigma^2}$	$\sqrt{e^{\sigma^2}-1}$	$e^{\sigma^2}-1$	$a_m^n\, e^{\frac{n^2\sigma^2}{2}}$

radius $\langle a \rangle$, effective radius a_{ef}, moments $\langle a^n \rangle$, effective variance Δ_{ef}, and coefficient of variance $(CV)\Delta$, defined as follows:

$$\langle a \rangle = \int_0^\infty af(a)\,da, \qquad a_{ef} = \frac{\int_0^\infty a^3 f(a)\,da}{\int_0^\infty a^2 f(a)\,da}, \qquad \langle a^n \rangle = \int_0^\infty a^n f(a)\,da \tag{1.29}$$

$$\Delta_{ef} = \frac{\int_0^\infty (a-a_{ef})^2 a^2 f(a)\,da}{a_{ef}^2 \int_0^\infty a^2 f(a)\,da}, \qquad \Delta = \frac{\sqrt{\int_0^\infty (a-\langle a \rangle)^2 f(a)\,da}}{\int_0^\infty af(a)\,da} \tag{1.30}$$

It was common to use the Junge particle size distribution (Junge, 1963) in earlier papers on atmospheric aerosols (Junge, 1963) and oceanic suspensions (Shifrin, 1988). This distribution can be represented by the following equation:

$$f(a) = \begin{cases} \text{const}\, a^{-\upsilon}, & a \in [a_1, a_2] \\ 0, & a \notin [a_1, a_2] \end{cases} \tag{1.31}$$

where the value of υ depends on the type of aerosol. Note that $\ln(f(a))$ is a linear function of the particle radius in this case.

Hansen and Travis (1974) found that disperse media with different particle size distributions but the same values of a_{ef}, Δ_{ef} have approximately the same light scattering and absorption characteristics. We can see that the same applies to the distributions with the same values of a_{ef}, Δ (see Table 1.2). The coefficient of variance Δ (see equation (1.30)), however, has a more transparent meaning. The gamma and lognormal particle size distributions (see Table 1.2) with the same values of a_{ef}, Δ, which are equal to 6 μm and 0.38, respectively, are presented for convenience in Figure 1.1. These PSDs are typical for water clouds.

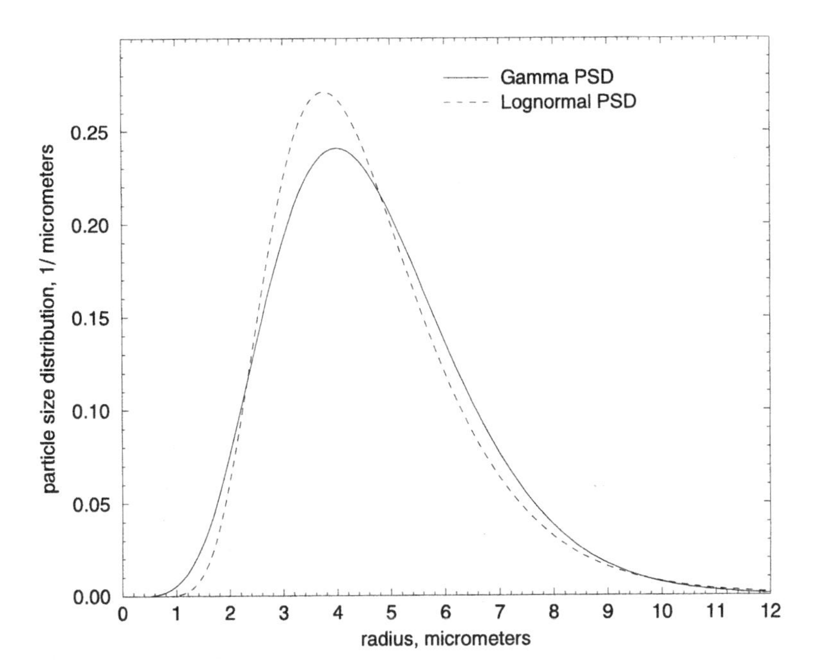

Figure 1.1. Lognormal and gamma particle size distributions with the effective radius $a_{ef} = 6\,\mu\text{m}$ and the coefficient of variance $\Delta = 0.38$.

We can see that the maximum of the equivalent lognormal distribution is shifted to smaller sizes when compared with the gamma distribution. The value of the particle size distribution at the maximum is smaller for the gamma particle size distribution.

For narrow particle size distributions ($\Delta \rightarrow 0$), only the parameter a_{ef}, which is proportional to the volume/surface area ratio (see equation (1.29)) , is important. Characteristic parameters of particle size distributions of some natural disperse media, which are often used to describe their optical properties, are presented in Table 1.3. It is worth pointing out that lognormal particle size distributions (see Table 1.3) are characterized by unrealistically large values of the coefficient of the variance Δ, which for atmospheric aerosols is usually smaller than 1.

Note that, according to the preliminary cloudless standard atmosphere model for radiation computations (WCP-112, 1986), the volumetric concentration of water soluble (WS), dust (D) , soot (S), and oceanic (OC) aerosols differs for continents (WS 29%, D 70%, S 1%), oceans (WS 5%, OC 95%), and urban areas (WS 61%, D 17%, S 22%).

A database of radiative characteristics of atmospheric aerosols is available via the Internet (Levoni *et al.*, 1997).

The size of a spherical homogeneous particle is uniquely defined by its diameter. For a cube the length along one edge is an important characteristic. With some regular nonspherical particles it may be necessary to specify more than one

Table 1.3. Parameters of particle size distributions of selected disperse media (WCP-112, 1986; Landolt-Bornstein, 1988).

Disperse medium	PSD	a_{ef} (μm)	Δ
Cloud C1 (Deirmendjian, 1969)	Gamma ($\mu = 6$, $a_0 = 4\,\mu m$)	6	$\frac{1}{\sqrt{7}} \approx 0.38$
Stratospheric aerosol	Gamma ($\mu = 2$, $a_0 = 0.1\,\mu m$)	0.25	$\frac{1}{\sqrt{3}} \approx 0.58$
Hailstones	Gamma ($\mu = 2$, $a_0 = 1000\,\mu m$)	2500	$\frac{1}{\sqrt{3}} \approx 0.58$
Water soluble aerosol (WS)	Lognormal ($\sigma = 1.09527$, $a_m = 0.05\,\mu m$)	0.1	1.52
Dust aerosol (D)	Lognormal ($\sigma = 1.09527$, $a_m = 0.5\,\mu m$)	10.0	1.52
Soot aerosol (S)	Lognormal ($\sigma = 0.69317$, $a_m = 0.0118\,\mu m$)	0.04	0.79
Oceanic aerosol (OC)	Lognormal ($\sigma = 0.92028$, $a_m = 0.3\,\mu m$)	2.5	1.15

dimension. For example: hexagonal cylinder, height and distance between opposite sides; plate, thickness and diameter; ellipsoid, three axes; circular cylinder, height and diameter.

It was found (Auer and Veal, 1970) that there are correlations between the height H and diameter D of ice crystals in natural clouds. This correlation could be expressed by the following approximate law in most of cases: $H = \nu D^{\mu}$, where constants ν and μ depend on the temperature and the type of crystals. For instance, it was found by Auer and Veal (1970) that $\nu = 2.506$ and $\mu = 0.398$ for plate crystals with values of $D \in [15\,\mu m, 8000\,\mu m]$ and $H \in [5\,\mu m, 90\,\mu m]$ at temperatures between -13 and -10°C. It follows that $\mu \approx \nu^{-1} \approx 2/5$ and $H \approx \frac{5}{2} D^{2/5}$ in this case. Thus, plates having larger diameters have also larger height. However, the ratio H/D is not constant. It is equal to $\frac{5}{2} D^{-3/5}$ and decreases with the thickness of a plate. The values ν and μ for some other types of crystals are presented in Table 1.4.

It follows for solid and hollow columns at temperatures between -10 and -8°C: $z = 0.866W$ and $W = 11.3L^{0.414}$ at $L > 200\,\mu m$ ($W = -8.479 + 1.002L - 0.00234L^2$ at $L \leq 200\,\mu m$), where L is the length of the column, z is the distance between opposite sides, and W is the width parameter (Auer and Veal, 1970).

Table 1.4. Parameters ν and μ for various crystals (Auer and Veal, 1970).

Crystal type	Temperature regime (°C)	ν	μ
Hexagonal plate	-10 to -13, -17 to -20	2.02	0.449
Stellar	-13 to -17	2.028	0.431
Fernlike crystal	-13 to -17	2.801	0.377

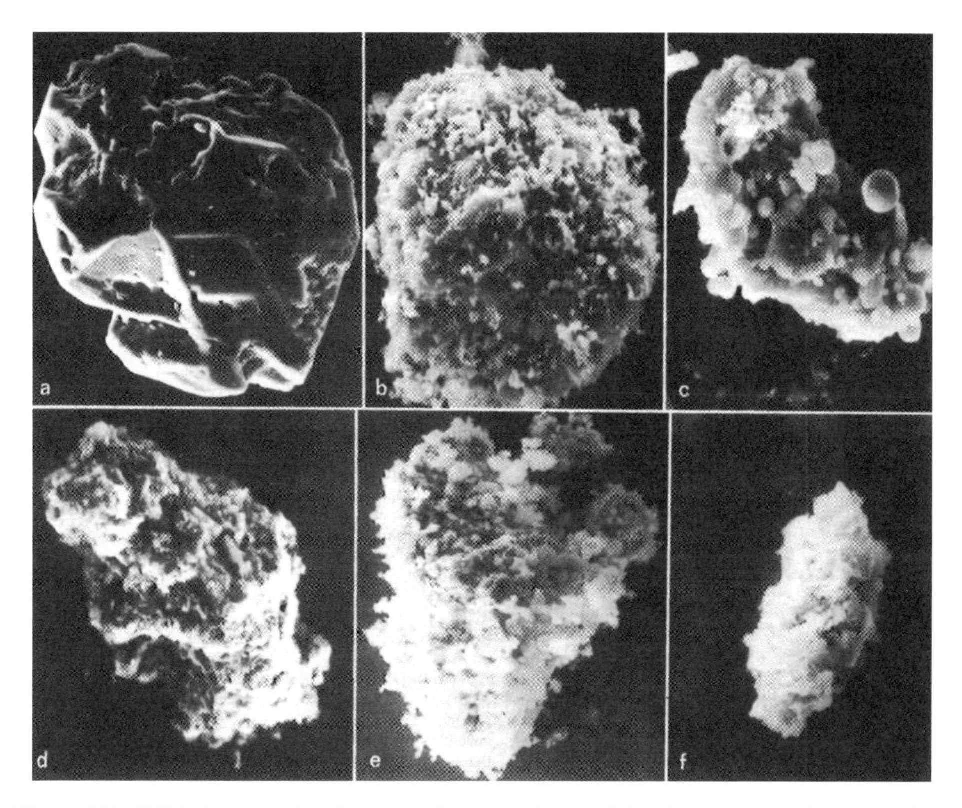

Figure 1.2. SEM photographs of some typical irregular particles: (a–c) terrestrial particles (a – quartz, b – activated charcoal, c – fly ash); (d–f) meteoritic samples (d – Meteorite Murchison, e – Meteorite Allende, matrix, f – Meteorite Allende, inclusion) (from Weiss-Wrana, 1983).

Note that, by using the correlation relations as presented above, such an approach allows reducing the number of variables from two or three to just one, which considerably simplifies the averaging procedures necessary in cloud optics. It is important, for instance, for calculations of the average extinction and absorption cross-sections of particles in crystalline clouds.

Regular nonspherical shapes occur infrequently. The most important cases are spherical and irregularly shaped scatterers (Shuerman, 1983). Spherical particles result from condensation and coagulation processes. They are in a liquid state in most cases. Characteristic examples are sulphate aerosols and water clouds.

Irregularly shaped particles (ISPs) can be classified in three categories: polyhedral solids, stochastically rough particles and stochastic aggregates (Lumme *et al.*, 1997). They come from the ocean, Earth surface or space (see Figure 1.2). ISPs are often compositions of different minerals. These particles can be generated inside the atmosphere as well. Soot particles, which also belong to this class, mostly result from biomass burning and human activity. They have a fractal structure. Particles can be not only irregular in shape, but also they can have an extremely complex internal structure. This is usually the case for bio-particles, for example (see Figure 1.3).

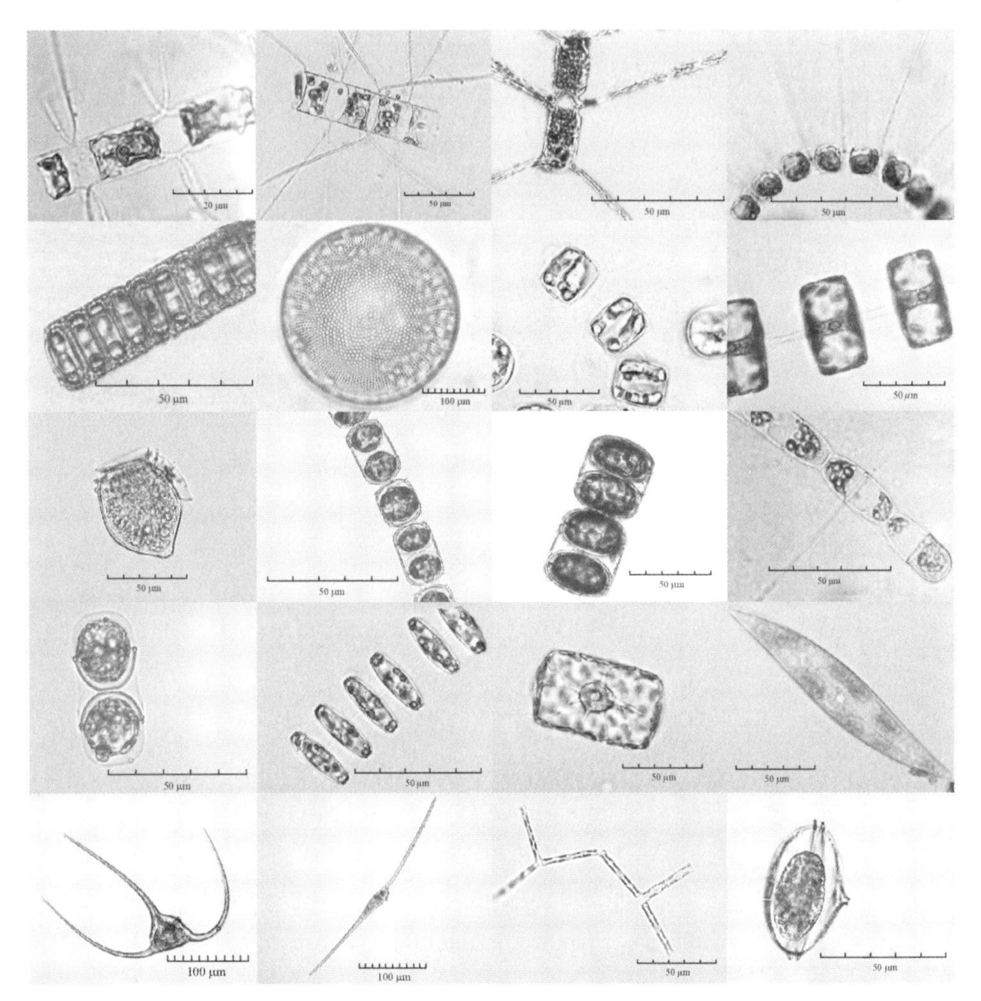

Figure 1.3. Images of living dominant phytoplankton taxa of the Barents and Kara Seas (from Smolyar, 2000).

Other examples of particulate media with irregularly shaped particles are snow fields and ice clouds (see Figure 1.4), which are formed due to crystallization processes at low temperatures. The Magono–Lee classification of natural snow crystals (Hobbs, 1974) includes eighty shapes, ranging from an elementary needle to an irregular germ (see Figure 1.5). The characterization of ice crystals by simple mathematical equations was proposed by Wang (1997). For instance, it was shown that the following expressions can be used to generate the combination of bullets:

$$r = [a \cos^{2b}(m\theta) + c]^d \, [\alpha \sin^{2\beta}(n\varphi) + \gamma]^\delta,$$

where (r, θ, φ) are spherical co-ordinates and $a, b, c, \alpha, \beta, \gamma, \delta, d, m, n$ are parameters that adjust to the size and shape of ice crystals. A four-branch combination of bullets

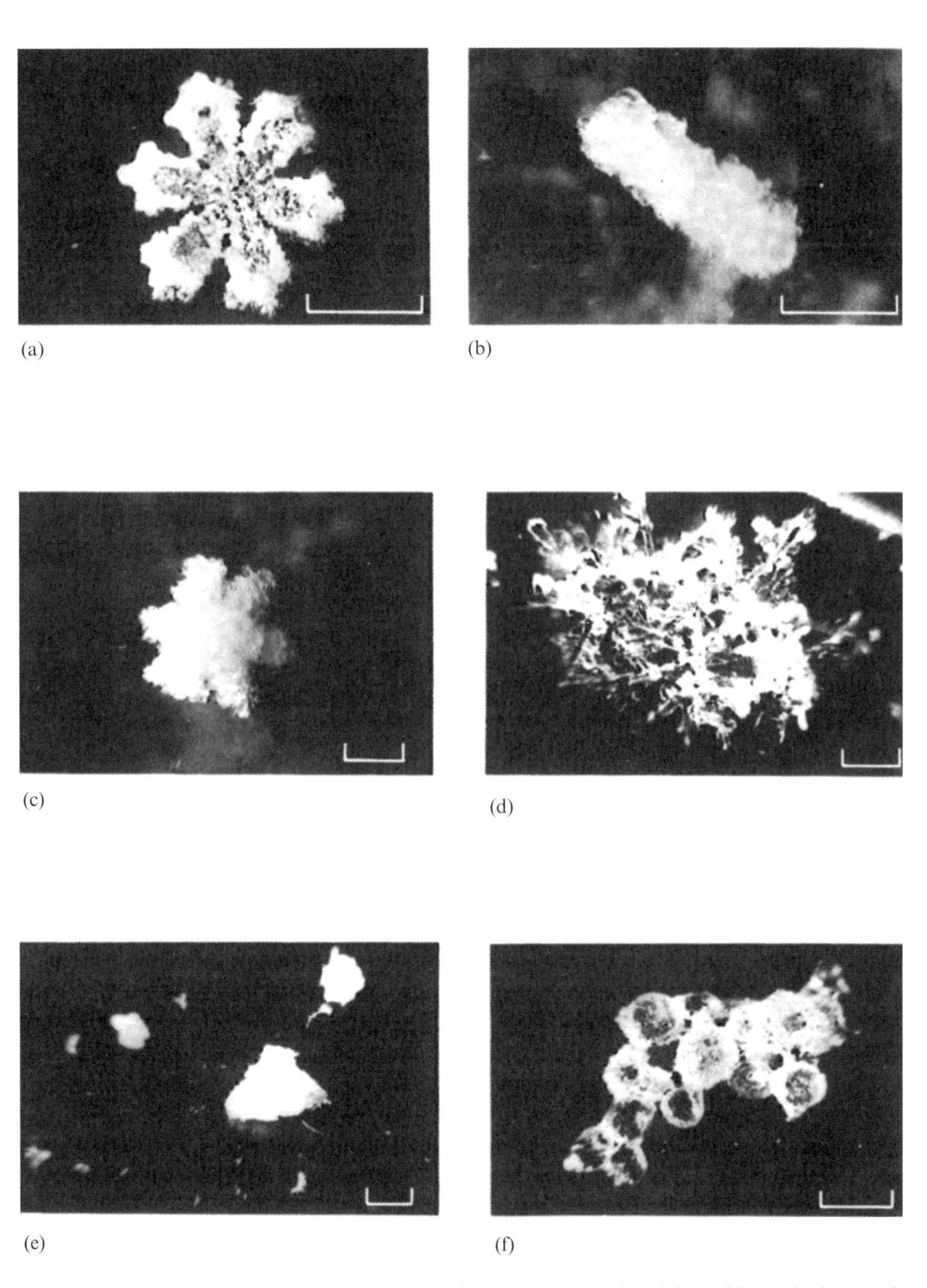

Figure 1.4. Examples of natural snow crystals. (a) Densely rimed broad-branched crystal. (b) Very densely rimed column. (c) Hexagonal graupel. (d) Aggregate of unrimed dendrites. (e) Conelike graupel. (f) Unrimed to densely rimed frozen water drops. The scaled lines represent 1 mm (from 'Ice Physics' by Peter V. Hobbs, Oxford University Press, 1974).

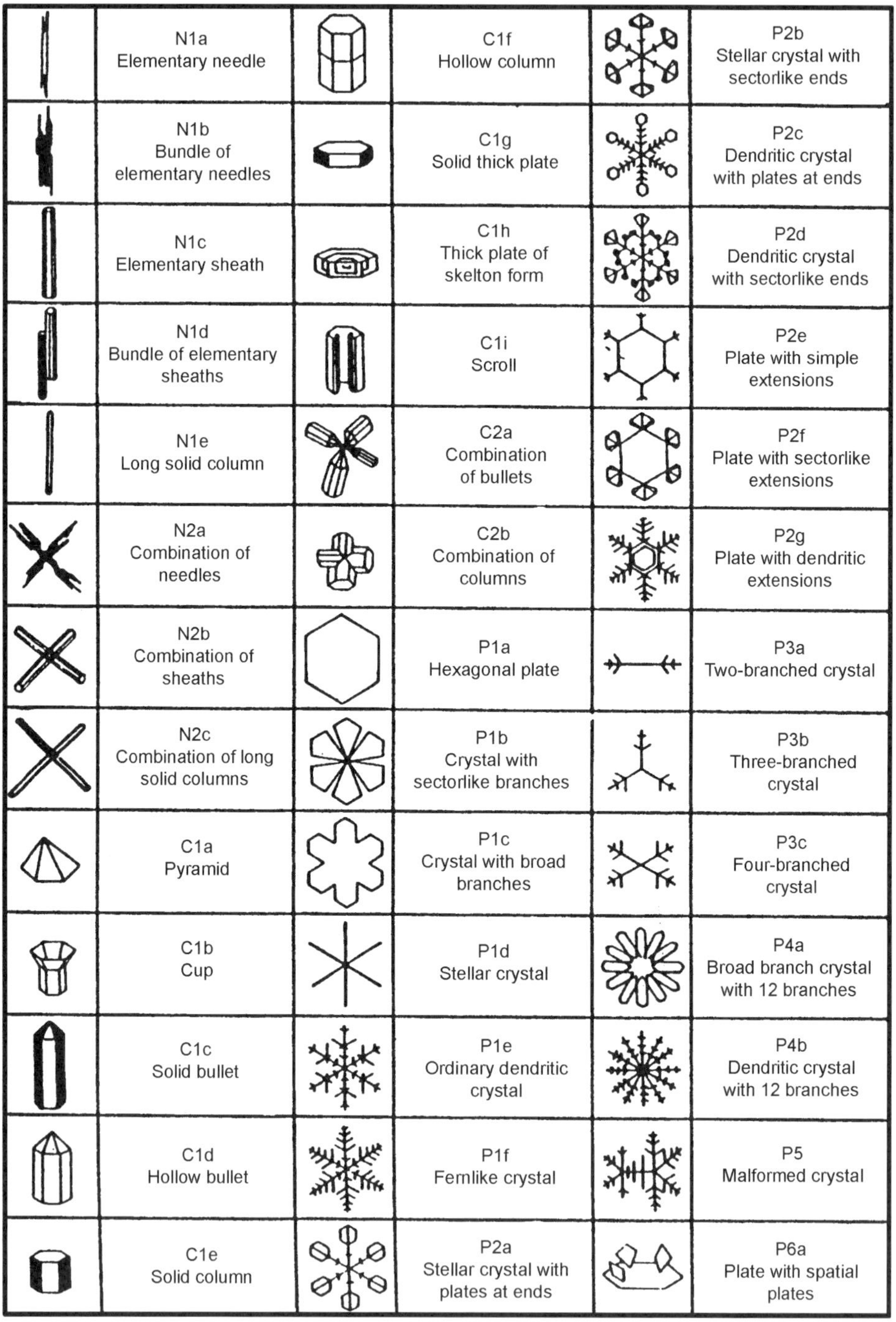

N1a Elementary needle	C1f Hollow column	P2b Stellar crystal with sectorlike ends
N1b Bundle of elementary needles	C1g Solid thick plate	P2c Dendritic crystal with plates at ends
N1c Elementary sheath	C1h Thick plate of skelton form	P2d Dendritic crystal with sectorlike ends
N1d Bundle of elementary sheaths	C1i Scroll	P2e Plate with simple extensions
N1e Long solid column	C2a Combination of bullets	P2f Plate with sectorlike extensions
N2a Combination of needles	C2b Combination of columns	P2g Plate with dendritic extensions
N2b Combination of sheaths	P1a Hexagonal plate	P3a Two-branched crystal
N2c Combination of long solid columns	P1b Crystal with sectorlike branches	P3b Three-branched crystal
C1a Pyramid	P1c Crystal with broad branches	P3c Four-branched crystal
C1b Cup	P1d Stellar crystal	P4a Broad branch crystal with 12 branches
C1c Solid bullet	P1e Ordinary dendritic crystal	P4b Dendritic crystal with 12 branches
C1d Hollow bullet	P1f Fernlike crystal	P5 Malformed crystal
C1e Solid column	P2a Stellar crystal with plates at ends	P6a Plate with spatial plates

Figure 1.5. Meteorological classification of snow crystals (Magono and Lee, 1966).

P6b Plate with spatial dendrites	CP3d Plate with scrolls at ends	R3c Graupellike snow with nonrimed extensions
P6c Stellar crystal with spatial plates	S1 Side plates	R4a Hexagonal graupel
P6d Stellar crystal with spatial dendrites	S2 Scalelike side planes	R4b Lump graupel
P7a Radiating assemblage of plates	S3 Combination of side planes, bullets and columns	R4c Conelike graupel
P7b Radiating assemblage of dendrites	R1a Rimed needle crystal	I1 Ice particle
CP1a Column with plates	R1b Rimcd columnar crystal	I2 Rimed partlcle
CP1b Column with dendrites	R1c Rimed plate or sector	I3a Broken branch
CP1c Multiple capped column	R1d Rimed stellar crystal	I3b Rimed broken branch
CP2a Bullet with plates	R2a Densely rimed plate or sector	14 Miscellaneous
CP2b Bullet with dendrites	R2b Densely rimed stellar crystal	G1 Minute column G2 Germ of skeleton form
CP3a Stellar crystal with needles	R2c Stellar crystal with rimed spatial branches	G3 Minute hexagonal plate
CP3b Stellar crystal with columns	R3a Graupellike snow of hexagonal type	G4 Minute stellar crystal G5 Minute assemblage of plates
CP3c Stellar crystal with scrolls at ends	R3b Graupellike snow of lump type	G6 Irregular germ

Table 1.5. Definitions of the size of nonspherical particles.

Symbol	Name	Definition	Formula
d_v	Volume diameter	Diameter of a sphere having the same volume V as the particle	$d_v = \sqrt[3]{\frac{6V}{\pi}}$
d_s	Surface diameter	Diameter of a sphere having the same surface area A as the particle	$d_s = \sqrt{\frac{A}{\pi}}$
d_{sv}	Surface volume diameter	Diameter of a sphere having the same volume to the surface area ratio as the particle	$d_{sv} = \frac{d_v^3}{d_s^2} = \frac{6V}{A}$
d_Σ	Projected area diameter	Diameter of a circle having the same area Σ as the projected area of the particle: (a) resting in a stable position; (b) randomly oriented	(a) $d_\Sigma = 2\sqrt{\frac{\Sigma}{\pi}}$ (b) mean value $\langle d_\Sigma \rangle$ for all possible orientations

can be obtained with this equation at $n = 1$, $m = 2$, $a = \alpha = -1$, $b = \beta = 2$, $c = \gamma = 1$, $d = \delta = 20$. It follows for dendrites:

$$r = a \sin^{2b}(m\theta) \sin^{2\beta}(n\varphi) + c$$

There are three well-known ways to handle the problem of light interaction with irregularly shaped particles. The first one is to neglect the shape of a particle and to replace it by a spherical one with the appropriate diameter (see Table 1.5). This is often used (Beddow and Meloy, 1980; Allen, 1990). For instance, hailstones, soot and dust particles in Table 1.3 are assumed to be spheres. But in reality, of course, this is not a case (Berry and Percival, 1986; Sorensen *et al.*, 1992; Mackowski, 1995; Markel and George, 2001). Thus, we can obtain only the approximate picture of a real process of light interaction with particles in the framework of this simplified approach. The correctness of results depends on the definition of the equivalent diameter used (see Table 1.5). For instance, Fraunhofer diffraction patterns can be analysed with the use of the projected area diameter d_Σ. The absorption of light by particles that are small compared with the wavelength λ ($d_v \ll \lambda$) should be described in terms of the volume diameter d_v. The extinction of light in dispersed media with randomly oriented convex particles ($d_v \gg \lambda$) is governed by the value of the surface volume diameter. However, some optical properties of nonspherical particles (e.g. polarization characteristics, phase functions) cannot be reproduced in the framework of this approximation at all (Weiss-Wrana, 1983).

The second way is to describe the optical characteristics of each nonspherical particle and to average the result obtained. This could be done only for a small number of particles, which is not the case for the majority of applications (e.g. see Figure 1.5).

Thus, the most promising is the third way: namely, the replacement of the optical characteristics of an elementary volume of a scattering medium by the corresponding optical characteristics of a single 'average' particle, whose shape is representative of the light scattering and absorption characteristics of an ensemble of scatterers. In the framework of this approach we should find the 'average' particle (or a finite number of such particles) with statistical properties of the surface (the mean diameter, covariance coefficient, correlation functions) close to the statistical properties of an ensemble of irregularly shaped particles.

This approach was mostly used for modelling the Fraunhofer patterns of irregularly shaped particles (Shifrin *et al.*, 1984; Jones, 1987, 1988; Al-Chalabi and Jones, 1994). The scattering of light by stochastically rough particles outside the diffraction peak was considered by Peltoniemi *et al.* (1989).

It was shown by Shifrin and Mikulinski (1982) that the intensity of the scattered light for an ensemble of irregularly shaped optically soft small particles is equal to the intensity of the scattered light of an 'average' particle. This result was obtained in the framework of the Rayleigh – Gans approximation. Statistical characteristics of an 'average' particle are represented by correlation functions of its transverse cross-sections (Shifrin, 1988). The Shifrin–Mikulinski approximation is extremely important for studies of light scattering by oceanic waters, where optically soft particles of various shapes are present (see Figure 1.3).

1.3 RADIATIVE TRANSFER

The characteristics of light fields in particulate media can be calculated in the framework of radiative transfer theory (Chandrasekhar, 1950; Sobolev, 1956, 1972; Ishimaru, 1978; Van de Hulst, 1980; Yanovitskij, 1997). This theory is based on solutions of the radiative transfer equation (RTE). The derivation of this equation for random media with discrete particles is simple and can be done as follows (Chandrasekhar, 1950).

Let us consider a light beam propagated in an absorbing, emitting, and scattering medium. The intensity of this beam will be constant in the absence of emittance, absorption, scattering, and refraction of light in a medium. It will change with distance for random media. The change in intensity after traversing the thickness dL in the linear optics approximation will be:

$$dI = -\sigma_{ext} I \, dL + \sigma_{em} \, dL \tag{1.32}$$

where σ_{ext} and σ_{em} are the extinction and emission coefficients. Nonlinear effects were considered in detail by Zege (1969). It is evident that the decrease in intensity is caused by the removal of photons, propagated in the fixed direction $\vec{\Omega} = (\theta, \phi)$, due to their absorption and scattering to other directions. This process is described by the first term in equation (1.32). Thus, the value of σ_{ext} can be divided into two parts:

$$\sigma_{ext} = \sigma_{abs} + \sigma_{sca} \tag{1.33}$$

where σ_{abs} and σ_{sca} are absorption and scattering coefficients. These values depend on the number concentration of particles N and their absorption C_{abs} and scattering C_{sca} cross-sections:

$$\sigma_{abs} = NC_{abs}, \qquad \sigma_{sca} = NC_{sca} \tag{1.34}$$

The extinction cross-section C_{ext} is defined as a sum of the scattering and absorption cross-sections, namely: $C_{ext} = C_{sca} + C_{abs}$. For particles of different sizes, shapes, or chemical compositions we should use mean values $\langle C_{sca} \rangle$ and $\langle C_{abs} \rangle$ in equations (1.34). In the case of uniform spheres, it follows that:

$$\langle C_{sca} \rangle = \int_0^\infty f(a) C_{sca}\, da \tag{1.35}$$

$$\langle C_{abs} \rangle = \int_0^\infty f(a) C_{abs}\, da \tag{1.36}$$

where $f(a)$ is the particle size distribution.

As a matter of fact, simple linear relationships between parameters of single particles (see e.g. equation (1.34)) and parameters of the radiative transfer equation (1.32) are valid only for disperse media with low concentrations of scatterers ($C_v < 0.01$–0.1). We must account for the correlation of particles positions in the case of their high concentrations (see Chapter 4 for more information).

Moreover, the whole approach, based on the RTE (1.32) becomes invalid for close-packed media (Apresyan and Kravtsov, 1983). For instance, spherical particles at $C_v \approx 0.74$ have an almost perfectly ordered structure. Scattering of light by such a structure is closer to the diffraction of light by crystalline media. For instance, Bragg maxima appear (Lock and Chiu, 1994). Clearly, highly ordered particulate media cannot be described in the framework of the radiative transfer equation.

Fortunately, almost all natural media (cosmic dust, atmospheric aerosols, clouds, oceanic suspensions) are characterized by extremely low values of volumetric concentration. Thus, the linear approximation (1.34) can be applied in this case. For instance, characteristic values of C_v are in the range 10^{-11}–10^{-7} for water clouds. This is not the case, however, for snow and soil, where the value C_v is in the range 0.3–0.4 or even higher. Note that the optics of close-packed media was considered in detail by Ivanov *et al.* (1988).

Let us return to equation (1.32). The emission coefficient σ_{em} in this equation consists of two parts:

$$\sigma_{em} = \sigma'_{em} + \sigma''_{em}, \tag{1.37}$$

where the value of σ'_{em} is due to the internal sources of radiation inside a medium and the value of σ''_{em} accounts for photons scattered from other directions to the direction $\vec{\Omega} = (\theta, \phi)$. It is evident that (Chandrasekhar, 1950):

$$\sigma'_{em} = B(T) \tag{1.38}$$

for media in the local thermodynamic equilibrium. Here:

$$B(T) = \frac{2hv^3}{c^2}\frac{1}{e^{hv/kT} - 1}$$

is the Planck function, v is the frequency, c is the speed of light, T is the temperature, h and k are the Boltzmann and Planck constants, respectively.

For the value of σ''_{em} it follows (Chandrasekhar, 1950) that:

$$\sigma''_{em} = \int_{4\pi} \sigma^d_{sca}(\Omega, \Omega') I(\Omega')\, d\Omega' \tag{1.39}$$

where $\sigma^d_{sca}(\Omega, \Omega')$ is the differential scattering cross-section from the direction $\vec{\Omega}' = (\theta', \phi')$ to the direction $\vec{\Omega} = (\theta, \phi)$. Thus, the RTE (1.32) can be represented as follows:

$$\frac{dI(\vartheta, \phi)}{dL} = -\sigma_{ext} I(\vartheta, \phi) + \int_0^{2\pi} d\phi' \int_0^{\pi} d\vartheta' \sin\vartheta' \sigma^d_{sca}(\vartheta, \vartheta', \phi, \phi') I(\vartheta', \phi') \tag{1.40}$$

where we neglected the term σ'_{em}, which is often not important in the visible region of the electromagnetic spectrum ($B(T) \to 0$ at $v \to \infty$). For isotropic media the value of $\sigma^d_{sca}(\vartheta, \vartheta', \phi, \phi')$ depends only on the scattering angle θ:

$$\theta = \arccos\left(uu' + \sqrt{1-u^2}\sqrt{1-u'^2}\cos(\phi - \phi')\right)$$

where $u = \cos\vartheta, u' = \cos\vartheta'$. We can see that the microstructure of a particulate medium enters the RTE (1.40) via the extinction coefficient σ_{ext} and the differential scattering coefficient σ^d_{sca} or alternatively (see equation (1.40)) via values of the optical thickness $\tau = \sigma_{ext} L$ and the ratio $\Xi = \sigma^d_{sca}/\sigma_{ext}$. Thus, media with the same values of τ and Ξ are characterized by exactly the same radiative characteristics. This allows, for example, the radiative transfer in cloudy media of different shapes to be modelled in the laboratory, avoiding expensive field measurements. For this, we should choose the value $\Phi = NL$ in the laboratory experiment, close to the value of this product for a cloudy medium. It could be done, for example, by reducing the value of the geometrical thickness L, which could be 1 km or higher for natural clouds, and increasing the number of particles N correspondingly. The source of light and receiver should also be modelled to simulate an experimental set-up in the real atmosphere. Of course, the same approach can be applied to other types of media (e.g. oceanic water). Some results in this direction were summarized by Ivanov (1969, 1975).

The scattering coefficient σ_{sca} is defined as an integral of the differential scattering cross-section σ^d_{sca}, namely:

$$\sigma_{sca} = \int_0^{2\pi} d\varphi \int_0^{\pi} \sigma^d_{sca}(\theta, \varphi) \sin\theta\, d\theta$$

or

$$\sigma_{sca} = 2\pi \int_0^{\pi} \sigma^d_{sca}(\theta) \sin\theta\, d\theta \tag{1.41}$$

for media with the azimuth independent local light scattering law. The value of σ^d_{sca} can be calculated if the differential cross-section $C_{sca}(\theta)$ per particle is known. In particular, it follows:

$$\sigma^d_{sca}(\theta) = N\langle C_{sca}(\theta)\rangle$$

where

$$C_{sca}(\theta) = \int_0^\infty f(a) C_{sca}(\theta, a)\, da$$

for spherical polydispersions. It follows from equation (1.41):

$$\int_0^\pi \frac{2\pi\sigma^d_{sca}(\theta)}{\sigma_{sca}} \sin\theta\, d\theta = 1 \tag{1.42}$$

The value

$$p(\theta) = \frac{4\pi\sigma^d_{sca}(\theta)}{\sigma_{sca}} \tag{1.43}$$

is called the phase function or the scattering indicatrix. The phase function is normalized by the following condition (see equations (1.42), (1.43)):

$$\frac{1}{2}\int_0^\pi p(\theta) \sin\theta\, d\theta = 1 \tag{1.44}$$

Note that the probability of photon scattering from the direction $\vec{\Omega}' = (\vartheta', \phi')$ to the direction $\vec{\Omega} = (\vartheta, \phi)$ is equal to $p(\theta)(d\Omega)/4\pi$. Thus, the notion of the phase function can be used for a probabilistic interpretation of the RTE. The differential scattering cross-section does not depend on the scattering angle in so-called isotropically scattering media. Thus, it follows in this case: $\sigma^d_{sca} = C$, where $C = \text{const}$. The total scattering cross-section can be obtained from equation (1.41): $\sigma_{sca} = 4\pi C$. The phase function is equal to 1 in this case (see equation (1.43)) and the probability of photon scattering within solid angle $d\Omega$ is equal simply to $d\Omega/4\pi$.

The characteristic phase function of a water cloud with the effective radius of droplets equal to 6 µm and different values of the half-width parameter μ are presented in Figure 1.6. Results were obtained at the wavelength 0.55 µm and $m = 1.333$, using the Mie theory (Van de Hulst, 1957). The particle size distribution was given by:

$$f(a) = Ba^\mu \exp\left(-\frac{(\mu+3)a}{a_{ef}}\right)$$

where $B = \text{const}$ (see Table 1.2). We can see that the phase function almost does not vary with the value of $\mu \in [2, 6]$. Note that this range is characteristic for water clouds.

Thus, it is a difficult task to measure the half-width parameter of particle size distributions by optical methods. The dependence of the phase function on the size of water droplets (see Figure 1.7) in a cloud is important mostly at small ($\theta \to 0$) and large ($\theta \to \pi$) scattering angles. The size of droplets is of importance also for rainbow scattering near the scattering angle of 140°. In particular, the rainbow scattering

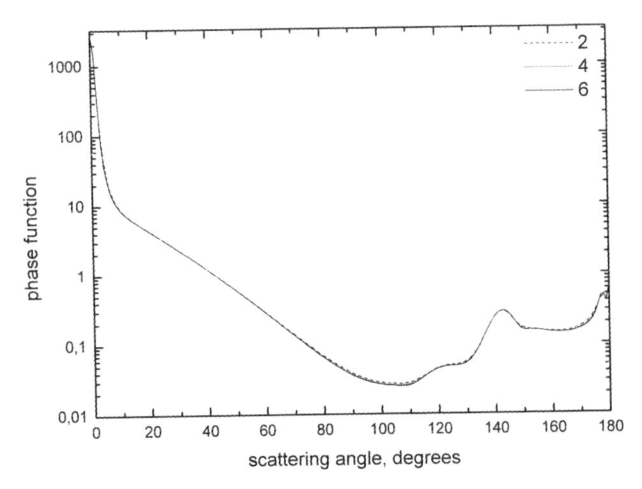

Figure 1.6. The phase function of water clouds with the gamma particle size distribution (see Table 1.2) at the wavelength 0.55 μm, the effective radius $a_{ef} = 6$ μm and different values of the half-width parameter $\mu = 2, 4, 6$.

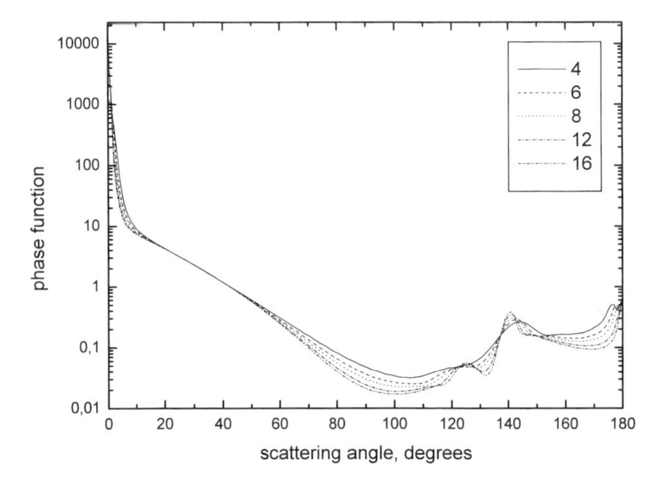

Figure 1.7. The same as in Figure 1.6 but for different values of the effective radius $a_{ef} = 4, 6, 8, 12, 16$ μm at $\mu = 6$.

angle θ_r depends on the average size of droplets in the cloud and their refractive index. The colour of the rainbow is also size-dependent.

Thus, angular scattering by water droplets at the diffraction ($\theta \to 0$), glory ($\theta \to \pi$), and rainbow ($\theta \to \theta_r(n)$) scattering regions can be used for optical particle sizing. Rainbow scattering can also be used for the solution of spectroscopic problems, namely for finding both real and imaginary parts of the refractive index of particles.

Only the radiative transfer in plane-parallel homogeneous light-scattering media is considered in this book. Thus, linear distances can be measured along the normal to the plane of the stratification. It follows from equation (1.40) for plane-parallel layers of disperse media (see equation (1.43)):

$$\cos\vartheta\frac{dI(\vartheta,\phi)}{dZ} = -\sigma_{ext}I(\vartheta,\phi) + \frac{\sigma_{sca}}{4\pi}\int_0^{2\pi}d\phi'\int_0^{\pi}d\vartheta'\sin\vartheta' p(\cos\theta)I(\vartheta',\phi') \tag{1.45}$$

where axis $\overrightarrow{OZ}$ (directed downward) is perpendicular to the boundary of a medium, $Z = L\cos\vartheta$ and $\cos\theta = \cos\vartheta\cos\vartheta' + \sin\vartheta\sin\vartheta'\cos(\phi - \phi')$ is the cosine of the scattering angle. The values of ϑ, ϑ' denote zenith angles with downward vertical, ϕ and ϕ' are azimuths referred to a chosen x-axis and measured clockwise when looking upward. This equation can be applied to radiative transfer problems in plane-parallel media with discrete scatterers. To account for the polarization of the scattered light beam equation (1.40) should be rewritten in the vector form (Chandrasekhar, 1950; Papanicolaou and Burridge, 1975; Mischenko *et al.*, 2002):

$$\frac{d\mathbf{I}(\vartheta,\phi)}{dL} = -\hat{\sigma}_{ext}\mathbf{I}(\vartheta,\phi) + \int_0^{2\pi}d\phi'\int_0^{\pi}d\vartheta'\sin\vartheta'\,\hat{\sigma}^d_{sca}(\vartheta,\vartheta',\phi,\phi')\mathbf{I}(\vartheta',\phi') \tag{1.46}$$

where $\mathbf{I} = (I, Q, U, V)$ is the Stokes vector (see equations (1.3)–(1.6)), $\hat{\sigma}_{ext}$ is the extinction matrix and $\hat{\sigma}^d_{sca}$ is the differential scattering matrix. Note that it follows for plane-parallel media (axis $\overrightarrow{OZ}$ is perpendicular to a boundary of a medium) (see equation (1.45)):

$$\cos\vartheta\frac{d\mathbf{I}(\vartheta,\phi)}{dZ} = -\hat{\sigma}_{ext}\mathbf{I}(\vartheta,\phi) + \int_0^{2\pi}d\phi'\int_0^{\pi}d\vartheta'\sin\vartheta'\,\hat{\sigma}^d_{sca}(\vartheta,\vartheta',\phi,\phi')\mathbf{I}(\vartheta',\phi') \tag{1.47}$$

where the Stokes vector $\mathbf{I}(\vartheta,\phi)$ is defined relative to the meridional plane, which holds directions $\overrightarrow{OZ}$ and $\vec{\Omega} = (\vartheta,\phi)$. Scattering matrices $\hat{\sigma}_{sca}(\theta)$ in the single light scattering theory (see Section 1.4) are usually calculated relative to the scattering plane, which holds directions $\vec{\Omega}' = (\vartheta',\phi')$ and $\vec{\Omega} = (\vartheta,\phi)$. Thus, we need to make rotations of the matrix $\hat{\sigma}_{sca}$, defined in the scattering theory, to obtain the matrix $\hat{\sigma}^d_{sca}$ in equation (1.47). Namely, it follows that:

$$\hat{\sigma}^d_{sca} = \hat{M}(\pi - i_2)\hat{\sigma}_{sca}(\theta)\hat{M}(-i_1) \tag{1.48}$$

where according to simple calculations (Hovenier, 1971):

$$\left.\begin{aligned}
&\hat{M}(-i_1) = \begin{pmatrix} 1 & 0 & 0 & 0 \\ 0 & \cos 2i_1 & -\sin 2i_1 & 0 \\ 0 & \sin 2i_1 & \cos 2i_1 & 0 \\ 0 & 0 & 0 & 1 \end{pmatrix}, \\
&\hat{M}(\pi - i_2) = \begin{pmatrix} 1 & 0 & 0 & 0 \\ 0 & \cos 2i_2 & -\sin 2i_2 & 0 \\ 0 & \sin 2i_2 & \cos 2i_2 & 0 \\ 0 & 0 & 0 & 1 \end{pmatrix} \\
&\cos 2i_j = 2\cos^2 i_j - 1, \qquad \sin 2i_j = 2\sqrt{1-\cos^2 i_j}\cos i_j \\
&\cos i_1 = \frac{-u + u'\cos\theta}{s\sqrt{(1-\cos^2\theta)(1-u'^2)}}, \qquad \cos i_2 = \frac{-u' + u\cos\theta}{s\sqrt{(1-\cos^2\theta)(1-u^2)}}
\end{aligned}\right\} \tag{1.49}$$

Table 1.6. Special cases for computation of the matrix $\hat{M}$; $\langle\pi, 2\pi\rangle$ signifies the interval from π to 2π with both end points included; $(0, \pi)$ stands for the interval from 0 to π with both end points excluded; arb means arbitrary within $\langle -1, +1\rangle$; $\Delta\phi = \phi - \phi'$ (from Hovenier, 1971).

u	u'	$\phi - \phi'$	Compute $\hat{M}(\pi - i_2)$ and $\hat{M}(-i_1)$ by utilization of
$\neq \pm 1$	$\neq \pm 1$	$0, \pi$	$\cos 2i_1 = 1, \cos 2i_2 = 1$
-1	arb	$\langle\pi, 2\pi\rangle$	$\cos 2i_1 = 1, \cos i_2 = -\cos(\Delta\phi)$
-1	arb	$(0, \pi)$	$\cos 2i_1 = 1, \cos i_2 = \cos(\Delta\phi)$
$+1$	arb	$\langle\pi, 2\pi\rangle$	$\cos 2i_1 = 1, \cos i_2 = \cos(\Delta\phi)$
$+1$	arb	$(0, \pi)$	$\cos 2i_1 = 1, \cos i_2 = -\cos(\Delta\phi)$
arb	-1	$\langle\pi, 2\pi\rangle$	$\cos i_1 = -\cos(\Delta\phi), \cos 2i_2 = 1$
arb	-1	$(0, \pi)$	$\cos i_1 = \cos(\Delta\phi), \cos 2i_2 = 1$
arb	$+1$	$\langle\pi, 2\pi\rangle$	$\cos i_1 = \cos(\Delta\phi), \cos 2i_2 = 1$
arb	$+1$	$(0, \pi)$	$\cos i_1 = -\cos(\Delta\phi), \cos 2i_2 = 1$

and $s = \text{sgn}(\phi - \phi' - \pi)$, $j = 1, 2$. If the denominator in equations (1.49) becomes zero, we should use the results presented in Table 1.6.

The matrix $\hat{\sigma}_{ext}$ is reduced to scalar for symmetric isotropic media. It follows from equation (1.47) in this case that:

$$\cos\vartheta \frac{d\mathbf{I}(\vartheta, \phi)}{d\tau} = -\mathbf{I}(\vartheta, \phi) + \frac{\omega_0}{4\pi} \int_0^{2\pi} d\phi' \int_0^{\pi} d\vartheta' \sin\vartheta' \, \hat{p}(\vartheta, \phi, \vartheta', \phi') \mathbf{I}(\vartheta', \phi') \tag{1.50}$$

where

$$\hat{p}(\vartheta, \phi, \vartheta', \phi') = \frac{4\pi\hat{\sigma}^d_{sca}(\vartheta, \vartheta', \phi, \phi')}{\sigma_{sca}} \tag{1.51}$$

is the phase matrix, $\tau = \sigma_{ext} Z$ is the optical depth, Z is the geometrical depth, $\omega_0 = \sigma_{sca}/\sigma_{ext}$ is the single scattering albedo.

Note that the phase matrix $\hat{p}(u, \phi, u', \phi')$ of randomly oriented particles with a plane of symmetry does not change if we perform one of the following operations:

1. interchanging ϕ and ϕ';
2. interchanging u and u';
3. changing the sign of u and u' at the same time.

These operations can be performed successively in an arbitrary order. Thus, the number of operations is equal to seven in total. This allows seven symmetry relationships to be obtained for the phase matrix (Hovenier, 1969). Their number, in the case of randomly oriented particles having no plane of symmetry, is reduced to just three.

Symmetry relationships can be used to find the symmetry in light that emerges at the top and bottom of a plane-parallel light scattering slab.

Equation (1.47) is a generalization of equation (1.45) for polarized radiation. This equation has the same form as equation (1.45), but it is much richer in terms of possible derived results. For instance, this equation allows effects of the circular and linear dichroism and birefringence in the multiple light scattering regime to be considered. Unfortunately solutions of this equation have not been studied in detail so far (except for the case when the matrix $\hat{\sigma}_{ext}$ is a scalar (Hansen and Travis, 1974; Hansen and Hovenier, 1974; De Rooij, 1985; De Haan, 1987; Wauben, 1992)).

Thus, the main object of radiative transfer theory is to solve equation (1.45) or the more general equations (1.46), (1.47). Equation (1.45) follows from equation (1.50) if we neglect polarization effects. Different numerical and approximate methods of the solution of these equations were proposed (Lenoble, 1985). In this book we will mostly consider equation (1.45), written in the following form:

$$u\frac{dI(\tau, u, \phi)}{d\tau} = -I(\tau, u, \phi) + \frac{\omega_0}{4\pi}\int_0^{2\pi} d\phi' \int_{-1}^{1} du' \, p(\cos\theta) I(u', \phi') \tag{1.52}$$

where $\tau = \sigma_{ext} Z$, $\cos\theta = uu' + \sqrt{(1-u^2)(1-u'^2)}\cos(\phi - \phi')$, Z is the geometrical depth. Main parameters of this equation are the optical depth τ, the single scattering albedo ω_0, and the phase function $p(\cos\theta)$.

Phase functions of water clouds in the visible are presented in Figures 1.6 and 1.7. The value of the single scattering albedo of clouds is close to 1 with a high accuracy due to a small absorption of light by water in the visible (see Appendix 1.1). The optical thickness of water clouds $\tau_0 = \sigma_{\text{ext}} Z_0$ is usually in the range 5–50. Here Z_0 is the cloud geometrical thickness.

Useful approximate solutions of equation (1.52) can be obtained as $\tau \to 0$, $\tau \to \infty$, $\omega_0 \to 0$, $\omega_0 \to 1$ and some special types of phase functions $p(\cos\theta)$ (e.g. for isotropic ($p(\cos\theta) = 1$) or highly anisotropic (see Figures 1.6 and 1.7) scattering, etc.). These approximations are presented in Chapter 3 of this book. They can easily be generalized to account for the polarization of light (Hansen and Travis, 1974; Van de Hulst, 1980; De Rooij, 1985; Domke, 1978a, b; Wauben, 1992).

Chapter 2 is devoted to approximate methods of the calculation of parameters σ_{ext}, ω_0, $p(\cos\theta)$ in equation (1.52) (or $\hat{\sigma}_{ext}$, $\hat{\sigma}_{sca}(\theta)$ for the vector case (1.47)) in the framework of scattering theory.

It should be pointed out that there are no longer problems in the solution of equations (1.47) and (1.52) nowadays. A lot of numerical and approximate techniques have been developed (Lenoble, 1985). The main difficulty lies now in accounting for the shape and morphology of particles in calculations of local optical characteristics (see Figures 1.2–1.5). Light propagation in inhomogeneous disperse media of different shapes is another ‘hot’ topic in radiative transfer theory (Rogovtsov, 1991, 1999).

1.4 SCATTERING THEORY

Single light scattering and absorption characteristics of an elementary volume of light-scattering media (e.g. matrices $\hat{\sigma}_{ext}$ and $\hat{\sigma}^d_{sca}$ in equation (1.46)) can be found in the framework of Maxwell's electromagnetic theory. Let us consider a particle which is illuminated by a plane harmonic wave:

$$\vec{E}_i = \vec{E}_0 \, e^{-i(\vec{k}\vec{r}-\omega t)} \tag{1.53}$$

where $\vec{E}_0$ is the vector amplitude, $|\vec{k}| = 2\pi/\lambda$, $\omega = kc$, λ is the wavelength, c is the light speed, t is time, $\vec{r}$ is the radius-vector. Any particle is composed of electric charges. These charges oscillate in the electromagnetic field and produce secondary radiation. This radiation from different places inside a particle in turn produces, at a large distance, an outgoing spherical electromagnetic wave:

$$\vec{E}_s = \vec{f}_s \frac{e^{-i(kr-\omega t)}}{r} \tag{1.54}$$

where $\vec{f}_s$ is the vector scattering amplitude, r is the distance from a particle to the observation point. The main task of scattering theory is calculation of the vector $\vec{f}_s$, which describes the polarization, amplitude, and phase of thc scattering wave as a function of the physical characteristics of particles and directions of the incidence and observation. Note that both the extinction matrix $\hat{\sigma}_{ext}$ and the differential scattering matrix $\hat{\sigma}^d_{sca}$ can be expressed via the vector scattering amplitude $\vec{f}_s$.

Both the vector $\vec{f}_s$ and $\vec{E}_0$ (see equation (1.53)) can be represented as a linear combination of electric vectors, parallel (E_{01}, f_{S1}) and perpendicular (E_{02}, f_{S2}) to the scattering plane, defined as a plane which holds the incident and scattered light beams:

$$\vec{E}_0 = E_{01}\vec{e}_1 + E_{02}\vec{e}_2 \tag{1.55}$$

$$\vec{f}_s = f_{S1}\vec{e}\,'_1 + f_{S2}\vec{e}\,'_2 \tag{1.56}$$

where vectors $\vec{e}_1$ and $\vec{e}\,'_1$ are unit vectors in the scattering plane and unit vectors $\vec{e}_2$ and $\vec{e}\,'_2$ are perpendicular to the scattering plane. It follows that $\vec{e}_1\vec{e}\,'_1 = \cos\theta$ and $\vec{e}_2\vec{e}\,'_2 = 1$, where θ is the scattering angle (the angle between the scattered and incident light beams).

Let us suppose that an incoming wave propagates along the positive direction of the co-ordinate axis OZ. Thus, it follows from equation (1.53) that:

$$\vec{E}_i = \vec{E}_0 \, e^{-i(kz-\omega t)} \tag{1.57}$$

and

$$\vec{e}_2 \times \vec{e}_1 = \vec{e}_3 \tag{1.58}$$

where $\vec{e}_3 \| OZ$. Vectors $\vec{e}_1, \vec{e}_2$, can be expressed as linear combinations of vectors $\vec{e}_x, \vec{e}_y$ along the co-ordinate axes OX, OY:

$$\vec{e}_1 = \cos\varphi\vec{e}_x + \sin\varphi\vec{e}_y \tag{1.59}$$

$$\vec{e}_2 = \sin\varphi\vec{e}_x - \cos\varphi\vec{e}_y \tag{1.60}$$

where $\cos\varphi = \vec{e}_1\vec{e}_x$. It follows for vectors $\vec{e}\,'_1$, $\vec{e}\,'_2$ that:

$$\vec{e}\,'_1 = \vec{e}_\theta, \qquad \vec{e}\,'_2 = -\vec{e}_\phi \tag{1.61}$$

where unit vectors $\vec{e}_\phi$, $\vec{e}_\theta$, $\vec{e}_r = \vec{e}_\theta \times \vec{e}_\varphi$ are determined in the spherical co-ordinate system. Thus, vectors $\vec{f}_s$ and $\vec{E}_0$ are defined in different systems of basis vectors. It should be pointed out that vectors $\vec{e}_1$ and $\vec{e}_2$ in equations (1.59), (1.60) can also be related to unit vectors $\vec{e}_r$, $\vec{e}_\theta$, $\vec{e}_\varphi$ of the spherical co-ordinate system. The correspondent relations are (Bohren and Huffman, 1983):

$$\vec{e}_1 = \cos\theta\,\vec{e}_\theta + \sin\theta\,\vec{e}_r, \qquad \vec{e}_2 = -\vec{e}_\varphi$$

We can see that $\vec{e}_1\vec{e}\,'_1 = \cos\theta$ and $\vec{e}_2\vec{e}\,'_2 = 1$, as it should be.

Components f_{S1} and f_{S2}, defined in equation (1.56), can be represented as a linear combination of amplitudes E_{01}, E_{02} (see equation (1.55)):

$$f_{S1} = S'_{11}E_{01} + S'_{12}E_{02} \tag{1.62}$$

$$f_{S2} = S'_{21}E_{01} + S'_{22}E_{02} \tag{1.63}$$

This follows from the linearity of Maxwell's equations. Introducing the amplitude scattering matrix $\hat{S}' = \begin{pmatrix} S'_{11} & S'_{12} \\ S'_{21} & S'_{22} \end{pmatrix}$, we obtain

$$\begin{pmatrix} f_{S1} \\ f_{S2} \end{pmatrix} = \begin{pmatrix} S'_{11} & S'_{12} \\ S'_{21} & S'_{22} \end{pmatrix} \begin{pmatrix} E_{01} \\ E_{02} \end{pmatrix} \tag{1.64}$$

Thus, the problem of determination of the vector scattering amplitude $\vec{f}_s$ in the framework of scattering theory is reduced to the calculation of four complex scalar numbers S'_{11}, S'_{12}, S'_{13}, S'_{14}. Note that scattering matrix $\hat{S}'$ in (1.64) in our definition is proportional to the dimensionless matrix, introduced by Van de Hulst (1957):

$$\hat{S} = \begin{pmatrix} S_{11} & S_{12} \\ S_{21} & S_{22} \end{pmatrix} \tag{1.65}$$

where

$$S_{11} = ikS'_{11}, \qquad S_{12} = ikS'_{12}, \qquad S_{21} = ikS'_{21}, \qquad S_{22} = ikS'_{22} \tag{1.66}$$

Indeed, it follows from equations (1.54)–(1.57), (1.64) that:

$$\begin{pmatrix} E_1 \\ E_2 \end{pmatrix} = \frac{e^{ik(z-r)}}{ikr} \begin{pmatrix} S_2 & S_3 \\ S_4 & S_1 \end{pmatrix} \begin{pmatrix} E_{i1} \\ E_{i2} \end{pmatrix} \tag{1.67}$$

where $S_1 \equiv S_{22}$, $S_2 \equiv S_{11}$, $S_3 \equiv S_{12}$, $S_4 \equiv S_{21}$ are Van de Hulst amplitude functions. This equation is identical to that presented by Van de Hulst (1957). Note that values E_{i1}, E_{i2}, E_1, E_2 in equation (1.67) are components of vectors $\vec{E}_i$ (1.53) and $\vec{E}_s$ (1.54):

$$\vec{E}_i = E_{i1}\vec{e}_1 + E_{i2}\vec{e}_2 \tag{1.68}$$

$$\vec{E}_s = E_1\vec{e}'_1 + E_2\vec{e}'_2 \tag{1.69}$$

Thus, we can see that the process of light scattering by a single particle is completely defined if the elements S_{ij} of the amplitude scattering matrix (S-matrix) are known. The extinction of the radiation by a small particle is determined by scattering and absorption processes simultaneously.

It follows for isotropic spherical particles (Van de Hulst, 1957):

$$S_{12}(0) = S_{21}(0) = 0 \tag{1.70}$$

and

$$C_{ext} = \frac{4\pi}{k^2}\mathrm{Re}(S(0)) \tag{1.71}$$

where $S(0) \equiv S_{11}(0) = S_{22}(0)$. Expression (1.71) is called the optical theorem (Van de Hulst, 1957). It relates the extinction of light by a particle with the amplitude function in the forward scattering direction. It is interesting to note that the S-matrix can be used not only for finding light scattering characteristics, it can also be applied to calculation of extinction and, therefore, absorption cross-sections.

It follows for the Stokes vector of the scattered light field (Van de Hulst, 1957):

$$\mathbf{I} = \frac{\hat{C}_{sca}}{r^2}\mathbf{I}_0 \tag{1.72}$$

where $\mathbf{I}_0$ is the Stokes vector of the incident light, $\mathbf{I}$ is the Stokes vector of the scattered radiation and $\hat{C}_{sca} \equiv k^{-2}\hat{F}_{sca}$ is the Stokes matrix. Equation (1.72) allows the scattering event to be considered not in terms of electromagnetic field components but in terms of observables, namely in terms of Stokes vector elements. This is consistent with the spirit of radiative transfer theory.

The structure of the dimensionless matrix $\hat{F}_{sca}$ depends on the definition of the Stokes vector and the plane of reference. This matrix can be found from equations (1.67), (1.72). It has the following form for the (I, Q, U, V) representation of the Stokes vector (see equations (1.3)–(1.6)):

$$\hat{F}_{sca} = \begin{pmatrix} F_{sca}^{11} & F_{sca}^{12} & F_{sca}^{13} & F_{sca}^{14} \\ F_{sca}^{21} & F_{sca}^{22} & F_{sca}^{23} & F_{sca}^{24} \\ F_{sca}^{31} & F_{sca}^{32} & F_{sca}^{33} & F_{sca}^{34} \\ F_{sca}^{41} & F_{sca}^{42} & F_{sca}^{43} & F_{sca}^{44} \end{pmatrix} \tag{1.73}$$

where (Bohren and Huffman, 1983):

$$\left.\begin{aligned}
F_{sca}^{11} &= \tfrac{1}{2}(|S_{11}|^2 + |S_{12}|^2 + |S_{21}|^2 + |S_{22}|^2) \\
F_{sca}^{12} &= \tfrac{1}{2}(|S_{11}|^2 - |S_{22}|^2 + |S_{21}|^2 - |S_{12}|^2) \\
F_{sca}^{13} &= \mathrm{Re}(S_{11}S_{12}^* + S_{22}S_{21}^*) \\
F_{sca}^{14} &= \mathrm{Im}(S_{11}S_{12}^* - S_{22}S_{21}^*) \\
F_{sca}^{21} &= \tfrac{1}{2}(|S_{11}|^2 - |S_{22}|^2 + |S_{12}|^2 - |S_{21}|^2) \\
F_{sca}^{22} &= \tfrac{1}{2}(|S_{11}|^2 + |S_{22}|^2 - |S_{12}|^2 - |S_{21}|^2) \\
F_{sca}^{23} &= \mathrm{Re}(S_{11}S_{12}^* - S_{22}S_{21}^*) \\
F_{sca}^{24} &= \mathrm{Im}(S_{11}S_{12}^* + S_{22}S_{21}^*) \\
F_{sca}^{31} &= \mathrm{Re}(S_{11}S_{21}^* + S_{22}S_{12}^*) \\
F_{sca}^{32} &= \mathrm{Re}(S_{11}S_{21}^* - S_{22}S_{12}^*) \\
F_{sca}^{33} &= \mathrm{Re}(S_{11}^*S_{22} + S_{12}S_{21}^*) \\
F_{sca}^{34} &= \mathrm{Im}(S_{11}S_{22}^* + S_{21}S_{12}^*) \\
F_{sca}^{41} &= \mathrm{Im}(S_{11}^*S_{21} + S_{12}^*S_{22}) \\
F_{sca}^{42} &= \mathrm{Im}(S_{11}^*S_{21} - S_{12}^*S_{22}) \\
F_{sca}^{43} &= \mathrm{Im}(S_{22}S_{11}^* - S_{12}S_{21}^*) \\
F_{sca}^{44} &= \mathrm{Re}(S_{22}S_{11}^* - S_{12}S_{21}^*)
\end{aligned}\right\} \tag{1.74}$$

The matrix (1.73) is defined on the basis of unit vectors $\vec{e}_1$, $\vec{e}_2$, and $\vec{e}_3$, where the plane of reference is the scattering plane.

It follows for spherical isotropic particles that $S_{12} = S_{21} = 0$ and

$$\hat{F}_{sca} = \begin{pmatrix} \dfrac{|S_{11}|^2 + |S_{22}|^2}{2} & \dfrac{|S_{11}|^2 - |S_{22}|^2}{2} & 0 & 0 \\ \dfrac{|S_{11}|^2 - |S_{22}|^2}{2} & \dfrac{|S_{11}|^2 + |S_{22}|^2}{2} & 0 & 0 \\ 0 & 0 & \mathrm{Re}(S_{11}S_{22}^*) & \mathrm{Im}(S_{11}S_{22}^*) \\ 0 & 0 & -\mathrm{Im}(S_{11}S_{22}^*) & \mathrm{Re}(S_{11}S_{22}^*) \end{pmatrix} \tag{1.75}$$

Thus, we can see that $F_{sca}^{11} = F_{sca}^{22}$, $F_{sca}^{33} = F_{sca}^{44}$, $F_{sca}^{12} = F_{sca}^{21}$, $F_{sca}^{34} = -F_{sca}^{43}$ in this specific case.

It follows for the intensity of scattered light from equations (1.72) and (1.75):

$$I = \frac{i_1 + i_2}{2k^2r^2} I_0$$

where I_0 is the intensity of incident light. The differential scattering cross-section in the solid angle Ω is defined as:

$$C_{sca}(\Omega) = \frac{r^2 I}{I_0}$$

or

$$C_{sca}(\Omega) = \frac{i_1 + i_2}{2k^2}$$

Therefore, for the total scattering cross-section we have:

$$C_{sca} = \int_{4\pi} C_{sca}(\Omega)\, d\Omega$$

or

$$C_{sca} = \frac{\pi}{k^2} \int_0^{\pi} (i_1 + i_2) \sin\theta\, d\theta$$

The phase function $p(\theta)$ for a single particle is defined as follows (see equation (1.51)):

$$p(\theta) = \frac{4\pi C_{sca}(\theta)}{C_{sca}}$$

and, therefore

$$p(\theta) = \frac{2\pi(i_1 + i_2)}{k^2 C_{sca}}$$

The extinction matrix determines the transformation of the coherent part of the electromagnetic field and can be written as follows:

$$\hat{C}_{ext} = \begin{pmatrix} C_{ext}^{11} & C_{ext}^{12} & C_{ext}^{13} & C_{ext}^{14} \\ C_{ext}^{21} & C_{ext}^{22} & C_{ext}^{23} & C_{ext}^{24} \\ C_{ext}^{31} & C_{ext}^{32} & C_{ext}^{33} & C_{ext}^{34} \\ C_{ext}^{41} & C_{ext}^{42} & C_{ext}^{43} & C_{ext}^{44} \end{pmatrix} \tag{1.76}$$

where, for the (I, Q, U, V) representation of the Stokes vector, we obtain (Ishimaru and Yeh, 1984; Mishchenko, 1990; Mishchenko, 1994a; Paramonov, 1995):

$$C_{ext}^{11} = C_{ext}^{22} = C_{ext}^{33} = C_{ext}^{44} = \frac{2\pi}{k^2} \mathrm{Re}(S_{11}(0) + S_{22}(0))$$

$$C_{ext}^{12} = C_{ext}^{21} = \frac{2\pi}{k^2} \mathrm{Re}(S_{11}(0) - S_{22}(0))$$

$$C_{ext}^{13} = C_{ext}^{31} = \frac{2\pi}{k^2} \mathrm{Re}(S_{12}(0) + S_{21}(0))$$

$$C_{ext}^{14} = C_{ext}^{41} = -\frac{2\pi}{k^2} \mathrm{Im}(S_{12}(0) - S_{21}(0))$$

$$C_{ext}^{23} = -C_{ext}^{32} = \frac{2\pi}{k^2} \mathrm{Re}(S_{12}(0) - S_{21}(0))$$

$$C_{ext}^{24} = -C_{ext}^{42} = -\frac{2\pi}{k^2} \mathrm{Im}(S_{12}(0) + S_{21}(0))$$

$$C_{ext}^{34} = -C_{ext}^{43} = \frac{2\pi}{k^2} \mathrm{Im}(S_{11}(0) - S_{22}(0))$$

It follows for isotropic spheres that:

$$S(0) \equiv S_{11}(0) = S_{22}(0), \qquad S_{12}(0) = S_{21}(0) = 0 \tag{1.77}$$

and the extinction matrix reduces to the scalar value (1.71). Note that for chiral spheres we obtain (Bohren and Huffman, 1983) that:

$$S_{11}(0) = S_{22}(0), \qquad S_{12}(0) = -S_{21}(0) \tag{1.78}$$

and

$$\hat{C}_{ext} = \varepsilon \begin{pmatrix} 1 & 0 & 0 & \beta \\ 0 & 1 & \alpha & 0 \\ 0 & -\alpha & 1 & 0 \\ \beta & 0 & 0 & 1 \end{pmatrix} \tag{1.79}$$

where

$$\varepsilon = \frac{4\pi}{k^2}\mathrm{Re}[S_{11}(0)], \qquad \alpha = \frac{\mathrm{Re}[S_{12}(0)]}{\mathrm{Re}[S_{11}(0)]}, \qquad \beta = -\frac{\mathrm{Im}[S_{12}(0)]}{\mathrm{Re}[S_{11}(0)]} \tag{1.80}$$

The Stokes matrix in this case ($S_{21} = -S_{12}$) can be written as follows:

$$\hat{C}_{sca} = k^{-2}\begin{pmatrix} \frac{|S_{11}|^2 + |S_{22}|^2}{2} + |S_{12}|^2 & \frac{|S_{11}|^2 - |S_{22}|^2}{2} & \mathrm{Re}[(S_{11} - S_{22})S_{12}^*] & \mathrm{Im}[(S_{11} + S_{22})S_{12}^*] \\ \frac{|S_{11}|^2 - |S_{22}|^2}{2} & \frac{|S_{11}|^2 + |S_{22}|^2}{2} - |S_{12}|^2 & \mathrm{Re}[(S_{11} + S_{22})S_{12}^*] & \mathrm{Im}[(S_{11} - S_{22})S_{12}^*] \\ -\mathrm{Re}[(S_{11} - S_{22})S_{12}^*] & -\mathrm{Re}[(S_{11} + S_{22})S_{12}^*] & \mathrm{Re}(S_{11}S_{22}^*) - |S_{12}|^2 & \mathrm{Im}(S_{11}S_{22}^*) \\ \mathrm{Im}[(S_{11} + S_{22})S_{12}^*] & \mathrm{Im}[(S_{11} - S_{22})S_{12}^*] & -\mathrm{Im}(S_{11}S_{22}^*) & \mathrm{Re}(S_{11}S_{22}^*) + |S_{12}|^2 \end{pmatrix} \tag{1.81}$$

Exact solutions for elements of the S-matrix for isotropic, chiral, and two-layered spheres are presented in Appendix 2.

We obtain for a unit volume of a light scattering medium (see equations (1.47), (1.48), (1.72)):

$$\hat{\sigma}_{sca} = N\langle\hat{C}_{sca}\rangle \tag{1.82}$$

$$\hat{\sigma}_{ext} = N\langle\hat{C}_{ext}\rangle \tag{1.83}$$

where N is the number concentration of particles (see Table 1.1) and:

$$\langle\hat{C}_{sca}\rangle = \int_0^\infty \hat{C}_{sca}(\vec{a})f(\vec{a})\,d\vec{a} \tag{1.84}$$

$$\langle\hat{C}_{ext}\rangle = \int_0^\infty \hat{C}_{ext}(\vec{a})f(\vec{a})\,d\vec{a} \tag{1.85}$$

are average values of the differential scattering and extinction matrices. The matrix $\hat{\sigma}_{sca}^d$ in equation (1.47) can be found from equations (1.48), (1.82).

The function $f(\vec{a})$ describes the probability density for a particle to have geometrical parameters (the size, shape, internal structure) characterized by a vector parameter $\vec{a}$. It follows by definition that $\int_0^\infty f(\vec{a})\,d\vec{a} = 1$. Note that for spherical particles $f(a)$ is the particle size distribution, defined in Section 1.2.

Equations (1.82), (1.83) can be used to define the absorption matrix (Karam, 1998):

$$\hat{\sigma}_{abs} = \hat{\sigma}_{ext} - \hat{\sigma}'_{sca} \tag{1.86}$$

where (see equation (1.41)):

$$\hat{\sigma}'_{sca} = \int_{4\pi} \hat{\sigma}_{sca}\,d\Omega. \tag{1.87}$$

Note that the dimensionless coefficients of variance ν_{ij} of elements σ_{ij} of all these matrices can be found from the following relation:

$$\nu_{ij} = \frac{\sqrt{\int_0^\infty (\sigma_{ij} - \langle\sigma_{ij}\rangle)^2 f(a)\,da}}{\langle\sigma_{ij}\rangle} \tag{1.88}$$

where $\langle\sigma_{ij}\rangle = \int_0^\infty \sigma_{ij} f(a)\,da$ is the average value of the element of the correspondent matrix (see equations (1.82), (1.83), (1.86), (1.87)). It follows for monodispersed media that $\nu_{ij} = 0$. Obviously, the coefficient of variance (1.88) increases with broadening of the particle size distribution.

As stated in Section 1.3, the radiative transfer theory considers light transfer in an 'effective' continuous medium with matrices, defined by equations (1.82), (1.83).

Exact results for these matrices in the case of spherical particles can be obtained from formulae which are presented in Appendix 2. They were derived from the solution of the following vector wave equation, which follows from Maxwell's equations:

$$\vec{\nabla} \times \vec{\nabla} \times \vec{E}(\vec{r}) - k^2 m^2\,\vec{E}(\vec{r}) = 0 \tag{1.89}$$

with tangential components of the electric vector $\vec{E}$ and magnetic vector $\vec{H} = (i/k)\vec{\nabla} \times \vec{E}$, which are continuous at the boundary of a particle. Note that equation (1.89) can be written in the integral form (see equations (1.53), (1.54) (Saxon, 1955)):

$$\left.\begin{aligned} \vec{E}(\vec{r}) &= \vec{E}_0(\vec{r}) + \vec{f}_s(\vec{r})\frac{e^{-i(kr-\omega t)}}{r} \\ \vec{f}_s(\vec{r}) &= \frac{k^2}{4\pi}\int_V (m^2(\vec{r}\,') - 1)\,e^{ik\vec{r}\,'\vec{j}}\,\vec{E}'(\vec{r}\,')\,d^3\vec{r}\,' \end{aligned}\right\} \tag{1.90}$$

where V is the volume of a particle, $\vec{E}_0(\vec{r})$ is the incident electric field, $\vec{E}'(\vec{r}\,') = -\vec{j} \times [\vec{j} \times \vec{E}(\vec{r}\,')]$. The unit vector $\vec{j}$ is along the scattering direction and $\vec{E}'(\vec{r}\,') \perp \vec{E}(\vec{r}\,')$. Equation (1.90) can be used to find the vector scattering amplitude, extinction, scattering and absorption matrices for disperse media with particles of different shapes. For isotropic particles, these matrices reduce to scalar values. It follows for cross-sections (Ishimaru, 1978; Klett and Sutherland, 1992; Markel and Polyakov, 1997) that:

$$C_{ext} = \frac{-4\pi}{k|\vec{E}_0|} \operatorname{Im}(\vec{f}_s(\vec{e}_3, \vec{e}_3)\vec{e}_p) \tag{1.91}$$

$$C_{sca} = \frac{1}{|\vec{E}_0|^2} \int_{4\pi} |\vec{f}_s|^2 \, d\Omega \tag{1.92}$$

$$C_{abs} = \frac{k}{|\vec{E}_0|^2} \int_V \varepsilon''(\vec{r}')|\vec{E}(\vec{r}')|^2 \, d^3\vec{r}' \tag{1.93}$$

where $\varepsilon'' = 2n\chi$ is the imaginary part of the dielectric permittivity, and the direction $\vec{e}_p$ coincides with the direction of the polarization of the incident linearly polarized electromagnetic wave. It should be stressed that the integral equation (1.90) is more convenient for the development of approximate techniques than the differential equation (1.89) (Ishimaru, 1978).

Generally speaking, the solution of the differential equation (1.89) is more complex than the solution of the integro-differential equation (1.52). This is related mostly to the fact that we must consider electric vector $\vec{E}$ instead of the Stokes vector $\mathbf{I}$ in the problem of light diffraction by a particle of a certain shape. Also boundary conditions for both equations differ considerably. The Stokes vector of singly scattered light can be easily found with equation (1.72) if a full solution of the electromagnetic scattering problem for a particle of a given shape is known. Thus, the solution of Maxwell's equations cannot be avoided in calculations of local optical characteristics of disperse media.

On the other hand, the concept of the light field (Rozenberg, 1977) allows us to substitute Maxwell's equations for the electromagnetic field by the far simpler RTE in the problem of light propagation on scales much larger than the wavelength of the incident radiation. We can see how the RTE is derived from the Maxwell theory in the book by Apresyan and Kravtsov (1983). Clearly, such a derivation cannot be performed for all types of disperse media. For instance, it is not valid for close-packed media. This is the real reason behind the slow progress in the field of disperse media with densely distributed particles (Tsang *et al.*, 1985; Ivanov *et al.*, 1988). However, the solution of this more complex problem is of great importance to many applications (e.g. for the calculation of reflectivity of planetary surfaces (Hapke, 1993)).

The next chapter is devoted to calculation of parameters of the RTE (σ_{ext}, σ_{abs}, $p(\theta)$ or correspondent matrices) within the framework of approximate techniques. We can find a description of exact methods for calculating these values for spheres, spheroids, and circular cylinders in books written by Barber and Hill (1990) and Mishchenko *et al.* (2002). Barber and Hill (1990) present not only exact analytical results but the computer codes as well. Kuik *et al.* (1992) published benchmark results for spheroidal particles.

Latest developments in the field were presented at the Conference on Light Scattering by Nonspherical Particles (1998) and summarized in a book edited by Mishchenko *et al.* (2000).

New effects in wave propagation and scattering in random media have been reviewed by Kravtsov (1993), Lagendijk and Van Tiggelen (1996), and the POAN Research Group (1998).

2

Single light scattering

2.1 PARTICLES THAT ARE SMALL COMPARED WITH THE WAVELENGTH

Light scattering and absorption characteristics of particles that are small compared with the wavelength λ of the incident radiation outside and inside of a particle can be studied using the Rayleigh approximation (Rayleigh, 1871). It is assumed within the framework of this approximation that:

$$x \ll 1, \qquad x|m| \ll 1$$

where $x = ka$ is the size parameter, $k = 2\pi/\lambda$, a is the size of a particle (e.g. the radius of a spherical scatterer), $m = n - i\chi$ is the complex refractive index of a particle, relative to that of the surrounding nonabsorbing medium.

For such a small particle the electric field within a particle is constant and it is in phase with the external field. Therefore, a particle can be replaced by a single oscillating dipole with the polarizability tensor $\hat{\alpha}$ and a simple theory of the dipole scattering can be applied for determination of light scattering and absorption characteristics. The components of the polarizability tensor depend on the shape and the internal structure of particles.

For randomly oriented particles it follows (Van de Hulst, 1957; Bohren and Huffman, 1983) that:

$$C_{abs} = \frac{-4\pi k}{3} \mathrm{Im}(\alpha_1 + \alpha_2 + \alpha_3), \qquad C_{sca} = \tfrac{8}{9}\pi k^4(|\alpha_1|^2 + |\alpha_2|^2 + |\alpha_3|^2) \tag{2.1}$$

$$\hat{p}(\theta) = \frac{3}{3A + B} \begin{pmatrix} A + B\cos^2\theta & B(\cos^2\theta - 1) & 0 & 0 \\ B(\cos^2\theta - 1) & B(\cos^2\theta + 1) & 0 & 0 \\ 0 & 0 & 2B\cos\theta & 0 \\ 0 & 0 & 0 & (12 - 10A)\cos\theta \end{pmatrix} \tag{2.2}$$

where α_1, α_2, α_3 are components of the diagonalized polarizability tensor of a particle:

$$\hat{\alpha} = \begin{pmatrix} \alpha_1 & 0 & 0 \\ 0 & \alpha_2 & 0 \\ 0 & 0 & \alpha_3 \end{pmatrix}$$

C_{sca} and C_{abs} are the average scattering and absorption cross-sections per particle, $\hat{p}(\theta)$ is the phase matrix,

$$p(\theta) \equiv p_{11}(\theta) = \frac{3(A + B\cos^2\theta)}{3A + B}$$

is the phase function, and

$$A = \frac{6 - M}{5}, \qquad B = \frac{2 + 3M}{5}, \qquad M = \frac{\mathrm{Re}(\alpha_1^*\alpha_2 + \alpha_1^*\alpha_3 + \alpha_2^*\alpha_3)}{|\alpha_1|^2 + |\alpha_2|^2 + |\alpha_3|^2} \tag{2.3}$$

It follows for spheres: $A = B = 1$. The Rayleigh phase matrix $\hat{p}(\theta)$ (2.2) is defined in the co-ordinate system, attached to the particle. Thus, it should be transformed according to the general rule (1.48) to be used in the radiative transfer equation (RTE) (1.50).

The matrix (2.2) depends on the internal structure of particles, their refractive index and shape. However, it does not depend on the size of particles. The Rayleigh phase function for randomly oriented particles can be presented in the following form:

$$p(\theta) = \frac{y + \cos^2\theta}{y + 1/3}$$

where $y = (6 - M)/(2 + 3M)$. We can see that $y = 1$ for spherical scatterers. It should be pointed out that the value of M is in the range $[-0.5, 1]$ (Bohren and Huffmen, 1983) and, correspondingly, $y \in [1, 13]$. The phase function at different values of y is presented in Figure 2.1. We can see that the nonsphericity of particles leads to more isotropic phase functions. However, it does not move the position of the minimum. Interestingly, all phase functions have two common intersection points, where the phase function is equal to 1. It takes place at scattering angles, which are equal to $\arccos(1/\sqrt{3})$ and $2\pi - \arccos(1/\sqrt{3})$ or 54.7° and 125.3° respectively. Phase functions are identical in forward and backward directions $(p(0) = p(\pi))$.

We can see that the components of the Stokes vector $\mathbf{I}_s = (I_s, Q_s, U_s, V_s)$ of the singly scattered light are proportional to the following values in the case of Rayleigh scattering (the incident light with the intensity I_0 is unpolarized) $I_s \sim (A + B\cos^2\theta)I_0$, $Q_s \sim B(\cos^2\theta - 1)I_0$, $U_s = 0$, $V_s = 0$. Therefore, it follows for the degree of polarization $P = -Q_s/I_s$, of the singly scattered light that:

$$P(\theta) = \frac{1 - \cos^2\theta}{y + \cos^2\theta}$$

The degree of polarization, obtained with this equation, is presented in Figure 2.2.

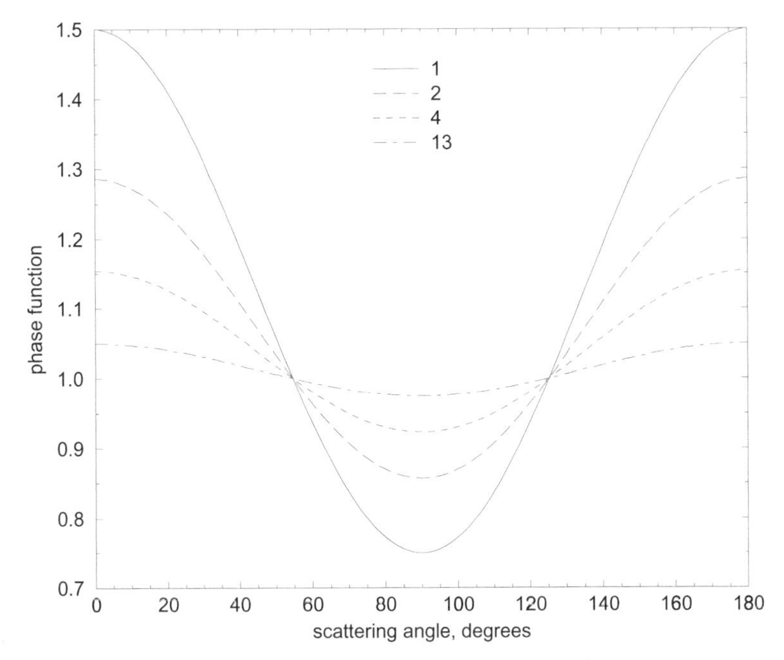

Figure 2.1. The phase function at different values of $y = 1, 2, 4, 13$.

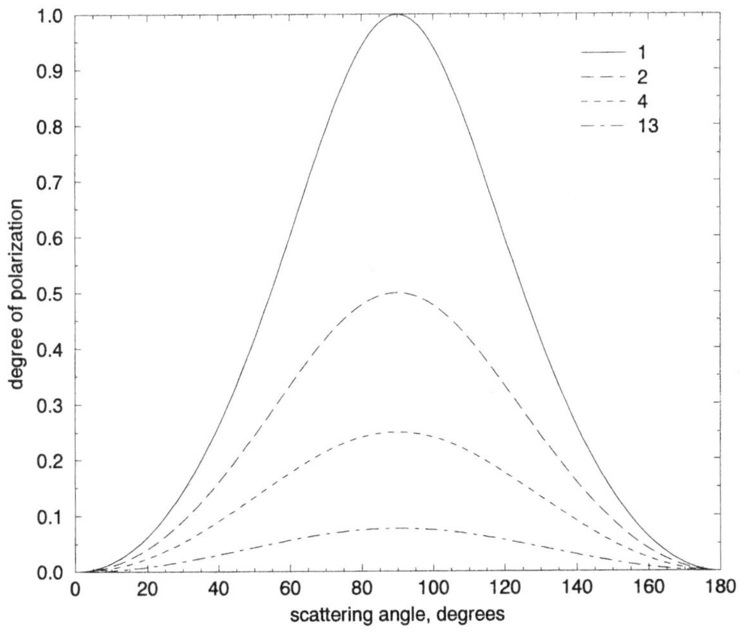

Figure 2.2. The degree of polarization at different values of $y = 1, 2, 4, 13$.

The nonsphericity of particles reduces the degree of polarization considerably. The maximum degree of polarization P at the scattering angle 90° is inversely proportional to the parameter y: $P(90^\circ) = y^{-1}$. For instance, it follows at $y = 2$: $P(90^\circ) = 0.5$.

The angle between the direction of the polarization and the plane containing the initial and final photon directions (the scattering plane) $\Psi = \frac{1}{2}\arctan(U_s/Q_s)$ is equal to 0 or $\pi/2$. The choice of the angle $\Psi = 0$ or $\Psi = \pi/2$ depends on the sign of the element p_{12} ($\Psi = 0$ for $p_{12} > 0$ and $\Psi = \pi/2$ for $p_{12} < 0$). For the Rayleigh scattering it follows that $p_{12}(\theta) \leq 0$ and $\chi = \pi/2$. Thus, the singly scattered light is partially linearly polarized in the plane perpendicular to the scattering plane (the ellipticity $\varepsilon = 0$).

We can see that the scattering process is a source of the polarization of natural light and the degree of polarization P depends on the scattering angle. The function $P(\theta)$ has the maximum at $\theta = \pi/2$ for randomly oriented particles in the framework of the Rayleigh approximation. Note that it follows for spherical particles that $\alpha_1 = \alpha_2 = \alpha_3$, $M = 1$, and $y = 1$. The light is completely polarized at the scattering angle $\theta = \pi/2$ in the direction perpendicular to the scattering plane in this case. The phase matrix $\hat{p}(\theta)$ does not depend on the refractive index for spherical Rayleigh particles ($A = B = 1$).

The Rayleigh phase function p_{11} and normalized elements p_{12}/p_{11} and p_{44}/p_{11} for spherical particles are presented in Figure 2.3 as functions of the scattering angle.

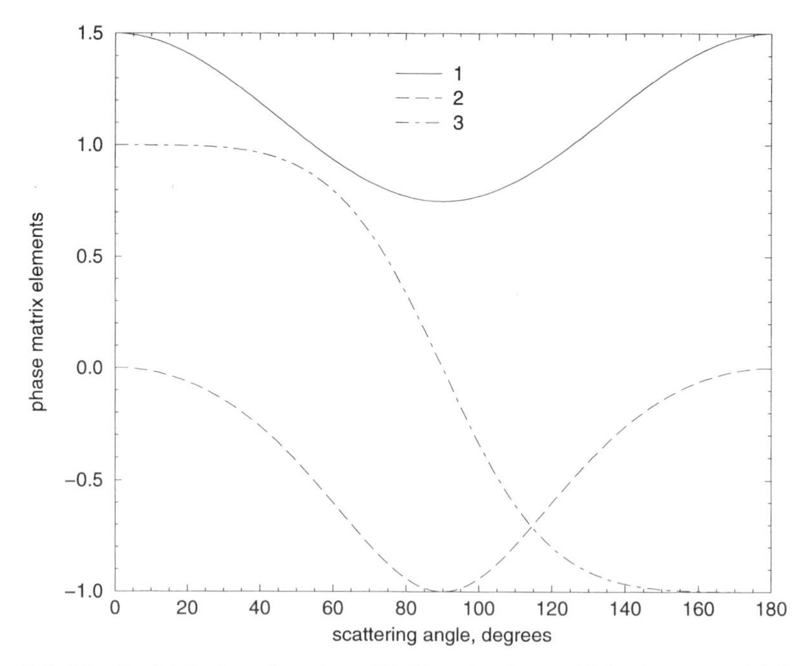

Figure 2.3. The Rayleigh phase function $p(\theta)$ (1) and ratios $p_{12}(\theta)/p_{11}(\theta)$ (2), $p_{44}(\theta)/p_{11}(\theta)$ (3), for spherical particles.

These angular dependencies are much simpler than in the case of larger particles, which results in simplification of radiative transfer problems for disperse media with Rayleigh particles (Chandrasekhar, 1950).

It is worth pointing out that the asymmetry parameter:

$$g = \frac{1}{2}\int_0^\pi p(\theta)\sin\theta\cos\theta\,d\theta$$

for Rayleigh particles of any shape is zero and the light pressure cross-section $C_{pr} = C_{ext} - gC_{sca}$ coincides with the extinction cross-section.

To apply the formulae for the values of C_{sca}, C_{abs} and $p(\theta)$ given above to radiative transfer problems we should specify the shape and the internal structure of particles and calculate the values of α_j. Analytical results for the components of the polarizability tensor $\hat{\alpha}$ of ellipsoids and soft ($|m-1| \ll 1$) Rayleigh particles of any shape are given in this section. It should be pointed out that the ellipsoidal shape is a very general one. An ellipsoid transforms to a sphere, prolate or oblate spheroid for a particular choice of axes lengths.

According to the definition, the induced dipole moment of a particle $\vec{d}$ can be described by the following equation (Van de Hulst, 1957):

$$\vec{d} = \hat{\alpha}\vec{E}_0$$

where $\hat{\alpha}$ is the polarizability tensor and $\vec{E}_0$ is the incident electric field. On the other hand, the dipole moment of a particle in an electric field is (Van de Hulst, 1957):

$$\vec{d} = \int_V \frac{m^2(\vec{r}) - 1}{4\pi}\vec{E}(\vec{r})\,d\vec{r}$$

where $\vec{E}(\vec{r})$ is the electric field and $m(\vec{r})$ is the refractive index of a particle in the point with the radius vector $\vec{r}$, V is the volume of a particle. It follows for soft ($|m-1| \ll 1$) Rayleigh particles of any shape that $\vec{E}(\vec{r}) \approx \vec{E}_0(\vec{r})$ and

$$\alpha_j = \frac{\langle m^2\rangle - 1}{4\pi}V \tag{2.4}$$

where $\langle m^2\rangle = 1/V\int_V m^2(\vec{r})\,d\vec{r}$. For example, we can obtain $\langle m^2\rangle = m^2$ for uniform particles with refractive index m. Thus, for optically soft isotropic uniform Rayleigh particles of any shape it follows (see equations (2.1), (2.2), (2.4)) that:

$$\left.\begin{aligned} &\hat{p}(\theta) = \frac{3}{4}\begin{pmatrix} 1+\cos^2\theta & \cos^2\theta - 1 & 0 & 0 \\ \cos^2\theta - 1 & 1+\cos^2\theta & 0 & 0 \\ 0 & 0 & 2\cos\theta & 0 \\ 0 & 0 & 0 & 2\cos\theta \end{pmatrix} \\ &C_{abs} = \gamma V, \qquad C_{sca} = \frac{k^4V^2}{6\pi}|m^2 - 1|^2 \end{aligned}\right\} \tag{2.5}$$

where $\gamma = (4\pi\chi)/\lambda$. It is worth remarking that scattering and absorption cross-sections of an optically soft Rayleigh particle depend on the volume V of a particle irrespectively to a specific shape. Thus, spherical, cylindrical, and cubic particles of the same volume have the same values of cross-sections C_{abs} and C_{sca}. The elements of the Rayleigh phase matrix $\hat{p}(\theta)$ do not depend on the size, shape, and refractive index of particles in this particular case. They are presented in Figure 2.3.

It follows that the electric field inside the particle (Van de Hulst, 1957) is given by the following equation for the specific case of Rayleigh uniform spherical particles:

$$E(\vec{r}) = \frac{3}{m^2+2}\vec{E}_0$$

and

$$\alpha = \frac{3(m^2-1)}{4\pi(m^2+2)}V \tag{2.6}$$

if one assumes that the refractive index of a host medium is equal to 1. Thus, we can obtain:

$$C_{abs} = -3kV\,\mathrm{Im}\left(\frac{m^2-1}{m^2+2}\right), \qquad C_{sca} = \frac{3}{2\pi}k^4V^2\left|\frac{m^2-1}{m^2+2}\right|^2 \tag{2.7}$$

Equations (2.5) follow from equations (2.7) as $m \to 1$. The phase matrix $\hat{p}(\theta)$ in this case is the same as for soft Rayleigh particles of any shape ($\alpha_1 = \alpha_2 = \alpha_3$ and $M = 1$).

The imaginary part of the refractive index is extremely small in the optical band of the electromagnetic spectrum for most substances ($\chi \ll n$). Thus, it follows from equations (2.7) that:

$$C_{abs} = f(n)\gamma V, \qquad C_{sca} = \frac{3}{2\pi}k^4V^2\left(\frac{n^2-1}{n^2+2}\right)^2$$

where $\gamma = (4\pi\chi)/\lambda$, $f(n) = 9n/(n^2+2)^2$ and we used the condition $\chi \ll n$. We can see that the light absorption by a Rayleigh particle decreases with the refractive index n and vice versa for the value of C_{sca}.

Another case in which the polarizability tensor of a particle can be calculated by elementary means is that of ellipsoids. The main components α_j of the polarizability tensor of an ellipsoid are (Van de Hulst, 1957):

$$\alpha_j = \frac{V}{4\pi}\frac{m^2-1}{1+L_j(m^2-1)}$$

where it follows for geometrical factors L_j ($L_1 + L_2 + L_3 = 1$):

$$L_j = \frac{a_1a_2a_3}{2}\int_0^\infty \frac{ds}{(a_j^2+s)f(s)}$$

and $f(s) = \sqrt{(s+a_1^2)(s+a_2^2)(s+a_3^2)}$, $a_1 \geq a_2 \geq a_3$ are the semi-axes of an ellipsoid. It follows that $\alpha_j \to (m^2-1)V/4\pi$ as $m \to 1$ and the polarizability tensor is reduced to the scalar value, which does not depend on the shape of particles in this case. However, for large and intermediate values of $m^2 - 1$, the dependence of α_j on the geometrical factors L_j is really important.

The values of L_j can be represented by elliptic integrals of the first $F(\vartheta, t)$ and the second $E(\vartheta, t)$ kind (Osborn, 1945):

$$L_1 = \frac{\cos\varphi\cos\vartheta}{\sin^3\vartheta\sin^2\beta}\{F(\vartheta,t) - E(\vartheta,t)\}$$

$$L_2 = \frac{\cos\varphi\cos\vartheta}{\sin^3\vartheta\sin^2\beta\cos^2\beta}\left\{E(\vartheta,t) - F(\vartheta,t)\cos^2\beta - \frac{\sin^2\beta\sin\vartheta\cos\vartheta}{\cos\varphi}\right\}$$

$$L_3 = \frac{\cos\varphi\cos\vartheta}{\sin^3\vartheta\cos^2\beta}\left\{\frac{\sin\vartheta\cos\varphi}{\cos\vartheta} - E(\vartheta,t)\right\}$$

where

$$\cos\vartheta = \frac{a_3}{a_1}, \qquad \cos\varphi = \frac{a_2}{a_1}, \qquad \sin\beta = \sqrt{\frac{1-\left(\frac{a_2}{a_1}\right)^2}{1-\left(\frac{a_3}{a_1}\right)^2}} = \frac{\sin\varphi}{\sin\vartheta} = t$$

$$F(\vartheta,t) = \int_0^\vartheta \frac{d\vartheta}{\sqrt{1-t^2\sin^2\vartheta}}, \qquad E(\vartheta,t) = \int_0^\vartheta \sqrt{1-t^2\sin^2\vartheta}\,d\vartheta$$

It follows from these equations that $L_1 + L_2 + L_3 = 1$ as it should be. General formulas for the values of L_j are simplified for specific shapes (see Table 2.1).

The optical characteristics of small irregular particles are obtained by integrating the corresponding optical characteristics (e.g. scattering and absorption cross-sections) over all possible values of L_1 and L_2 weighted by a shape probability function $P(L_1, L_2)$. Such an approach was used by Nevitt and Bohren (1984) for calculation of the backscattering coefficients of irregularly shaped aerosol particles in the infrared region of the electromagnetic spectrum.

Scattering of a plane wave by a general anisotropic dielectric ellipsoid was considered by Papadakis *et al.* (1990).

We obtain for particles in random orientation:

$$C_{abs} = \frac{\gamma n V}{3}\sum_{j=1}^{3}\frac{1}{(1+\nu L_j)^2 + (\mu L_j)^2}$$

$$C_{sca} = \frac{k^4 V^2}{18\pi}\sum_{j=1}^{3}\frac{(\nu + (\nu^2+\mu^2)L_j)^2 + \mu^2}{[(1+\nu L_j)^2 + (\mu L_j)^2]^2}$$

where $\gamma = (4\pi\chi)/\lambda$, $\nu = n^2 - 1 - \chi^2$, $\mu = 2n\chi$.

There are exact solutions of light-scattering problems for spheres, spheroids, and infinitely long cylinders (Barber and Hill, 1990). The Rayleigh approximation can be obtained from these solutions, assuming that x, $mx \to 0$ (see Appendices 2 and 3).

Table 2.1. The values of L_j for different shapes of particles

$$\left(\zeta = \frac{a_1}{a_3}, \varsigma = \frac{\zeta + \sqrt{\zeta^2 - 1}}{\zeta - \sqrt{\zeta^2 - 1}}, \upsilon = \frac{\sqrt{\zeta^2 - 1}}{\zeta}\right)$$

Shape	L_1	L_2	L_3
Prolate spheroid ($a_1 > a_2 = a_3$)	$\frac{1}{\zeta^2 - 1}\left\{\frac{\zeta \ln \varsigma}{2\sqrt{\zeta^2 - 1}} - 1\right\}$	$\frac{1 - L_1}{2}$	$\frac{1 - L_1}{2}$
Very slender prolate spheroid ($a_1 \gg a_2 = a_3$)	$\frac{1}{\zeta^2}\{\ln 2\zeta - 1\}$	$\frac{1 - L_1}{2}$	$\frac{1 - L_1}{2}$
Needle	0	$\frac{1}{2}$	$\frac{1}{2}$
Oblate spheroid ($a_1 = a_2 > a_3$)	$\frac{1}{2(\zeta^2 - 1)}\left\{\frac{\zeta^2 \arcsin \upsilon}{\sqrt{\zeta^2 - 1}} - 1\right\}$	L_1	$1 - L_1 - L_2$
Disk	0	0	1
Very flat oblate spheroid ($a_1 = a_2 \gg a_3$)	$\frac{\pi}{4\zeta}\left\{1 - \frac{4}{\pi\zeta}\right\}$	L_1	$1 - L_1 - L_2$
Sphere ($a_1 = a_2 > a_3$)	$\frac{1}{3}$	$\frac{1}{3}$	$\frac{1}{3}$

For example, in this way the extinction efficiency factor $Q_{ext} = C_{ext}/\pi a^2$ of small spheres with regard to the terms of the order of x^4 was obtained by Pendorf (1960):

$$Q_{ext} = \frac{24n\chi}{F_1(n,\chi)}x + \left\{\frac{4n\chi}{15} + \frac{20n\chi}{3F_2(n,\chi)} + \frac{48n\chi[7(n^2+\chi^2)^2 + 4(n^2 - \chi^2 - 5)]}{F_1^2(n,\chi)}\right\}x^3$$
$$+ \frac{8}{3}\left\{\frac{[(n^2+\chi^2)^2 + (n^2-\chi^2-2)]^2 - 36n^2\chi^2}{F_1^2(n,\chi)}\right\}x^4 \tag{2.8}$$

where

$$F_1(n,\chi) = (n^2+\chi^2)^2 + 4(n^2 + 4(n^2-\chi^2) + 4$$
$$F_2(n,\chi) = 4(n^2+\chi^2)^2 + 12(n^2-\chi^2) + 9$$

The extinction efficiency factor Q_{ext} is the sum of the absorption efficiency factor $Q_{abs} = (C_{abs})/\pi a^2$ (terms $\sim x$ and x^3 in equation (2.8)) and the scattering efficiency factor $Q_{sca} = (C_{sca})/\pi a^2$ (term x^4 in equation (2.8)). The accuracy of this equation was discussed by Kerker (1969). Errors of the formula do not exceed 2% up to $m = 1.5$, $x = 1.4$.

The error of the Rayleigh approximation for the extinction cross-section initially decreases with the refractive index, reaches minimum at the refractive index in the range 1.6–1.8 and increases again with further growth of n (Kerker, 1969). Spectral extinction and scattering properties of small-size parameter particles, including soot and TiO_2, were studied in detail by Kim *et al.*(1996).

2.2 PARTICLES WITH THE SIZE COMPARABLE WITH THE WAVELENGTH

Optical properties of particles with sizes comparable with the wavelength of the incident light can be calculated within the framework of the discrete dipole approximation (DDA). In the framework of this approximation, a particle of any shape is replaced not just by a single dipole, as it was in the previous section, but by means of an array of N ($N \to \infty$) point dipoles, with the spacing between dipoles small compared with the wavelength. The replacement requires specification of both geometry (location $\vec{r}_j$ of the dipoles $j = 1, \ldots, N$) and the dipole polarizabilities α_j. Note that for a finite array of point dipoles the scattering problem may be solved exactly (Purcell and Pennypacker, 1973; Bohren and Singham, 1991). This approach makes it possible to consider spherical and nonspherical particles with the characteristic size $a \leq \lambda$. For larger particles numerical calculations become unstable.

Each dipole has an oscillating polarization in response to both an incident plane wave and electric fields due to all other dipoles in an array. This is why the DDA is also sometimes referred to as the coupled dipole approximation. The dipole at location $\vec{r}_i$ acquires a dipole moment:

$$\vec{d}_i = \alpha_i \vec{E}_i \tag{2.9}$$

where α_i is the polarizability of a dipole i. As mentioned before, the electric field $\vec{E}_i$ is due to the incident field $\vec{E}_{inc} = \vec{E}_0 \exp(-i(\vec{k}\vec{r}_i - \omega t))$ at the point with the radius-vector $\vec{r}_i$ and contributions from dipoles with amplitudes $\vec{d}_j$, located at points $\vec{r}_j$. Thus, according to the linear superposition principle, we obtain for the electric field $\vec{E}_i$ at the point $\vec{r}_i$, where the i dipole is located:

$$\vec{E}_i = \vec{E}_{inc} + \vec{E} \tag{2.10}$$

where $\vec{E} = \sum_{j \neq i}^{N} \vec{E}_j$ is the sum of fields from all dipoles except the dipole with the radius-vector $\vec{r}_i$. The electric field $\vec{E}_j$, in vacuum, of a dipole with amplitude $\vec{d}_j$, evaluated at the distance $r_{ij} = |\vec{r}_i - \vec{r}_j|$ has the following analytical form:

$$\vec{E}_j = \frac{\exp(-i(kr_{ij} - \omega t))}{r_{ij}} \left[\frac{1 + ikr_{ij}}{r_{ij}^2} \left(\frac{3(\vec{d}_j \cdot \vec{r}_{ij})\vec{r}_{ij}}{r_{ij}^2} - \vec{d}_j \right) - k^2 \left(\frac{(\vec{d}_j \cdot \vec{r}_{ij})\vec{r}_{ij}}{r_{ij}^2} - \vec{d}_j \right) \right] \tag{2.11}$$

where $\vec{r}_{ij} = \vec{r}_i - \vec{r}_j$.

It should be pointed out that the sum $\sum_{j\neq i}^{N} \vec{E}_j$ can be replaced by the integral (Rouleau and Martin, 1993):

$$\vec{E}(\vec{\rho}) = \int_V \frac{\exp(-i(\vec{k}\vec{r} - \omega t))}{r} \left[\frac{1 + ikr}{r^2} \left(\frac{3(\vec{P}\cdot\vec{r})\vec{r}}{r^2} - \vec{P} \right) - k^2 \left(\frac{(\vec{P}\cdot\vec{r})\vec{r}}{r^2} - \vec{P} \right) \right] d^3\vec{r}$$

where $\vec{r} = \vec{\rho} - \vec{\rho}'$, $\vec{\rho}' \neq \vec{\rho}$ and V is the volume of a particle. Equation (2.11) follows from this equation after replacement of a particle by a cubic array of point dipoles, each of which has the dipole moment $\vec{d} = \vec{P}\Delta V$, where ΔV is the small volume associated with each dipole.

Let us substitute equations (2.10), (2.11) into equation (2.9). Then it follows:

$$\vec{d}_i = \alpha_i \exp(i\omega t) \left[\vec{E}_0 \exp(-i\vec{k}\vec{r}_i) + \sum_{j\neq i}^{N} \frac{\exp(-ikr_{ij})}{r_{ij}^3} \right.$$
$$\left. \times \left(\frac{1 + ikr_{ij}}{r_{ij}^4} (3(\vec{d}_j\cdot\vec{r}_{ij})\vec{r}_{ij} - \vec{d}_j r_{ij}^2) - k^2 \left(\frac{(\vec{d}_j\cdot\vec{r}_{ij})\vec{r}_{ij}}{r_{ij}^2} - \vec{d}_j \right) \right) \right] \tag{2.12}$$

The system of $3N$ complex equations (2.12) allows us to find the dipole moments $\vec{d}_i$ of each single dipole in the array at points $\vec{r}_i$, which should be specified before calculations are performed. Generally speaking, the more points that are chosen, the more correct results that are obtained. However, the numerical solution of equations (2.12) becomes numerically unstable as $N \to \infty$. This does not allow consideration of values of N larger than 10^6 at the moment. This imposes limitations on the size of particles which can be studied in the framework of the discrete dipole approximation.

It should be pointed out that the calculation of light scattering and absorption characteristics of particles using dipole moments d_j is simple. For instance, the electric field $\vec{E}_s$ in the far field zone is given by:

$$\vec{E}_s = \vec{f}_s \frac{\exp(-i(kr - \omega t))}{r} \tag{2.13}$$

where (Draine, 1988):

$$\vec{f}_s = k^2 \sum_{j=1}^{N} e^{ik\hat{r}\vec{r}_j} (\hat{r}\cdot\hat{r} - \hat{I})\vec{d}_j \tag{2.14}$$

where $\hat{r} = \vec{r}/r$, $r = |\vec{r}|$, $\hat{I}$ is the unit matrix.

For the extinction and absorption cross-sections it follows in the framework of the DDA (Draine and Flatau, 1994):

$$C_{ext} = \frac{4\pi k}{|\vec{E}_0|^2} \sum_{j=1}^{N} \mathrm{Im}(\vec{E}_{0j}^* \vec{d}_j) \tag{2.15}$$

$$C_{abs} = \frac{4\pi k}{|\vec{E}_0|^2} \sum_{j=1}^{N} \{ \mathrm{Im}\, \vec{d}_j (\alpha_j^{-1})^* \vec{d}_j^* - \tfrac{2}{3} k^3 |\vec{d}_j|^2 \} \tag{2.16}$$

It is often assumed that the polarizability α in equation (2.12) is isotropic. However, the case of a tensor polarizability could also be considered in the framework of the DDA. The first coarse approximation for the value of α can be obtained from the Clausius–Mossotti relation:

$$\varepsilon - 1 = \frac{4\pi n\alpha}{1 - \dfrac{4\pi n\alpha}{3}} \tag{2.17}$$

where ε is the dielectic constant of a particle and n is the number density of dipoles. Thus, it follows from equation (2.17) that:

$$\alpha = \frac{3}{4\pi n}\frac{\varepsilon - 1}{\varepsilon + 2} \tag{2.18}$$

The value of n could be chosen equal to D^{-3}, where D is the distance between dipoles, which should be much smaller than the wavelength of the incident radiation. Clearly, it follows for inhomogeneous particles that:

$$\alpha(\vec{r}_j) = \frac{3D^3}{4\pi}\frac{\varepsilon_j - 1}{\varepsilon_j + 2} \tag{2.19}$$

where ε_j is the dielectric permittivity of a particle at the location $\vec{r}_j$. Both ε_j and $\alpha(\vec{r}_j)$ become tensors for anisotropic and chiral scatterers. Note that the DDA can also be used to study radiative torques on particles of different shapes (Draine and Weingarter, 1996). This underlines the power of the DDA, which can treat very complex particles in the framework of the same system of equations.

Equation (2.18) is exact for polarizable point dipoles located on a cubic latice in the zero-frequency limit, but it is not exact at finite frequencies. Thus, the choice of polatizability α is nontrivial (Draine and Goodman, 1993). Draine (1988) found that the extinction theorem can be satisfied, assuming that:

$$\alpha = \frac{3}{4\pi n}\frac{\varepsilon - 1}{\varepsilon + 2}\left[1 + \frac{2i}{3N}(ka_{ef})^3\frac{\varepsilon - 1}{\varepsilon + 2}\right]^{-1} \tag{2.20}$$

where $a_{ef} = \sqrt[3]{4\pi/3n}$. Draine and Goodman (1993) proposed the so-called lattice dispersion relation (LDR) for the value of α. This relation takes into account terms proportional to $(ka_{ef})^2$ and improves the accuracy of the DDA. It has the following form:

$$\alpha = \frac{\alpha^{cm}}{1 + \dfrac{\alpha^{cm}}{D^3}(b_1 + \varepsilon b_2 + \varepsilon b_3\, s)(kD)^2 - \dfrac{2i}{3}(kD)^3} \tag{2.21}$$

where α^{cm} is the Clausius–Mossotti polarizability (see equation (2.18)), $b_1 = -1.891\,531$, $b_2 = 0.164\,8469$, $b_3 = -1.770\,0004$, $s = \sum_{j=1}^{3}(\vec{n}_j\vec{e}_j)^2$, where $\vec{n}_j$ and $\vec{e}_j$ are unit vectors defining the incident direction and polarization, $D = a_{ef}\sqrt[3]{4\pi/3N}$ is the distance between dipoles.

As the number of dipoles $N \to \infty$, the DDA represents exact results with higher and higher accuracy. For instance, Draine and Flatau (1994) found that for spheres with the size parameter $x < 12$ and the refractive index $m = 1.33 - 0.01i$ the errors of calculations of values (2.15), (2.16) with the DDA are smaller than 3% at $N \leq 17{,}904$. About the same accuracy was obtained for the phase function. This accuracy is high enough for most radiative transfer applications. It should be pointed out that the number of dipoles needed depends on the size of the particle. At the size parameter $x = 10$ we should consider about 10^5 dipoles.

Different methods have been introduced to solve a set of coupled linear equations for dipole moments at large N (Draine, 1988; Draine and Flatau, 1994). However, they cannot be used at $N > 10^6$ because of the lack of stability. Therefore, this method cannot be applied for particles with the size parameter $x \geq 15$–20 at the moment.

The chief disadvantage of the method considered in comparison with the Rayleigh approximation is the necessity to perform complex numerical calculations. The advantage is the possibility of considering very complex shapes (including fractals and anisotropic particles) in the framework of the same algorithm. Note that the DDA and many other electromagnetic codes are available free via the Internet (see Appendix 4).

The application of the DDA to particles with real refractive index smaller than unity was studied by Laczik (1996).

The phase function and the normalized phase matrix elemets for ice cubic, hexagonal, and spherical particles calculated with the DDA are shown in Figure 2.4. It was assumed that particles ae monodispersed and randomly oriented. All scatterers have the same volume $V = 4\pi a^3/3$, where $a = 80\,\text{nm}$ is the radius of a volume-equivalent spherical particle. It was assumed that the length and the side of the hexagonal cross-section are equal and $\lambda = 270\,\text{nm}$, $n = 1.34$. Such small ice particles exist in noctilucent clouds (Gadsden and Schroder, 1989). However, information on their shape is highly uncertain at the moment.

It follows from Figure 2.4(a) that the shape of particles strongly influences the scattering in the backward hemisphere. Also polarization characteristics differ considerably depending on the shape. In particular, the backward linear depolarization factor

$$\delta(\pi) = \frac{\mathrm{I}_\perp(\pi)}{\mathrm{I}_\parallel(\pi)}$$

is equal to zero for spheres. It substantially deviates from zero for nonspherical scatterers considered here. Note that $\mathrm{I}_\perp$ is the cross-polarized component of the scattered signal. Correspondingly, $\mathrm{I}_\parallel$ is the co-polarized component of the scattered signal. Its polarization coincides with the polarization of an emitted light beam. It's easy to show (Kokhanovsky, 2003) that $\delta(\pi) = (1 - p_{22}(\pi))/(1 + p_{22}(\pi))$.

We present the angular dependence of the degree of polarizatin $(-p_{12})$ in Figure 2.4(b) both for nonspherical particles and for spheres. It follows that the degree of polarization by nonspherical particles is generally closer to that of Rayleigh scattering. It means in particular, that the interpretation of measurements using Mie theory may underestimate the size of particles. The possibility of such an underestimation was first found by Mishchenko (1991), who used the model of ellipsoidal particles. The largest differences are around the scattering angle of 140° (as in Figure 2.4(a)). Small differences take place at small scattering angles.

The element p_{22} is given in Figure 2.4(c). The difference $1 - p_{22}$ can be as small as 0.04 for nonspherical particles considered here. Such a small deviation from one is possible to measure, however. This was demonstrated by Baumgarten *et al.* (2002), who measured the value of δ in the exact backward scattering direction and found that it was an average equal to $(1.7 \pm 1.0)\%$ at the wavelength 532 mm.

Let us consider now the matrix element p_{34}. The deviations between different particle models are quite strong for the scattering angle of 140° if the value of p_{34} is measured (see Figure 2.4(d)). This can be used for the identification of the shape of particles. The element p_{34} gives the strength of the transition of the linearly polarized light to the circular polarization mode as it was outlined above. Therefore, such a transition is of importance around the scattering angle of 140°.

The elements p_{33} and p_{44} for hexagonal particles, cubes, and spheres are given in Figure 2.4(e). They coincide for spheres. They almost coincide for cubes and hexagonal cylinders, being closer to a correspondent Rayleigh curve ($p_{44} = 2\cos\theta/(1+\cos^2\theta)$).

Overall, our analysis shows that nonspherical particles can give large deviations in the scattering matrix (including phase function) compared with spheres of the same volume. In particular, non-spherical scattering produces more Rayleigh-like scattering behaviour, which ads to the underestimation of the size of particles if the spherical model is used in the retrieval procedure.

2.3 PARTICLES THAT ARE LARGE COMPARED WITH THE WAVELENGTH

2.3.1 Geometrical optics approximation

The sizes of particles in natural media (e. g. in water and ice clouds, fogs, dust aerosols, foams, snow, ocean water) are often much larger than the wavelength of incident light. Thus, the Rayleigh and discrete dipole approximations, described in previous sections, cannot be applied for such particles in the optical band. Approximations for soft ($|m-1| \ll 1$) particles can be very useful for biological applications (optics of blood and leaves, oceanic suspensions), but not for atmospheric optics, because the refractive index n of atmospheric aerosols and clouds in the visible range is not close to the refractive index of air. For water and ice clouds, sulphur and dust aerosols the value of n is around 1.3 and 1.5, respectively (see Appendix 1). Exact solutions of the wave equation should be applied in this case (Wickramasinghe, 1973; Kerker, 1988). They are well known for infinite cylinders (Rayleigh, 1881; Wait, 1955), spheres (Mie, 1908; Logan, 1965; Lilienfeld, 1991), and spheroids (Asano and Yamamoto, 1975; Farafonov, 1983, 1990). Solutions for layered and nonuniform particles were derived by Aden and Kerker (1951), Kerker and Matijevich (1961), Shah (1972), Onaka (1980), Farafonov *et al.* (1994), Johnson (1996), and Perelman (1996). Results for optically active particles were obtained by Bohren

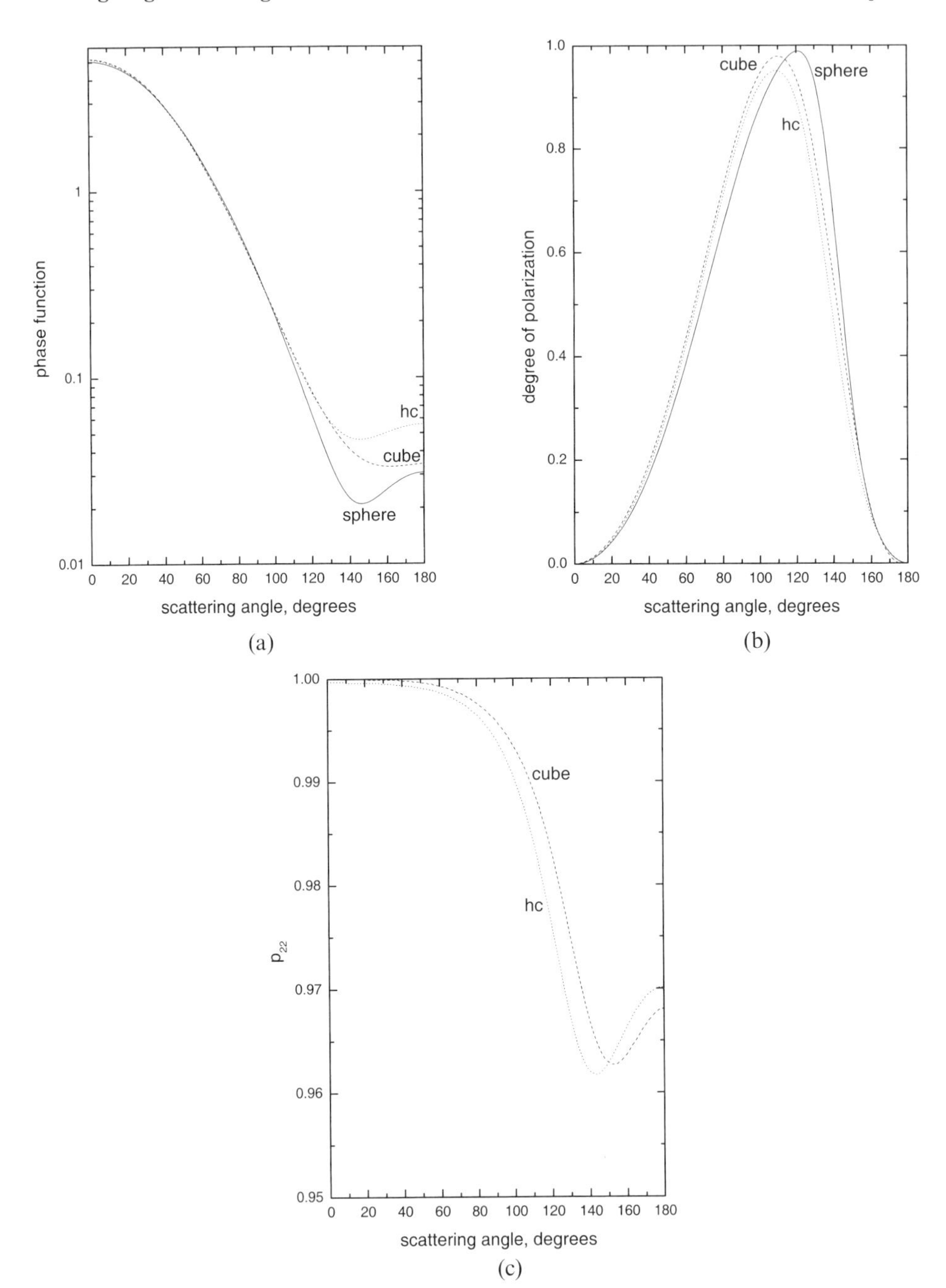

Figure 2.4. The phase function (a), degree of polarization (b), elements p_{22} (c), p_{34} (d), and p_{33}, p_{44} (e) for randomly oriented ice cubes, hexagonal cylinders, and spheres at the wavelength 270 nm. The volume of all particles is the same, corresponding to that of a sphere having a radius 80 nm. The length of the cylinder and the side of the hexagonal cross-section are equal. The refractive index is assumed to be equal to 1.34. The curves marked "hc" correspond to the case of a hexagonal cylinder.

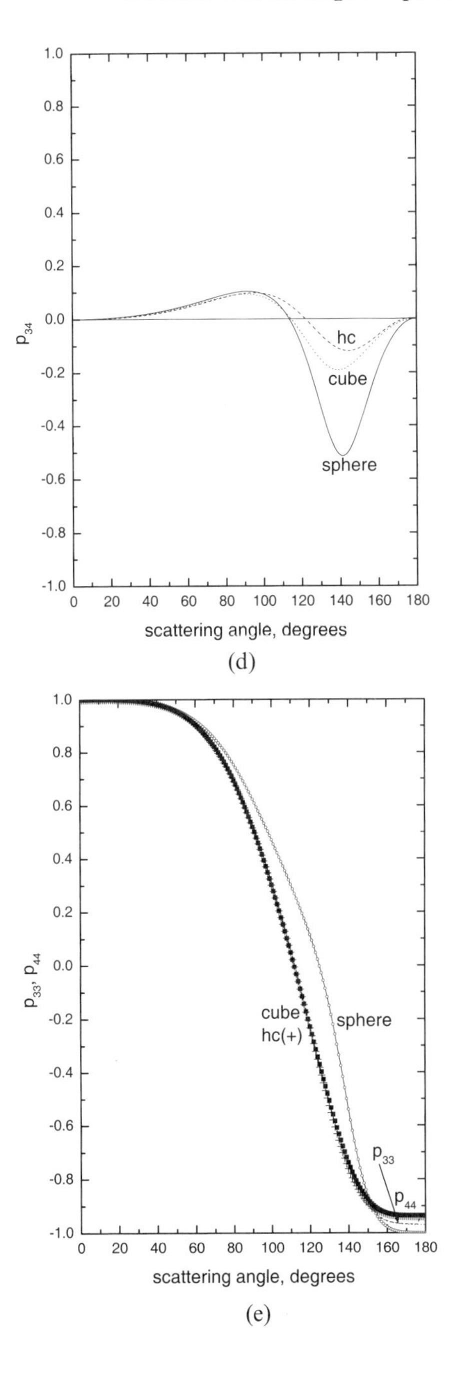

Figure 2.4. (*cont.*)

(1974, 1975, 1978), Kluskens and Newman (1991), Cooray and Ciric (1993). Unfortunately, these solutions are expansions on the parameter a/λ, where a is the size of a particle and λ is the wavelength. Thus, we need to account for a lot of rather complex terms in these general solutions for large values of the parameter a/λ (see Appendix 2). For water drops with radius 10 μm at wavelength $\lambda = 0.5$ μm, we should consider 150 terms for a general solution and for snow grains with size 1000 μm this number is 100 times larger. It is possible to handle these cases with modern computers in the case of spherical particles, but for spheroids there is still no progress with numerical computations if the number of terms needed is larger than 100. Other exact methods (Waterman, 1971; Barber and Yeh, 1975; Dau-Sing and Barber, 1979; Barber and Hill, 1990; Hage *et al.*, 1991; Yang and Liou, 1996a; Mischchenko *et al.*, 2000), including the T-matrix approach (Wiscombe and Mugnai, 1986; Mishchenko and Travis, 1994a,b; Paramonov, 1995), have the same trouble.

The solution of this problem lies in the use of the ray optics approximation. Indeed, in the framework of this approach the scattering characteristics can be represented as expansions on the parameter λ/a. This parameter is small for large particles ($a \gg \lambda$) and only a few terms should be considered.

The basis of the geometrical optics approximation (GOA) is very simple. Namely, it is assumed that the energy incident on a large particle is split into parts which are reflected or refracted according to the Fresnel formulae for the plane surfaces because all radii of curvature of a particle are large. Then the diffraction component with the intensity $I^D(\theta)$ is added to the geometrical optics solution. Sometimes the interference of geometrical optics and diffraction components is considered as well (Van de Hulst, 1957). This is of importance for small phase shifts $p = 2x|m - 1|$. However, we will consider only the case $p \gg 1$ in this section. The diffraction component is important in the small-angle region ($\theta \leq \lambda/a$) and it is essentially the same as for the diffraction of light on an aperture of the same area Σ and shape as the area and shape of the particle projection onto the plane perpendicular to the incident radiation (Born and Wolf, 1965).

It follows for the geometrical optics (GO) component $C_{sca}^{GO}(\theta)$ of the differential scattering cross-section:

$$C_{sca}^{GO}(\theta) = A\Sigma[R(\theta) + T(\theta)\, e^{-\gamma\alpha_1(\theta)}(1 + R(\theta)\, e^{-\gamma\alpha_2(\theta)} + R^2(\theta)\, e^{-\gamma\alpha_3(\theta)} + \cdots)T(\theta)]D(\theta)$$

where $A = \text{const}$, Σ is the geometrical cross-section, which is equal to the area of a geometrical projection of a particle into the plane, perpendicular to the direction of propagation, $R(\theta)$ and $T(\theta)$ are the reflected and refracted fluxes, determined by Fresnel equations (Born and Wolf, 1965), $\gamma = 4\pi\chi/\lambda$, χ is the imaginary part of the refractive index of a particle, $\alpha_j(\theta)$ is the path length of the j-ray inside the particle, $D(\theta)$ is the divergence factor (Van de Hulst, 1957). The first, second, and third terms represent the following physical processes respectively:

(1) the first term – reflection of a ray by a surface of a particle;
(2) the second term – refraction of a ray by a surface of a particle → absorption of

light on a pass of a ray inside a particle → refraction of a ray by a surface of a particle into a host medium;

(3) the third term – refraction of a ray by a surface of a particle → absorption of a light on a pass of a ray → reflection of a ray by a surface of a particle back into a particle → absorption of light in a pass of a ray inside a particle → refraction of a ray by a surface of a particle into a host medium.

Generally speaking, the scattering angle θ for a given incident ray depends on the refractive index, shape, internal structure of a particle, and the angle between the normal to the surface of a particle and incident ray direction. For instance, rays incident along the diameter of a spherical particle can travel only in the forward or backward scattering directions. Correspondingly, the scattering angle is equal to 0 or π, depending on the number of interactions of the ray with the surface of a particle.

It follows for the phase matrix $\hat{p}$ in the framework of the geometrical optics approach:

$$\hat{p} = B\Sigma[\hat{R} + \hat{T}e^{-\gamma\alpha_1(\theta)}(1 + \hat{R}e^{-\gamma\alpha_2(\theta)} + \hat{R}\hat{R}e^{-\gamma\alpha_3(\theta)} + \cdots)\hat{T}]D(\theta)$$

where B = constant, $\hat{R}$ and $\hat{T}$ are reflection and transmission matrices, multiplied by rotation matrices (Peltoniemi *et al.*, 1989; Macke, 1994).

The geometrical optics approach was used to calculate optical properties of different types of particles, including spheres (Shifrin, 1951; Van de Hulst, 1957; Kerker, 1969; Newton, 1982; Bohren and Huffman, 1983), spheroids (Liou *et al.*, 1983; Kokhanovsky and Macke, 1997; Kokhanovsky and Nakajima, 1998), hexagonal cylinders (Liou *et al.*, 1983; Liou, 1992; Yang and Liou, 1995, 1996b; Zhang and Xu, 1995; Mishchenko and Macke, 1998), stochastic particles (Peltoniemi *et al.*, 1989), and circular cylinders (Mischenko, *et al.*, 1996).

The Monte Carlo ray-tracing algorithm (MCRTA) was described by Takano and Liou (1989), Peltoniemi *et al.* (1989), Macke (1994), Kokhanovsky and Nakajima (1998), and Yang and Liou (1998). The deterministic Fresnelian interactions between rays and a particle surface are treated as a number of stochastic events in the framework of the MCRTA. When a ray hits a particle surface, a special criterion based on Fresnel coefficients is evaluated to determine whether the reflection or refraction takes place. This criterion has the following form (Yang and Liou, 1998):

$$\Upsilon = \frac{|R_1|^2|h_1|^2 + |R_2|^2|h_2|^2}{|h_1|^2 + |h_2|^2}$$

where $R_1 = (\sin\tau - m\sin\tau')/(\sin\tau + m\sin\tau')$ and $R_2 = (m\sin\tau - \sin\tau')/(m\sin\tau + \sin\tau')$ are Fresnel coefficients, $\cos\tau' = \cos\tau/n$, τ is the incidence angle ($\tau = \pi/2$ means the normal incidence of a ray on the surface of a particle), $m = n - i\chi$ is the relative refractive index of a particle, h_1 and h_2 are incident electromagnetic field components, which are parallel and perpendicular to the incident surface for a given incident polarization configuration. If a random number distributed uniformly in the region of (0,1) is larger than Υ, refraction is expected, otherwise reflection occurs. When a ray finally escapes from a particle, the contribution of this ray to the scattered light field is normalized according to the energy carried by the corresponding initial incident ray.

The attractivity of the MCRTA lies in the possibility of considering very complex shapes, including fractals, plates with sector-like branches, etc. The particle surface roughness can be accounted for by introducing the random distribution of tilt angles of facets (Macke *et al.*, 1996). The influence of internal inclusions (e.g. bubbles in ice crystals) and non- uniformity of particles can also be studied in the framework of the same scheme.

It is assumed that the extinction cross-section C_{ext} is equal to 2Σ in the framework of the geometrical optics approximation. This result is general and does not depend on the shape of a particle. The values of the scattering cross-section

$$C_{sca} = \int_{4\pi} C_{sca}(\Omega)\, d\Omega$$

and the asymmetry parameter

$$g = \frac{\int_{4\pi} \cos\theta C_{sca}(\Omega)\, d\Omega}{\int_{4\pi} C_{sca}(\Omega)\, d\Omega}$$

can be obtained by the integration of the differential cross-section $C_{sca}(\Omega) = C_{sca}^{GO}(\Omega) + C_{sca}^{D}(\Omega)$ on the entire sphere. Here the values of C_{sca}^{GO}, C_{sca}^{D} are contributions of the geometrical optics and diffraction terms to the full value of C_{sca}, respectively.

Let us consider now the optical properties of large particles of different shapes in more details.

2.3.2 Spheres

2.3.2.1 The differential scattering matrix

It follows for the elements of the amplitude matrix (1.65) of spherical particles of the radius a within the framework of the geometrical optics approximation (Van de Hulst, 1957; Newton, 1982):

$$S_{11}^{GO} = x \sum_{p=0}^{\infty} \varepsilon_{2p} \sqrt{D_p} \exp(i\beta_2 - \zeta),$$

$$S_{22}^{GO} = x \sum_{p=0}^{\infty} \varepsilon_{1p} \sqrt{D_p} \exp(i\beta_1 - \zeta),$$

$$S_{12}^{GO}(\theta) = S_{21}^{GO}(\theta) = 0 \tag{2.22}$$

where

$$\varepsilon_{jp} = \begin{cases} R_j, & p = 0 \\ (1 - R_j^2)(-R_j)^{p-1}, & p = 1, 2, 3, \ldots \end{cases}$$

$x = ka$ is the size parameter, $k = 2\pi/\lambda$, $R_1 = (\sin\tau - m\sin\tau')/(\sin\tau + m\sin\tau')$, $R_2 = (m\sin\tau - \sin\tau')/(m\sin\tau + \sin\tau')$, $\zeta = pc\sin\tau'/2$, $c = 4\chi x$, $\cos\tau' = (\cos\tau)/n$, τ is the incidence angle ($\tau = \pi/2$ means the normal incidence of a ray on the surface of a particle), $m = n - i\chi$ is the relative refractive index of a particle, β_1 and β_2 are phases. Expressions for β_1 and β_2 are given by van de Hulst (1957), who also showed that $\beta_1 = \beta_2$ for nonabsorbing particles. The divergency factor

$$D_p = \frac{\sin 2\tau\, d\tau\, d\varphi}{2\sin\theta\, d\theta\, d\varphi}$$

can be presented in the following form:

$$D_p = \frac{\sin 2\tau}{4\psi(\tau)\sin\theta_p}$$

where $\psi(\tau) = |1 - p\tan\tau/\tan\tau'|$, $\theta_p = q(\theta' - 2\pi N)$ is the scattering angle, $\theta' = 2(\tau - p\tau')$ is the deflection angle. The integer numbers N and $q = \pm 1$ define the scattering angle θ_p in the interval $[0, \pi]$.

To obtain the full amplitude matrix $\hat{S}(\theta) = \hat{S}^{GO}(\theta) + S^D(\theta)\hat{I}_2$, where $\hat{I}_2$ is the unity matrix, wc should add to equations (2.22) the diffraction contribution (Van de Hulst, 1957):

$$S^D(\theta) = \frac{xJ_1(x\theta)}{\theta} \tag{2.23}$$

where $J_1(x\theta)$ is the Bessel function.

The Fraunhofer diffraction patterns for other shapes of particles were studied by many authors (e.g. Shifrin *et al.*, 1984; Bohren and Koh, 1985; Jones, 1988; Uno *et al.*, 1993; Borovoi *et al.*, 1998; Muhlenweg and Hirleman, 1998).

Equations (2.22) can be written in the following form for S_{11}^{GO}, S_{22}^{GO}:

$$S_{jj}^{GO} = M_j \exp(it_j) \tag{2.24}$$

where M_j are real numbers and $t_1 = t_2$ for nonabsorbing particles. It follows for the elements of the Stokes matrix of nonabsorbing particles, neglecting the interference between diffracted and geometrical optics beams (see equation (1.75)), that:

$$\hat{C}_{sca} = \frac{(S^D)^2\hat{I}_4 + \hat{G}}{k^2}$$

where $\hat{I}_4$ is the 4×4 unity matrix and

$$\hat{G} = \begin{pmatrix} \dfrac{M_1^2 + M_2^2}{2} & \dfrac{M_1^2 - M_2^2}{2} & 0 & 0 \\ \dfrac{M_1^2 - M_2^2}{2} & \dfrac{M_1^2 + M_2^2}{2} & 0 & 0 \\ 0 & 0 & M_1M_2 & 0 \\ 0 & 0 & 0 & M_1M_2 \end{pmatrix} \tag{2.25}$$

Thus, it follows that $C_{sca}^{33} = C_{sca}^{44}$ and $C_{sca}^{34} = C_{sca}^{43} = 0$ in the approximation under study. Elements C_{sca}^{34}, C_{sca}^{43} are due to the interference of the diffracted and geometrical optics components and phase differences $t_1 - t_2$, which are not considered here.

It follows for the intensity of the scattered light outside the diffraction peak in the framework of the geometrical optics approximation:

$$I^{GO}(\theta) = \frac{I_0(i_1^{GO} + i_2^{GO})}{2k^2r^2} \tag{2.26}$$

where r is the distance to the point of observation ($r \gg \lambda$), I_0 is the intensity of the incident natural light and $i_j(\theta) = |S_{jj}(\theta)|^2$. Thus, we obtain from equation (2.22), neglecting the interference of the geometrical optics beams (Glautshing and Chen, 1981), that:

$$i_j^{GO}(\theta) = x^2\left[r_j + (1 - r_j)^2 \sum_{p=1}^{\infty} r_j^{p-1} D_p e^{-pc \sin \tau'}\right] \tag{2.27}$$

where $r_j = |R_j|^2$.

It follows from equation (2.23) for the diffraction contribution to the scattered light intensity:

$$i^D(\theta) = \frac{x^4}{4} F(x\theta)$$

The function

$$F(y) = \frac{4J_1^2(y)}{y^2}$$

is given in Figure 2.5. We have: $F(0) = 1$, $F(\infty) = 0$. It follows that the diffraction contribution to the scattered signal is restricted mostly to angles $\theta \ll 1/x$. Summing up, for the intensity of scattered light we have as $x \to \infty$, $2|(m-1)x| \to \infty$:

$$I = \frac{i_1^{GO} + i_2^{GO} + 2i^D}{2k^2r^2} I_0$$

Note that $i^D(0) \sim x^4$ and $i^{GO}(0) \sim x^2$, meaning that the scattered light intensity is dominated by the diffraction process at small angles in the framework of the approximation under study. The phase function obtained in the framework of this approximation as compared to Mie results is given in Figure 2.6(a).

We can also obtain for the degree of polarization of unpolarizecd incident light in the geometrical optics approximation:

$$P = \frac{i_1^{GO} - i_2^{GO}}{i_1^{GO} + i_2^{GO}}$$

Note that the subscript 1 indicates the component perpendicular to the scattering plane (see expression for R_1 above). Therefore, the degree of polarization is positive for $i_1^{GO} > i_2^{GO}$.

The dependence $P(\theta)$ is given in Figure 2.6(b) at $n = 1.333$. For comparison we also show $P(\theta)$, found using exact Mie theory at $a_{ef} = 15\,\mu\text{m}$. The differences mostly occur in the backward scattering region.

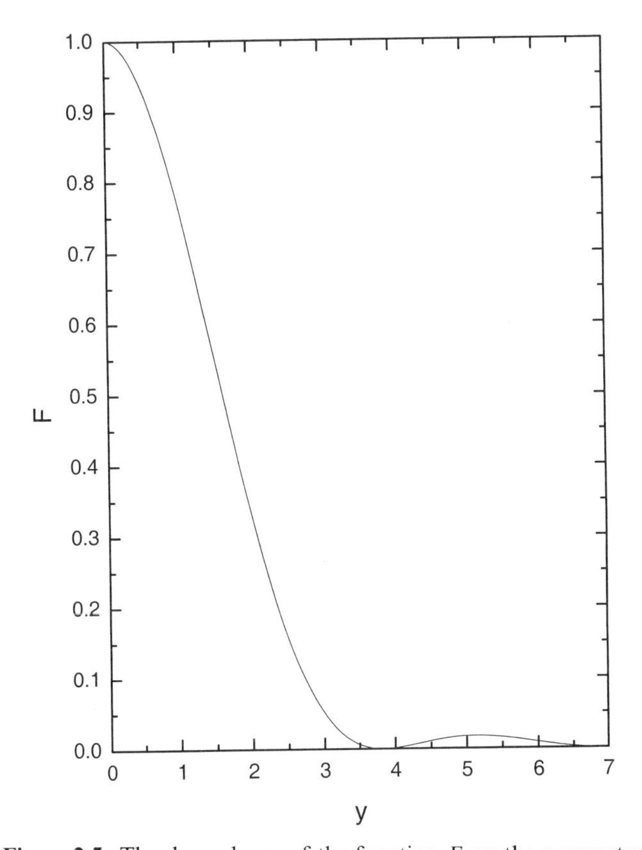

Figure 2.5. The dependence of the function F on the parameter y.

The normalized element $p_{33} = p_{44} = 2M_1M_2/(M_1^2 + M_2^2)$ of the matrix (2.25) is given in Figure 2.6(c). The physical meaning of this element is very simple. It gives the degree of circular polarization of scattered light under the assumption that incident light is completely right-hand circularly polarized (Kokhanovsky, 2003b).

2.3.2.2 *Cross-sections*

The geometrical optical scattering cross-section:

$$C_{sca}^{GO} = \int_0^{2\pi} d\varphi \int_0^{\pi} d\theta \sin\theta \frac{I^{GO}(\theta)}{I_0} r^2$$

can be calculated from the formula for $I^{GO}(\theta)$ (2.26):

$$C_{sca}^{GO} = \pi a^2 \omega \tag{2.28}$$

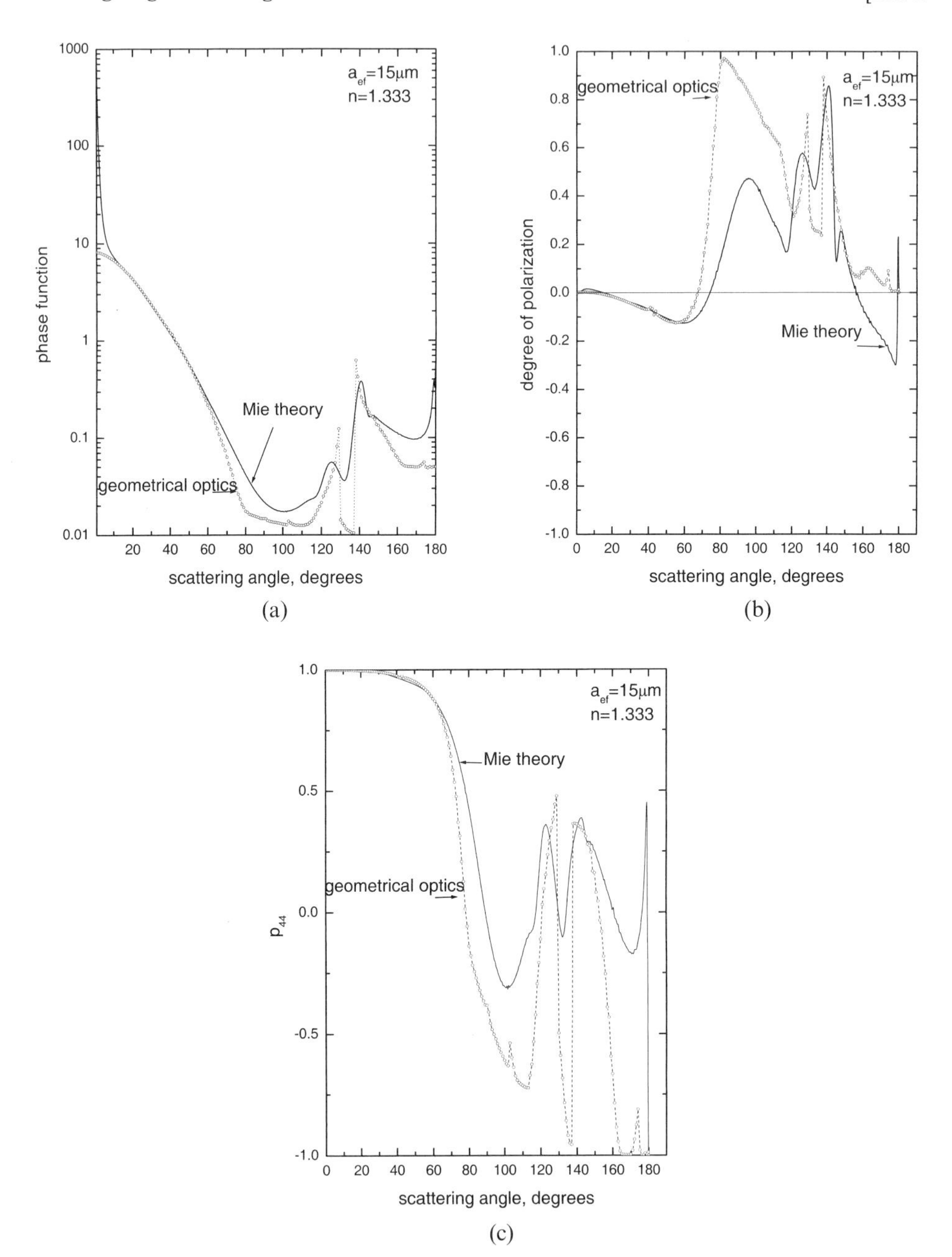

Figure 2.6. (a) The phase function, (b) the degree of polarization, and (c) the phase matrix element p_{44} calculated using the geometrical optics approach and Mie theory for spherical polydispersions of water droplets having an effective radius of 15 μm. The coefficient of variance of the droplet size distribution is equal to 38%. The refractive index is assumed to be equal 1.333 and the wavelength is equal to 0.55 μm.

where

$$\omega = \tfrac{1}{2}(\omega_1 + \omega_2) \tag{2.29}$$

$$\omega_j = \int_0^{\pi/2} \left(r_j + \frac{(1 - r_j)^2 e^{-c\xi(\tau)}}{1 - r_j e^{-c\xi(\tau)}} \right) \sin 2\tau \, d\tau \tag{2.30}$$

and $\xi(\tau) = \sqrt{1 - (\cos^2 \tau)/n^2}$. We have used the formula for the geometric infinite progression and the equality $\sin\theta \, d\theta = 2 \sin\theta_\rho \psi(\tau) \, d\tau$ to find the formulae for ω_j.

We can obtain $C_{sca}^{GO} = \pi a^2$ at $c = 0$ from equation (2.28). It follows as $c \to \infty$ (Cohen and Tirosh, 1990) that:

$$C_{sca}^{GO} = \pi a^2 W \tag{2.31}$$

where

$$W = \frac{1}{2} \int_0^{\pi/2} (r_1 + r_2) \sin 2\tau \, d\tau \tag{2.32}$$

Thus, the total scattering cross-section, including the diffraction term $C_{sca}^D = \pi a^2$, is:

$$C_{sca} = \pi a^2 (1 + W)$$

and, correspondingly, the absorption cross-section is given by:

$$C_{abs} = \pi a^2 (1 - W)$$

where we accounted for the fact that the extinction cross-section:

$$C_{ext} = 2\pi a^2$$

for spherical particles of any chemical composition at $x \gg 1$, $2x|m - 1| \gg 1$.

Integral (2.32) can be evaluated analytically at $\chi \ll n$ (Gershun, 1937; Zolotova and Shifrin, 1993):

$$W = W_1 \ln n + W_2 \ln\left(\frac{\Delta n}{n+1}\right) + W_3 \tag{2.33}$$

where at $n > 1$:

$$\left.\begin{aligned} &\Delta n = n - 1, \qquad W_1 = \frac{8n^4(n^4 + 1)}{(n^4 - 1)^2(n^2 + 1)}, \qquad W_2 = \frac{n^2(n^2 - 1)^2}{(n^2 + 1)^3} \\ &W_3 = \frac{\sum_{j=0}^{7} A_j n^j}{3(n^4 - 1)(n^2 + 1)(n + 1)}, \qquad A_j = (-1, -1, -3, 7, -9, -13, -7, 3) \end{aligned}\right\} \tag{2.34}$$

The total scattering cross-section C_{sca}, which includes the diffraction term, is determined by the following formula (Van de Hulst, 1957):

$$C_{sca} = \pi a^2 (1 + \omega) \tag{2.35}$$

where the well-known fact that the total diffraction cross-section of the aperture is equal to its area was used. It follows for the absorption cross-section (see equation 2.35)) that:

$$C_{abs} = \frac{\pi a^2}{2} \sum_{j=1}^{2} \int_0^{\pi/2} \frac{(1 - r_j)(1 - \exp(-c\xi))}{1 - r_j \exp(-c\xi)} \sin 2\, d\tau \tag{2.36}$$

where we accounted that $C_{ext} = 2\pi a^2$ for large particles. This expression was used by Sharma and Jones (2000) to study the absorption of light by highly absorbing inclusions in water droplets. Note that integral (2.36) can be transformed to:

$$C_{abs} = \frac{\pi a^2}{2} \sum_{j=1}^{2} \int_0^{1} \frac{(1 - r_j(\sigma))(1 - \exp(-c\xi(\sigma)))}{1 - r_j(\sigma) \exp(-c\xi(\sigma))} d\sigma$$

and evaluated using the Simpson quadrature formula. It follows for the functions $r_1(\sigma)$, $r_2(\sigma)$, $\xi(\sigma)$ that:

$$r_1(\sigma) = \left| \frac{\sqrt{1-\sigma} - m\sqrt{1-\sigma/m^2}}{\sqrt{1-\sigma} + m\sqrt{1-\sigma/m^2}} \right|^2, \qquad r_2(\sigma) = \left| \frac{m\sqrt{1-\sigma} - \sqrt{1-\sigma/m^2}}{m\sqrt{1-\sigma} + \sqrt{1-\sigma/m^2}} \right|^2,$$

$$\xi(\sigma) = \sqrt{1 - \sigma/n^2}$$

where we used the substitution $\sigma = \cos^2 \tau$.

This integral was approximately evaluated by Bohren and Nevitt (1983):

$$C_{abs} = \frac{4n\pi a^2 Q(c)}{(n+1)^2 - (n-1)^2 e^{-c}} \tag{2.37}$$

where

$$Q(c) = \frac{2n^2}{c^2}\{e^{-bc}(1+bc) - e^{-c}(1+c)\}, \qquad b = \sqrt{1 - n^{-2}} \tag{2.38}$$

Note, that it follows as $n \to 1$, $r_j \to 0$ that:

$$\omega_j = \int_0^{\pi/2} e^{-c\sqrt{1-\cos^2\tau}} \sin 2\tau \, d\tau \tag{2.39}$$

We can obtain from equation (2.39) after integration:

$$\omega_j = \frac{2}{c^2}[1 - (1+c)\, e^{-c}] \tag{2.40}$$

Thus, the formulae for the absorption C_{abs} and scattering C_{sca} cross-sections in the framework of this approximation are:

$$C_{sca} = \pi a^2 \left\{ 1 + \frac{2}{c^2}[1 - (1+c)\, e^{-c}] \right\}, \qquad C_{abs} = \pi a^2 \left\{ 1 - \frac{2}{c^2}[1 - (1+c)\, e^{-c}] \right\} \tag{2.41}$$

Note that it follows from equations (2.41) that:

$$C_{ext} \equiv C_{abs} + C_{sca} = 2\pi a^2 \tag{2.42}$$

as it should be. The formula for C_{abs} coincides with the result, obtained within the framework of the Van de Hulst approximation for optically soft particles (Van de Hulst, 1957). It follows that:

$$C_{abs} = \frac{4\pi\chi}{\lambda} V \tag{2.43}$$

as $c \to 0$ and $n \to 1$. Here V is the volume of a particle. It appears that this equation also holds for nonspherical particles as $n \to 1$. Simple formula (2.43) follows from general equation (1.93), assuming that $\vec{E}(\vec{r}') = \vec{E}_0$ (i.e. the electric field inside the particle is equal to that of an incident field), which is a good approximation as $c \to 0$ and $n \to 1$. The same equation can be obtained in the framework of the Rayleigh–Gans approximation (Van de Hulst, 1957).

The simple approximation for the absorption cross-section can be derived from the integral for C_{abs} at small values of c and any n. It follows in this case (Bohren and Huffman, 1983) that:

$$\frac{1 - r_j}{1 - r_j e^{-c\xi}} \approx 1 \tag{2.44}$$

and we can obtain (see equation (2.36)):

$$C_{abs} = \pi a^2 \int_0^{\pi/2} \left\{ 1 - \exp\left(-c\sqrt{1 - \left(\frac{\cos\tau}{n}\right)^2} \right) \right\} \sin 2\tau \, d\tau \tag{2.45}$$

or

$$C_{abs} = \pi a^2 \left\{ 1 - \frac{2n^2}{c^2} (e^{-bc}(1 + cb) - e^{-c}(1 - c)) \right\} \tag{2.46}$$

where $b = \sqrt{1 - n^{-2}}$. It appears as $c \to 0$ that:

$$C_{abs} = f(n) \frac{4\pi\chi}{\lambda} V \tag{2.47}$$

where $f(n) = n^2[1 - (1 - n^{-2})^{3/2}]$ (Bohren and Nevitt, 1983; Bohren and Huffman, 1983). Equation (2.43) follows from equation (2.47) as $n \to 1$. The multiplier $f(n)$ accounts for the enhancement of absorption due to the focusing effect of light by particles. The pass of a ray inside the particle increases due to this phenomenon. It should lead to higher values of $f(n)$ for larger values of the refractive index n. Note that it follows that $f(1.33) = 1.26$, $f(1.5) = 1.58$. Indeed, the value of f increases with n, as one may expect.

Thus, the change of the relative refractive index of particles n, will change their absorption characteristics. In particular, sand grains in air are more absorbing particles that the same grains in water.

It follows from equations (2.41), (2.46) that $C_{abs} \to \pi a^2$ as $c \to \infty$ instead of $C_{abs} \to (1 - W)\pi a^2$ as it should be. The approximate formula with correct limits at small and large values of c was proposed by Shifrin (1951):

$$C_{abs} = \pi a^2 (1 - W)(1 - e^{-c}) \tag{2.48}$$

and by Kokhanovsky and Zege (1995):

$$C_{abs} = \pi a^2 \{1 - Q(c) - W(1 - e^{-cb})^2\} \tag{2.49}$$

The last formula gives an accurate approximation of the geometrical optics integral for the value of C_{abs}. The error is less than 5% at any c, n. Note that it follows for the value of the scattering cross-section that:

$$C_{sca} = \pi a^2 \{1 + Q(c) + W(1 - e^{-cb})^2\} \tag{2.50}$$

within the framework of the same approximation.

Note that the function $Q(c)$ (2.38) can be approximated by the following simple equation (Shifrin and Tonna, 1992):

$$Q = \exp\left(-\tfrac{2}{3}c\right)$$

It follows from equation (2.49) under this assumption that:

$$C_{abs} = \pi a^2 \left\{1 - \exp\left(-\frac{2c}{3}\right) - W(1 - e^{-cb})^2\right\} \tag{2.51}$$

2.3.2.3 *The asymmetry parameter*

The next important characteristic of light scattering by large particles is the asymmetry parameter. It is defined by the following equation:

$$g = \frac{1}{2}\int_0^\pi p(\theta) \sin\theta \cos\theta \, d\theta \tag{2.52}$$

where $p(\theta)$ is the phase function $\left(\frac{1}{2}\int_0^\pi p(\theta)\sin\theta\, d\theta = 1\right)$. Within the framework of the geometrical optics approximation, taking account of diffraction, it follows that:

$$p(\theta) = \frac{C_{sca}^D p^D(\theta) + C_{sca}^{GO} p^{GO}(\theta)}{C_{sca}^D + C_{sca}^{GO}}, \qquad g = \frac{C_{sca}^D g^D + C_{sca}^{GO} g^{GO}}{C_{sca}^D + C_{sca}^{GO}} \tag{2.53}$$

$$p^D(\theta) = \frac{4\pi i^D}{k^2 C_{sca}^D}, \qquad g^D = \frac{1}{2}\int_0^\pi p^D(\theta)\sin\theta\cos\theta\, d\theta, \qquad C_{sca}^D = \pi a^2 \tag{2.54}$$

$$p^{GO}(\theta) = \frac{2\pi(i_1^{GO} + i_2^{GO})}{k^2 C_{sca}^{GO}}, \qquad g^{GO} = \frac{1}{2}\int_0^\pi p^{GO}(\theta)\sin\theta\cos\theta\, d\theta \tag{2.55}$$

where $i_j^{GO} = |S_j^{GO}(\theta)|^2$, $i^D = |S^D(\theta)|^2$. The function $p^{GO}(\theta)$ for different values of the refractive index n at $\chi = 0$ is presented in Appendix 5 and Figure 2.7. It does not depend on the size of particles. It follows that:

$$p^D(\theta) = F(\theta x)x^2 \tag{2.56}$$

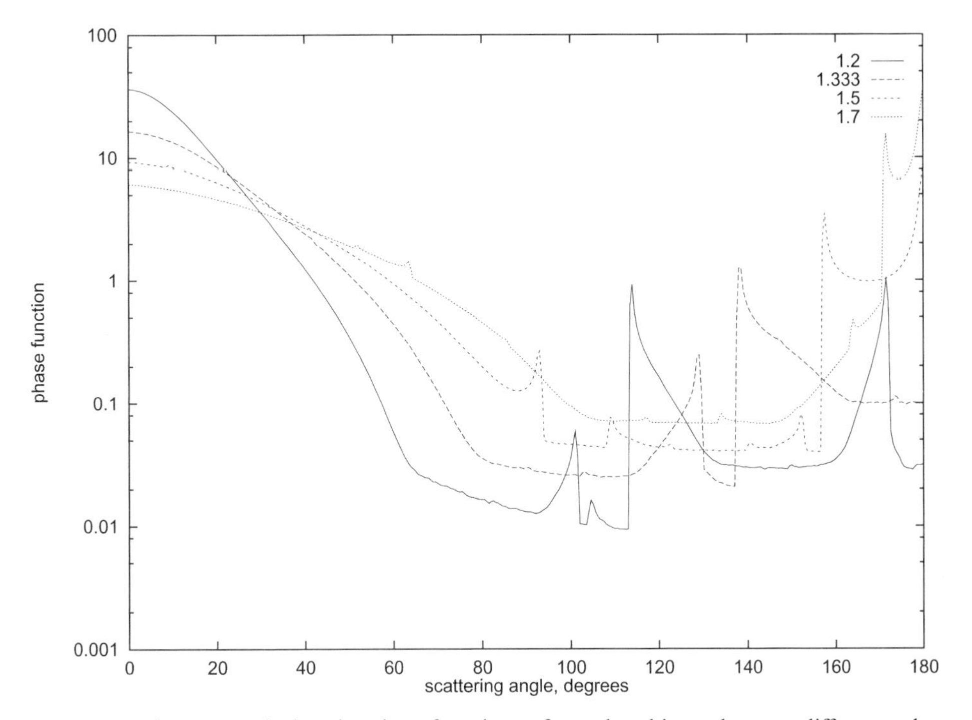

Figure 2.7. The geometrical optics phase functions of nonabsorbing spheres at different values of the refractive index $n = 1.2$, 1.333, 1.5, 1.7.

where $F(\theta x) = (4J_1^2(\theta x))/(\theta x)^2$. The vadility of this approximation for large spheres was studied by Jones (1977). We obtain for the asymmetry parameter g^D as $x \to \infty$:

$$g^D = \frac{1}{2}\int_0^{\pi} p^D(\theta)\sin\theta\cos\theta\,d\theta \approx \frac{x^2}{2}\int_0^{\infty} F(\theta x)\theta\left(1-\frac{\theta^2}{2}\right)d\theta$$
$$= \frac{1}{2}\int_0^{\infty} F(Z)Z\left(1-\frac{Z^2}{2x^2}\right)dZ \to \frac{1}{2}\int_0^{\infty} F(Z)Z\,dZ \to 1 \qquad (2.57)$$

where $Z = \theta x$. Thus, the asymmetry parameter of the diffracted light is close to 1. It follows from equation (2.53), accounting for this fact, that:

$$g = \frac{1+\omega g^{GO}}{1+\omega} \qquad (2.58)$$

where

$$g^{GO} = \frac{1}{2\omega}\sum_{j=1}^{2}\int_0^{\pi/2} \frac{d_j(\tau)\sin 2\tau\,d\tau}{1-2r_j\,e^{-c\xi}\cos 2\tau' + r_j^2\,e^{-2c\xi}} \qquad (2.59)$$

where

$$d_j(\tau) = e^{-c\xi}(1-r_j)^2 \cos 2(\tau - \tau') + r_j \cos 2\tau(1 - e^{-2c\xi}) + 2r_j^2(e^{-c\xi} - \cos 2\tau')\cos 2\tau\, e^{-c\xi}$$

and

$$\xi = \sqrt{1 - \frac{\cos^2\tau}{n^2}}, \qquad \tau' = \arccos\left(\frac{\cos\tau}{n}\right)$$

Equation (2.59) simplifies for nonabsorbing particles. Namely, it follows at $c = 0$ that:

$$g_0^{GO} \equiv g^{GO}(c=0) = \frac{1}{2}\sum_{j=1}^{2}\int_0^{\pi/2} \frac{[(1-r_j)^2\cos 2(\tau' - \tau)\sin 2\tau + r_j^2(1 - \cos 2\tau')\sin 4\tau]\, d\tau}{1 - 2r_j\cos 2\tau' + r_j^2}$$

Unfortunately, it is not possible to express the value of g^{GO} by means of elementary functions at any c. However, it follows as $c \to \infty$ that:

$$g^{GO} = \int_0^{\pi/2} (r_1 + r_2)\sin 4\tau\, d\tau \tag{2.60}$$

This integral can be evaluated analytically. It follows after the integration at $\chi \ll n$ (Barun and Gavrilovich, 1987) that:

$$g_\infty^{GO} = g^{GO}(c \to \infty) = \frac{1}{W}\left\{1 + g_1 \ln n - g_2 \ln\frac{n+1}{n-1} + g_3\right\} \tag{2.61}$$

where

$$\left.\begin{aligned} g_1 &= \frac{8n^4(n^6 - 3n^4 + n^2 - 1)}{(n^4-1)^2(n^2+1)^2}, \\ g_2 &= \frac{(n^2-1)^2(n^8 + 12n^6 - 54n^4 - 4n^2 + 1)}{16(n^2+1)^4} \\ g_3 &= \frac{\sum_{j=1}^{12} B_j n^j}{24(n^2+1)^2(n^4-1)(n+1)}, \\ B_j &= (-3, 13, -89, 151, 186, 138, -282, 22, 25, 25, 3, 3) \end{aligned}\right\} \tag{2.62}$$

Thus, the asymmetry parameter g_∞ can be represented by the following equation (see Table 2.2) as $c \to \infty$:

$$g_\infty = \frac{1 + W g_\infty^{GO}}{1 + W} \tag{2.63}$$

We can obtain at $c = 0$ (see Tables 2.2, 2.3):

$$g_0 = \frac{1 + g_0^{GO}}{2} \tag{2.64}$$

Table 2.2. Parameters W, g_0, g_∞, and y for different values of the refractive index n.

n	W	g_0	g_∞	y
1.1	0.0252	0.9731	0.9946	0.5180
1.2	0.0443	0.9341	0.9856	0.6528
1.25	0.0529	0.9147	0.9806	0.6948
1.3	0.0611	0.8961	0.9751	0.7280
1.333	0.0664	0.8843	0.9714	0.7468
1.34	0.0675	0.8818	0.9706	0.7505
1.35	0.0691	0.8783	0.9695	0.7555
1.4	0.0768	0.8613	0.9638	0.7785
1.45	0.0844	0.8542	0.9579	0.7985
1.5	0.0918	0.8299	0.9520	0.8160
1.55	0.0991	0.8154	0.9460	0.8315
1.60	0.1063	0.8015	0.9400	0.8453
1.65	0.1133	0.7884	0.9340	0.8580
1.70	0.1203	0.7759	0.9280	0.8695
1.90	0.1475	0.7314	0.9046	0.9080
2.00	0.1606	0.7121	0.8933	0.9340
2.10	0.1734	0.6945	0.8823	0.9383

Table 2.3. The dependence of the asymmetry parameter g_0 on the value of n at $n < 1$.

n^{-1}	1.1	1.2	1.3	1.333	1.4	1.5	1.6	1.7	2.0
g_0	0.9670	0.9139	0.8609	0.8444	0.8132	0.7718	0.7365	0.7064	0.6408

The values of W, g_0, g_∞, as functions of the real part of the refractive index n are presented in Tables 2.2, 2.3.

The formula for the value of g at any c was proposed by Kokhanovsky and Zege (1995):

$$g = g_\infty - (g_\infty - g_0)\, e^{-y(n)c} \tag{2.65}$$

where $y(n)$ is the fitting parameter (see Table 2.2).

Formulae presented here can be used to estimate other scattering characteristics; namely, the single scattering albedo $\omega_0 = C_{abs}/C_{ext}$, light pressure cross-section $C_{pr} = C_{ext} - gC_{sca}$, the average scattering angle $\langle\theta\rangle$:

$$\langle\theta\rangle = \frac{\int_0^\pi \theta p(\theta) \sin\theta \, d\theta}{\int_0^\pi p(\theta) \sin\theta \, d\theta} \tag{2.66}$$

Table 2.4. Asymptotic fluxes scattered by spherical particles between two angles at $c = 0(F_0)$ and $c \to \infty(F_\infty)$ as functions of the refractive index.

n	$F_0(\pi/2)$	$F_\infty(\pi/2)$	$F_0(\pi/4)$	$F_\infty(\pi/4)$
1.05	0.056	0.0004	0.0062	0.0016
1.10	0.0104	0.0013	0.0124	0.0051
1.15	0.0149	0.027	0.0211	0.0093
1.20	0.0192	0.0045	0.0359	0.0140
1.25	0.0235	0.0065	0.0554	0.0189
1.30	0.0277	0.0088	0.0775	0.0240
1.33	0.0302	0.0103	0.0913	0.0271
1.40	0.0359	0.0140	0.1234	0.0344
1.45	0.0363	0.0167	0.1451	0.0397
1.50	0.0401	0.0195	0.1656	0.0450
1.53	0.0424	0.0213	0.1772	0.0481
1.60	0.0483	0.0254	0.2020	0.0554
1.65	0.0528	0.0284	0.2181	0.0606
1.70	0.0577	0.0314	0.2327	0.0657
1.75	0.0628	0.0344	0.2461	0.0708
1.80	0.0681	0.0374	0.2584	0.0758
1.85	0.0736	0.0404	0.2696	0.0807
1.90	0.0793	0.0433	0.2799	0.0856
1.95	0.0850	0.0463	0.2894	0.0904
2.00	0.0907	0.0492	0.2981	0.0951

the mean square angle $\langle\theta^2\rangle$:

$$\langle\theta^2\rangle = \frac{\int_0^\pi \theta^2 p(\theta) \sin\theta \, d\theta}{\int_0^\pi p(\theta) \sin\theta \, d\theta} \tag{2.67}$$

and the light flux in the scattering angle range $\theta \in [0, \theta_0]$:

$$F(\theta_0) = \frac{1}{2}\int_0^{\theta_0} p(\theta) \sin\theta \, d\theta \tag{2.68}$$

For instance, results for values of $F(\pi/4)$, $F(\pi/2)$ of large nonabsorbing ($c \to 0$) and highly absorbing ($c \to \infty$) particles are presented in Table 2.4. They were obtained by numerical integration of the geometrical optics phase function $p^{GO}(\theta)$. Note that we can obtain the following analytical result as $c \to \infty$:

$$F(\theta_0) = 1 - \frac{1 + A(\theta_0)}{1 + W} \tag{2.69}$$

where

$$A(\theta_0) = (n^2-1)^{-2}[\tfrac{4}{3}z^6 + 2\alpha z^4 + \gamma^2\beta^2 z^2 - 2n^2\beta\gamma^2 z\nu - \tfrac{4}{3}z^3\nu^3 - 4n^6 z\alpha\gamma^2(n^2 z+\nu)^{-1}$$
$$+ 8n^4\beta\gamma^3\ln((n^2 z+\nu)/\mu) - 2n^2(n^8+6n^4+1)\gamma^3\ln((z+\nu)/\mu)]$$

$z = \sin\theta/2$, $\alpha = n^2-1$, $\beta = n^4+1$, $\gamma = (n^2+1)^{-1}$, $\nu = \sqrt{n^2+n^2-1}$, $\mu = \sqrt{n^2-1}$. Values of $F(\pi/2)$ are important parameters of various approximations of the radiative transfer theory (Wiscombe and Grams, 1976; Zege *et al.*, 1991; Marshall *et al.*, 1995).

2.3.2.4 *Expansion of the phase function on Legendre polynomials*

Let us find approximate equations for the values of the coefficients of the expansion of the phase function on the Legendre polynomials:

$$p(\theta) = \sum_{l=0}^{\infty} x_l P_l(\cos\theta) \tag{2.70}$$

This expansion is often used in radiative transfer problems (Sobolev, 1972; Garcia and Siewert, 1985). The coefficients of this expansion:

$$x_l = \frac{2l+1}{2}\int_0^{\pi} p(\theta)P_l(\cos\theta)\sin\theta\,d\theta \tag{2.71}$$

can be calculated in the geometrical optics limit as well. Let us consider this problem in detail.

We can obtain in the framework of the geometrical optics approximation (see equation (2.53)):

$$x_l = \frac{C^D_{sca}x^D_l + C^{GO}_{sca}x^{GO}_l}{C^D_{sca} + C^{GO}_{sca}} \tag{2.72}$$

where

$$x^D_l = \frac{2l+1}{2}\int_0^{\pi} p^D(\theta)P_l(\cos\theta)\sin\theta\,d\theta \tag{2.73}$$

$$x^{GO}_l = \frac{2l+1}{2}\int_0^{\pi} p^{GO}(\theta)P_l(\cos\theta)\sin\theta\,d\theta \tag{2.74}$$

It is straightforward to find analytical equations for values of x^D_l and x^{GO}_l. First of all it should be noted that the function $p^D(\theta) \ll 1$ at large values of θ. Thus, only the small region of the integration at $\theta \approx 0$ is important in the expression for x^D_l. It follows as $\theta \to 0$ that:

$$P_l(\theta) \approx J_0((l+\tfrac{1}{2})\theta), \qquad p^D(\theta) = \frac{4J_1^2(\theta x)}{\theta^2}, \qquad \sin\theta \approx \theta \tag{2.75}$$

and (see equation (2.73))

$$x^D_l = (2l+1)F(l) \tag{2.76}$$

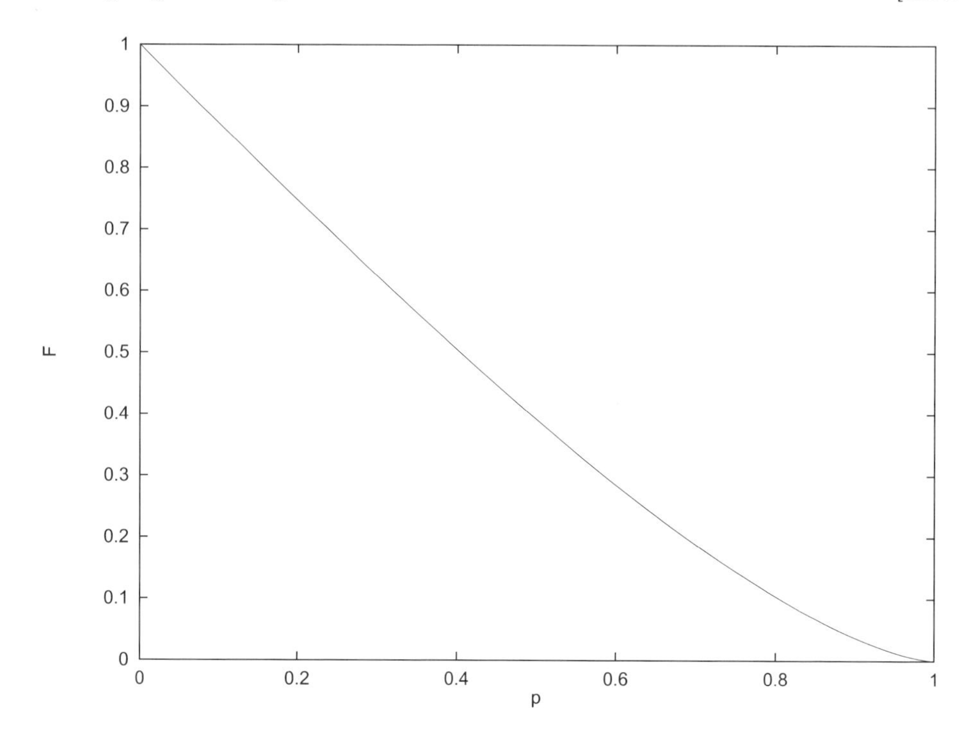

Figure 2.8. Dependence of the function F on the parameter $p = l/(2x)$.

where the integral

$$F(l) = 2\int_0^{\pi} J_0(\theta(l+\tfrac{1}{2}))J_1^2(\theta x)\theta^{-1}\,d\theta$$

can be evaluated analytically, extending the upper limit to infinity. This is possible due to a sharp pike of the integrand at small scattering angles. Thus, it follows that:

$$F(l) = \begin{cases} \dfrac{2}{\pi}\left\{\arccos\left(\dfrac{l}{2x}\right) - \left(\dfrac{l}{2x}\right)\sqrt{1-\left(\dfrac{l}{2x}\right)^2}\right\}, & l \le 2x \\ 0, & l > 2x \end{cases} \tag{2.77}$$

We can see that the function F depends on the only parameter $p = l/(2x)$. This dependence is presented in Figure 2.8.

To obtain the analytical solution for the values of x_l^{GO} the following expansion can be used:

$$P_l(\cos\theta) = \sum_{s=0}^{l} \varphi_l(s)\exp(i(2s-l)\theta) \tag{2.78}$$

where

$$\varphi_l(s) = \frac{\Gamma(s+\frac{1}{2})\Gamma(l-s+\frac{1}{2})}{\pi\Gamma(s+1)\Gamma(l-s+1)} \tag{2.79}$$

Here $\Gamma(a)$ is the gamma function.

After substitution of this formula into the integral for the value of x_l^{GO} it follows that:

$$x_l^{GO} = (2l+1)\sum_{s=0}^{l} \varphi_l(s) I(|2s-l|) \tag{2.80}$$

where

$$I(\alpha) = \frac{1}{2\omega}\sum_{j=1}^{2}\int_0^{\pi/2} \frac{D_j(\alpha,\tau)\sin 2\tau\, d\tau}{1-2r_j e^{-c\xi}\cos(2\alpha\tau') + r_j^2 e^{-2c\xi}} \tag{2.81}$$

and

$$D_j(\alpha,\tau) = r_j(1-e^{-2c\xi})\cos 2\alpha\tau + (1-r_j)^2 e^{-c\xi}\cos 2\alpha(\tau-\tau') + 2r_j^2 e^{-c\xi}(e^{-c\xi} - \cos 2\alpha\tau')\cos(2\alpha\tau)$$

Note that values of x_l^D are much larger than x_l^{GO} at $l > 10$. Thus, it follows at $1 \ll l < 2x$ that:

$$x_l = \frac{2(2l+1)}{\pi(1+\omega)}\left\{\arccos\left(\frac{l}{2ka}\right) - \left(\frac{l}{2ka}\right)\sqrt{1-\left(\frac{l}{2ka}\right)^2}\right\} \tag{2.82}$$

We can see that values of x_l at large values l are determined mostly by the diffraction part of the scattered field. The geometrical optics scattering is important only at small values of l.

Note that equation (2.59) follows from equations (2.80), (2.81) at $l = 1$ ($g^{GO} = x_1^{GO}/3$). The accuracy of equations (2.72), (2.76), (2.80) was studied by Kokhanovsky (1998).

2.3.2.5 *The glory*

As mentioned before, the geometrical optics approximation cannot be used at scattering angles, where there is a high concentration of light scattered energy. In this case other approximations should be used (Van de Hulst, 1957). For instance, it was shown by Kokhanovsky and Zege (1997b) that for the phase function of polydispersions of large transparent particles at large scattering angles $\theta(u = \pi - \theta \ll 1)$ it follows that:

$$p(\theta) = p(\pi)\frac{\langle a^2 J_0^2(kau)\rangle}{\langle a^2\rangle} + q\frac{\langle a^2 J_2^2(kau)\rangle}{\langle a^2\rangle}$$

where J_0 and J_2 are Bessel functions, k is the wavenumber, a is the radius of particles and the constant q does not depend on the scattering angle. Brackets mean averaging over the particle size distribution. It was found that this simple formula holds for natural water clouds in visible and near-infrared regions of the electromagnetic spectrum, where, from comparisons with the Mie theory for polydispersed media, it follows that:

$$q \approx 0.5(7 - 0.015ka_{ef})(1 - 1.1/(ka_{ef})^{2/3}), \qquad p(\pi) \approx \tfrac{2}{3}$$

where $a_{ef} = \langle a^3 \rangle / \langle a^2 \rangle$ is the effective radius of particles. The accuracy of this approximation is better than 10%. This simple equation for the phase function proved to be helpful in lidar cloud remote-sensing problems (Barun, 2000). It describes the oscillations of light intensity at scattering angles close to π. This phenomenon, which is called glory, is often observed in natural clouds. It produces the coloured rings around the shadows of aircraft flying over cloud when special conditions on the viewing geometry are satisfied ($\theta \approx \pi$).

2.3.3 Nonspherical particles

Optical properties of nonspherical particles can be considered in the framework of the geometrical optics approximation if all curvature radii of these particles are much larger than the wavelength of the incident radiation (Hedkinson, 1963; Hovenac, 1991). Unlike the case of scattering by spheres it is a difficult task to derive analytical formulae (Lock, 1996a, b). Thus, numerical computations should be used to solve the problem (Liou *et al.*, 1983; Peltoniemi *et al.*, 1989; Macke, 1993, 1994; Mitchell and Arnott, 1994; Macke *et al.*, 1995; Macke and Mishchenko, 1996; Kokhanovsky and Nakajima, 1998; Yang *et al.*, 2001; Kokhanovsky, 2003a).

Results of calculations of phase functions of randomly oriented prolate and oblate spheroids with the Monte Carlo ray-tracing code (MCRTC) are presented in Figures 2.9, 2.10 at different values of the shape parameter $\xi = c/a$, where c and $a = b$ are semiaxes of spheroids. The same results are given in tabular form in Appendix 5, where data for randomly oriented hexagonal cylinders are also presented. It is interesting that the main rainbow position moves towards the forward hemisphere with increasing nonsphericity of particles.

Phase functions of irregularly shaped particles can be modelled by distributions of spheroids with different values of the shape parameter. In this case phase functions become featureless (such as, e.g., the phase function presented in Figure 2.11). This phase function was obtained with the ray-tracing code for fractal particles (Macke and Tzschihholz, 1992; Macke, 1996) having refractive index 1.5. We can see that the logarithm of the phase function in Figure 2.11 can be represented as a linear function of the logarithm of the scattering angle at $\theta > 10°$. This is not the case for smaller angles, where the diffraction of light should be taken into account.

It was found (Kokhanovsky and Macke, 1997) that the following simple parametrizations of average absorption cross-sections and asymmetry parameters for large randomly oriented nonspherical particles can be used:

$$C_{abs} = \frac{A}{4}(1 - W(n))(1 - e^{-\psi(n,s)c}) \tag{2.83}$$

$$g = g_\infty(n) - (g_\infty(n) - g_0(n,s))\, e^{-y(n,s)c} \tag{2.84}$$

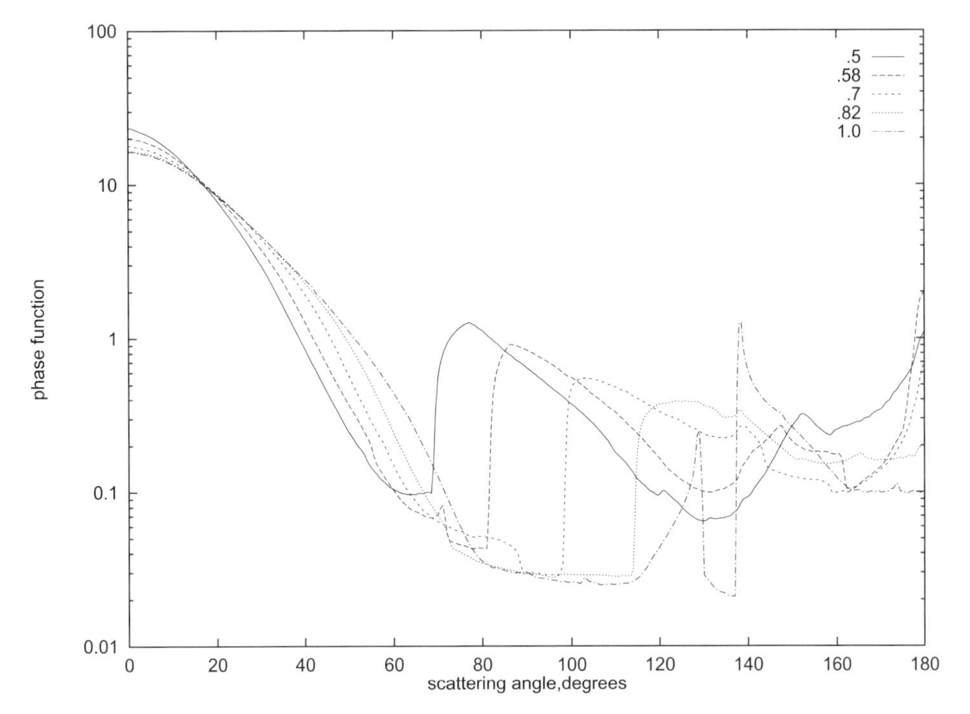

Figure 2.9. The geometrical optics phase functions of nonabsorbing oblate spheroids at $n = 1.333$, $\xi = 0.5, 0.58, 0.7, 0.82, 1.0$.

where $c = 4\chi x$, $x = 2\pi a_{ef}/\lambda$, $a_{ef} = (3V)/A$, V is the volume of particles and A is their surface area, $\psi(n, s)$ and $y(n, s)$ are phenomenological constants depending on the refractive index n and shape s, and $g_0(n, s)$ is the asymmetry parameter of a nonabsorbing particle. The values $W(n)$, $g_\infty(n)$ determined by equations (2.33), (2.61), respectively. Formulae (2.83), (2.84) are very similar to equations (2.48), (2.65) which were obtained for spherical particles. It follows from equation (2.83) as $c \to 0$ that:

$$C_{abs} = f(n, s)\alpha V$$

where $\alpha = (4\pi\chi)/\lambda$, $f(n, s) = \frac{3}{2}(1 - W(n))\psi(n, s)$. Thus, the absorption coefficient $\sigma_{abs} = C_{abs}/V$ of weakly absorbing particles does not depend on their size. Only the shape of particles is important in this case.

It should be pointed out that the spherical particles are characterized by smaller values of absorption than randomly oriented nonspherical particles with the same effective size a_{ef}.

Note that for the value of C_{ext} of convex randomly oriented particles it follows that:

$$C_{ext} = A/2 \tag{2.85}$$

where A is the surface area of a particle. It is equal to $4\pi a^2$ for spherical particles with the radius a.

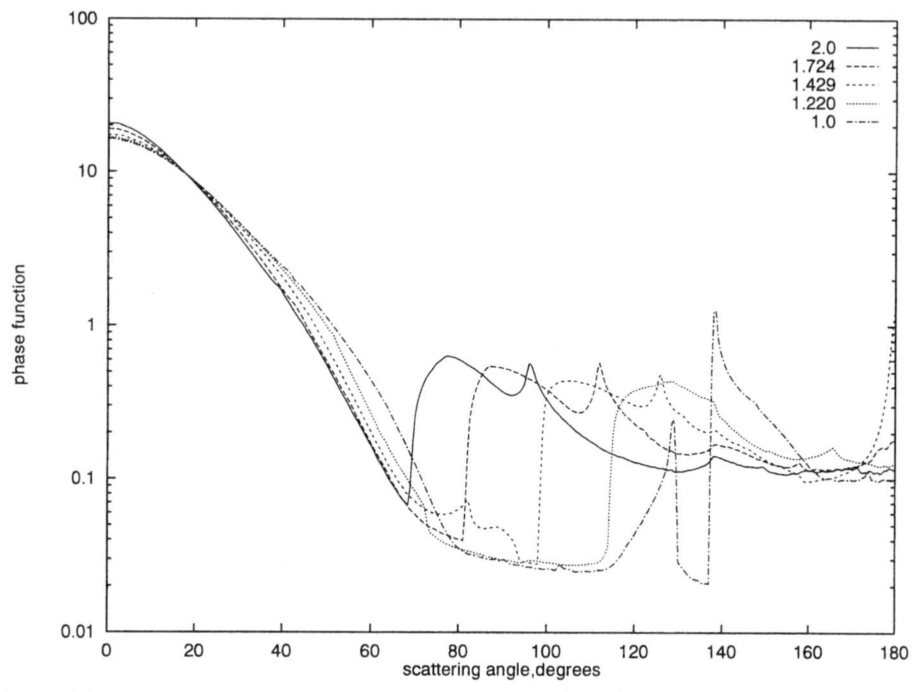

Figure 2.10. The same as in Figure 2.6 but for prolate spheroids at $\xi = 1.0, 1.22, 1.429, 1.724, 2.0$.

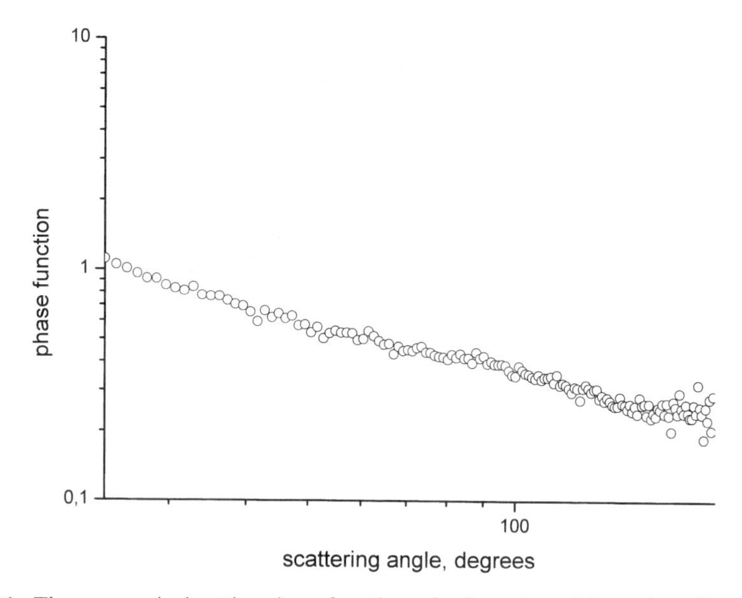

Figure 2.11. The geometrical optics phase function of a fractal particle at the refractive index $m = 1.5$.

Table 2.5. Value of g_0 for randomly oriented spheroidal particles at different values of the shape parameter s and real part of the refractive index n.

s/n	1.1	1.2	1.333	1.4	1.5	1.6	1.7
0.3	0.9582	0.9123	0.8774	0.8590	0.8257	0.7873	0.7518
0.5	0.9583	0.8976	0.8129	0.7718	0.7135	0.6677	0.6407
0.7	0.9764	0.9127	0.8428	0.8041	0.7522	0.7065	0.6665
1.0	0.9731	0.9341	0.8843	0.8613	0.8299	0.8015	0.7759
1.5	0.9652	0.9154	0.8510	0.8208	0.7747	0.7322	0.6934
2.0	0.9642	0.9158	0.8556	0.8298	0.7965	0.7685	0.7479
3.5	0.9711	0.9413	0.9085	0.8911	0.8624	0.8320	0.8020
∞	0.9817	0.9551	0.9201	0.9029	0.8795	0.8577	0.8360

Table 2.6. Values of $y(n, s)$ for randomly oriented spheroidal particles at different values of the shape parameter s and real part of the refractive index n.

s/n	1.1	1.2	1.333	1.4	1.5	1.6	1.7
0.3	2.49	2.21	3.09	3.21	3.53	3.70	3.83
0.5	1.32	1.59	1.59	1.66	1.73	1.78	1.84
0.7	0.9	0.83	1.16	1.26	1.29	1.37	1.43
1.0	0.47	0.54	0.76	0.82	0.83	0.86	0.89
1.5	0.86	1.05	1.09	1.13	1.18	1.23	1.28
2.0	1.04	1.10	1.27	1.32	1.32	1.35	1.40
3.5	1.43	1.37	1.40	1.48	1.65	1.83	1.98

The values $g_0(n, s)$, $y(n, s)$, and $f(n, s)$ for randomly oriented particles of different shapes (spheroids, hexagonal particles, and infinite circular cylinders) are presented in Tables 2.5–2.8 (see Figures 2.12–2.14 as well). They were obtained using the MCRTC. The definition of parameter s depends on the shape of the particles in question. For instance, the shape parameter s for spheroids is equal to the ratio c/a ($a = b, c$ are semiaxes of spheroids). It is equal to infinity for infinite circular

Table 2.7. Values of $f(n, s)$ for randomly oriented spheroidal particles at different values of the shape parameter s and real part of the refractive index n.

s/n	1.1	1.2	1.33	1.4	1.5	1.6	1.7
0.3	2.46	2.78	3.16	3.35	3.64	3.93	4.21
0.5	1.66	1.84	2.09	2.21	2.40	2.59	2.76
0.7	1.32	1.45	1.60	1.67	1.78	1.88	2.00
1.0	1.11	1.18	1.24	1.26	1.29	1.31	1.33
1.5	1.14	1.25	1.36	1.42	1.50	1.57	1.64
2.0	1.18	1.29	1.43	1.49	1.58	1.68	1.76
3.5	1.24	1.37	1.52	1.59	1.70	1.80	1.89

Table 2.8. Values of ψ, y, and g_0 for hexagonal cylinders at $n = 1.333$ and different values of the ratio $\nu = L/l$. Here L is the length of a hexagonal cylinder and l is the side of a hexagonal cylinder cross-section.

ν	ψ	y	g_0
0.2	1.9	4.8	0.9031
0.4	1.8	1.7	0.8607
1	1.6	1.1	0.7847
2	1.6	0.9	0.7601
4	1.8	0.7	0.7971
10	1.9	1.5	0.8442
20	1.9	1.5	0.8665

cylinders and to the ratio of the length L to the side l of a transverse cross-section for hexagonal cylinders.

The results in Table 2.5 at $s = 1$ (sphere) coincide with those in Table 2.2. This confirms both calculations.

Generally speaking, deviations from sphericity reduce the asymmetry parameter of large particles. However, this is not a general rule. For instance, randomly oriented nonabsorbing infinite cylinders ($s \to \infty$) are characterized by higher values of asymmetry parameters than nonabsorbing spherical particles with the same chemical composition. The same is true for thin discs ($s \to 0$). The value of g_0 decreases with refractive index n for particles of all shapes studied. This result

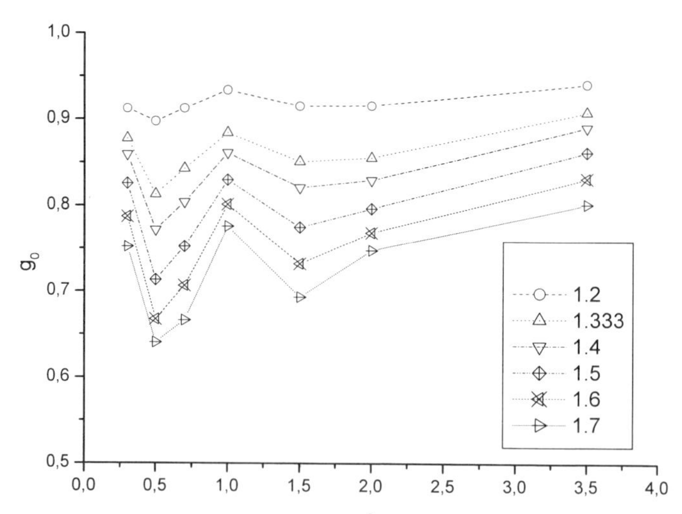

Figure 2.12. The dependence of the asymmetry parameter g_0 of nonabsorbing spheroids on the value of s at different values of the refractive index of particles $n = 1.2, 1.333, 1.4, 1.5, 1.6, 1.7$.

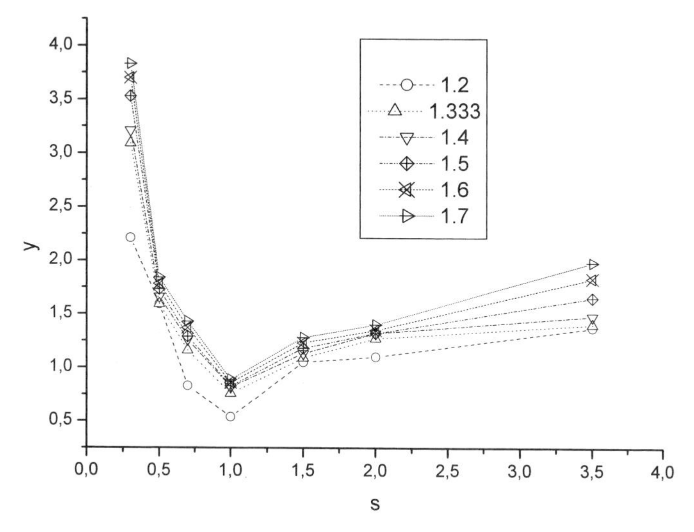

Figure 2.13. The same as in Figure 2.9 for the value of y.

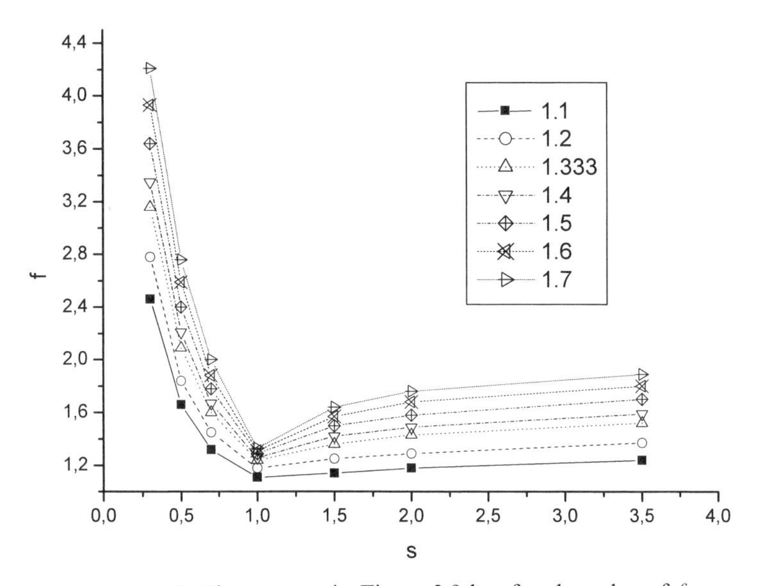

Figure 2.14. The same as in Figure 2.9 but for the value of f.

could be understood on physical grounds. Indeed, rays are almost not deflected by particles at $n \approx 1$. Thus, the value of g is close to 1 in this case. Naturally, the asymmetry parameter g_0 decreases with the refractive index n. Note that the parameter y is smaller for spheres than for nonspherical particles at $n = \text{const}$ (see Table 2.6).

2.3.4 Edge effects

The geometrical optics results can be improved if we account for wave optics corrections (Jones, 1957; Nussenzveig and Wiscombe, 1980; Nussenzveig, 1992). It is especially important for scattering angles, where the intensity of the scattered light is characterized by the considerable increase in magnitude (e.g. rainbow and glory scattering). Glory scattering is important at scattering angles near backward direction ($\theta \to \pi$). The primary rainbow for spherical particles is located at the scattering angle θ_r, which can be obtained from the following equation:

$$\theta_r = \pi + 2\arccos\varsigma - 4\arcsin\upsilon \tag{2.86}$$

where

$$\varsigma = \sqrt{\frac{n^2 - 1}{3}} \tag{2.87}$$

and

$$\upsilon = \sqrt{\frac{4 - n^2}{3n^2}} \tag{2.88}$$

For instance, this angle is equal to 137.9 degrees at $n = 1.333$. It should be pointed out that the value of the refractive index n depends on the temperature of a particle. Thus, the same is true for the primary rainbow scattering angle θ_r. This fact is used for the measurements of the temperature of droplets.

Asymptotic methods to study wave corrections for spheres were developed by Nussenzveig (1992). Unfortunately, these results, which have great importance from the general point of view, are too complex for applications (Slingo and Shrecker, 1982; Slingo, 1989; Chou, 1998). A simpler method for treatment of wave corrections is parametrization (Evans and Fournier, 1990, 1996; Kokhanovsky and Zege, 1995; Van de Bosch *et al.*, 1996; Baran and Havemann, 1999). For parametrization purposes the following general equation can be used:

$$F = F^G + F^E \tag{2.89}$$

Here F^G is the geometrical optical solution of a problem, F is the result of the solution of Maxwell equations for the given shape of a particle (e.g. the Mie solution for spheres), and F^E is an unknown function which describes wave corrections. The so-called edge term F^E can easily be found from equation (2.89). Namely, it follows that: $F^E = F - F^G$. The values F^E can be parametrized, using the exact solution for the electromagnetic scattering problem for a particle of given shape. Thus, the exact form of the edge term will depend on the shape of the particle and the optical characteristic in question.

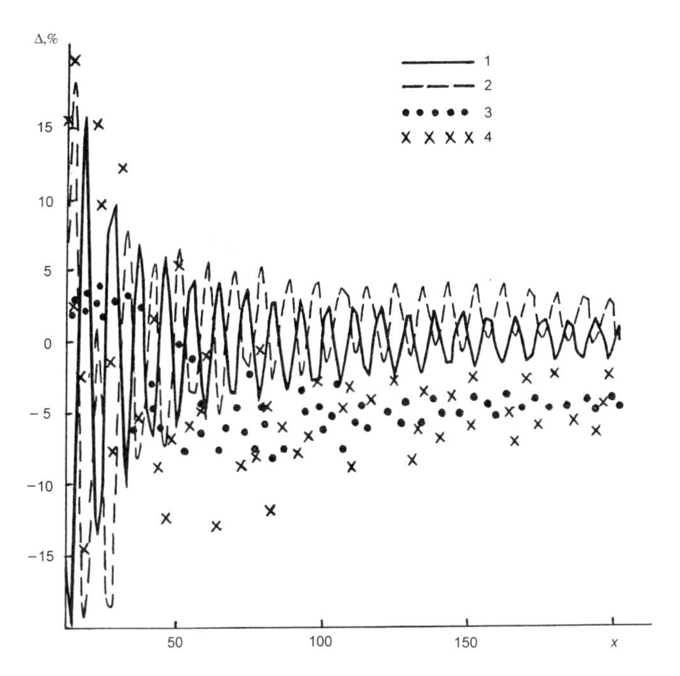

Figure 2.15. The relative error Δ of equations (2.90) (1), (2.92) (2) for the value of $(1-g)$, (2.91) (3), and the error of finding the probability of photon absorption C_{abs}/C_{ext} (4) at $n = 1.34$, $\chi = 0.0001$ as the function of size parameter x. The results were obtained by comparison with Mie theory calculations for monodipersed spheres.

For instance, we obtain for spherical particles (Kokhanovsky and Zege, 1995):

$$C_{ext} = 2\left(1 + \frac{1}{x^{2/3}}\right)\pi a^2 \tag{2.90}$$

$$C_{abs} = \left[1 - \frac{2n^2}{c^2}(e^{-cb}(1+cb) - e^{-c}(1+c)) - W(1-e^{-cb})^2\right]\Upsilon\pi a^2 \tag{2.91}$$

$$g = g_\infty - \left(g_\infty - g_0 + \frac{\gamma_1 + \gamma_2 c}{x^{2/3}}\right)e^{-cy} \tag{2.92}$$

where $\Upsilon = 1 + (n-1)(1 - exp(-1/\beta\rho))$, $\rho = 2x(n-1)$, $\beta = (21.2 + 20.1\lg\chi + 11.1\lg^2\chi + \lg^3\chi)$, $\gamma_1 = 0.5$, $\gamma_2 = 0.2$, $x = 2\pi a/\lambda$, $c = 4\chi x$, $b = \sqrt{1-n^{-2}}$, and $m = n - i\chi$ is the relative refractive index of particles. It should be pointed out that equations (2.42), (2.51), (2.65) follow from equations (2.90)–(2.92) as $x \to \infty$. These parametrizations can be used at $n \in [1.2, 1.6]$, $\chi > 10^{-5}$, $x > 45$ with relative errors less than 15% (see Figures 2.15–2.17). The error reduces considerably if we take into account the polydispersity of a scattering medium.

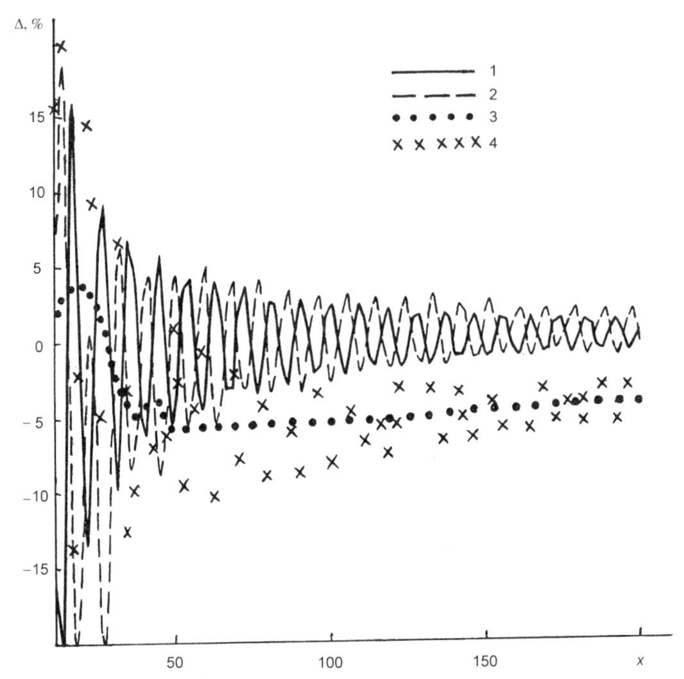

Figure 2.16. The same as in Figure 2.15 but at $\chi = 0.001$.

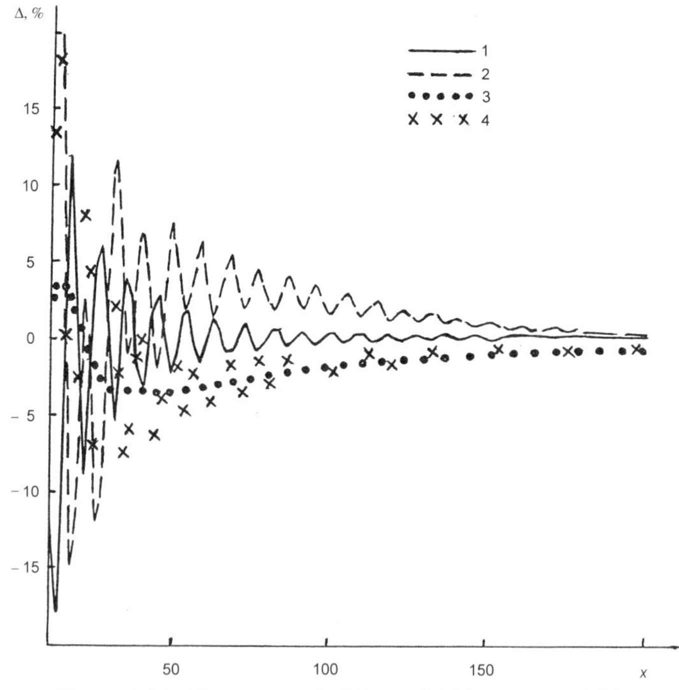

Figure 2.17. The same as in Figure 2.15 but at $\chi = 0.01$.

Similar results for spheroids were obtained by Fournier and Evans (1991, 1993) and Evans and Fournier (1994). Extinction cross-sections for randomly oriented circular and elliptical cylinders were studied by Fournier and Evans (1996). Baran and Havemann (1999) investigated wave corrections for hexagonal cylinders, which are of importance for studies of optical characteristics of crystalline clouds.

Equations (2.90)–(2.92) can be used for calculation of the local optical characteristics of water clouds, fogs, and mists. Droplets in these media are almost perfect spheres. They can be characterized by the particle size distribution. It follows from equations (2.90)–(2.92) for the extinction coefficient σ_{ext}, absorption coefficient σ_{abs}, and co-asymmetry parameter $1-g$ of spherical polydispersions (see equations (1.82), (1.83) and Appendix 2) that:

$$\sigma_{ext} = \frac{1.5C_\nu}{a_{ef}}\left\{1+\frac{1.1}{(ka_{ef})^{2/3}}\right\} \tag{2.93}$$

$$\sigma_{abs} = \frac{5\pi\chi(1-\alpha a_{ef})}{\lambda}C_\nu\left\{1+0.34\left(1-\exp\left(-\frac{8\lambda}{a_{ef}}\right)\right)\right\} \tag{2.94}$$

$$1-g = 0.12+0.5(ka_{ef})^{-2/3}-0.15\alpha a_{ef} \tag{2.95}$$

where $\alpha=(4\pi\chi)/\lambda$, $k=(2\pi)/\lambda$, C_ν is the volumetric concentration of droplets. Note that we use the constant value of the refractive index $n=1.34$ in equations (2.93)–(2.95). This is due to the fact that the variation of this value for water and ice at $\lambda < 2.2\,\mu m$ is weak (see Figure 2.18(a)). Also it follows from Figure 2.18(b) that the value of χ is small for water droplets and ice crystals at $\lambda < 2.2\,\mu m$. Thus, it was possible to assume that the value $\alpha a_{ef} \to 0$ in derivations of equations (2.93)–(2.95).

Equations (2.93)–(2.95) allow us to find the probability of photon absorption by the elementary volume of a cloudy medium:

$$\beta = \frac{\sigma_{abs}}{\sigma_{ext}} \tag{2.96}$$

and the similarity parameter:

$$K = \sqrt{\frac{\beta}{3(1-g)}} \tag{2.97}$$

These parameters are of great importance in radiative transfer theory (Van de Hulst, 1980). For instance, semi-infinite weakly absorbing disperse media with different phase functions but the same values of parameter K have very close values of radiative characteristics. Spectral dependencies of parameters K and β are presented in Figures 2.19, 2.20 for water clouds with $a_{ef}=4\,\mu m$ and $16\,\mu m$ at $\lambda=0.55\,\mu m$. They are obtained with exact calculations using the Mie theory (see Appendix 2) and approximate equations (2.93)–(2.95). The approximate agreement and exact results is excellent at $\lambda < 2.2\,\mu m$.

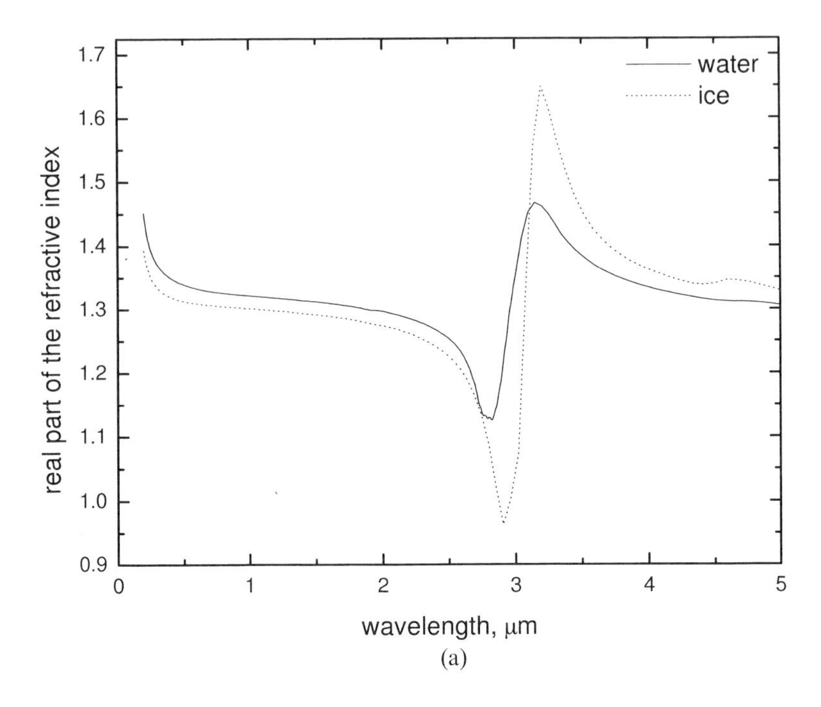

(a)

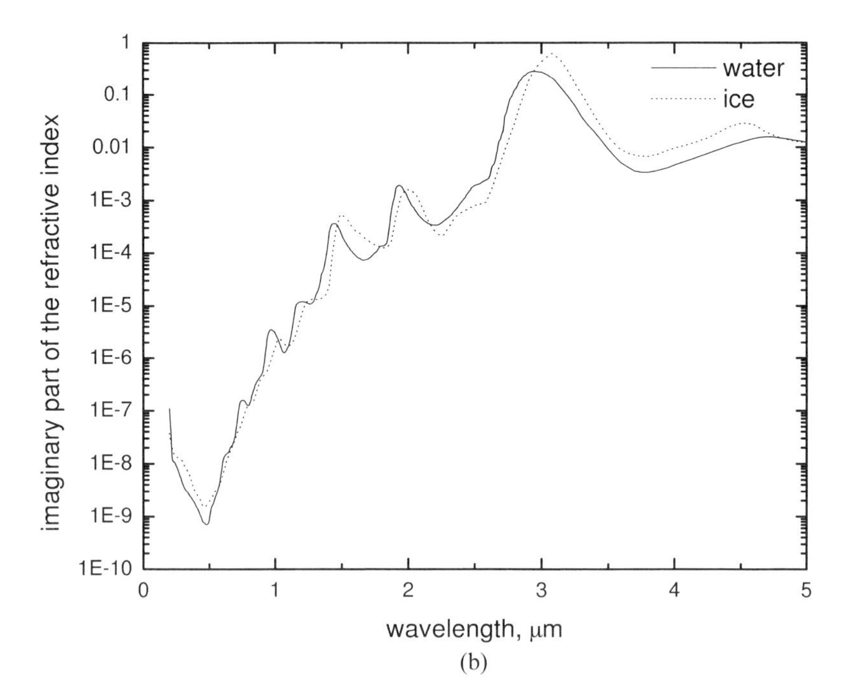

(b)

Figure 2.18. (a) The real and (b) the imaginary parts of the refractive index of water and ice.

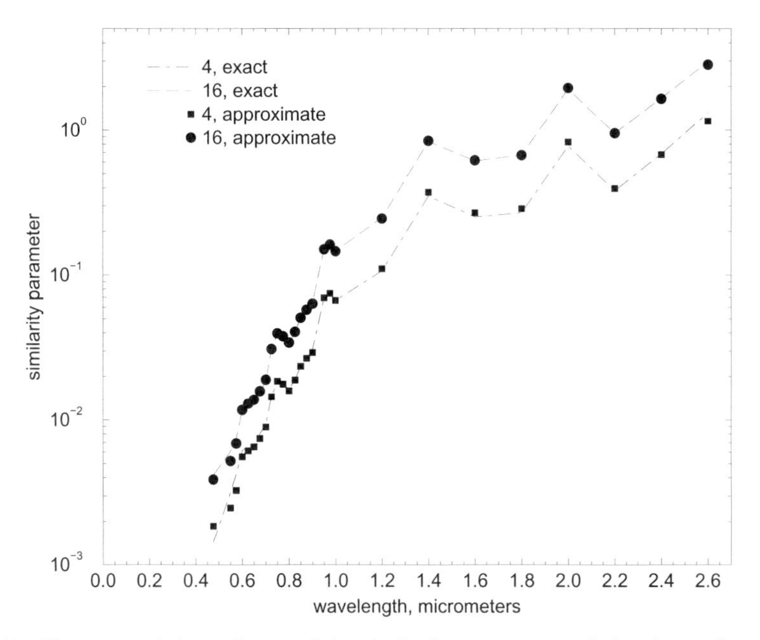

Figure 2.19. The spectral dependence of the similarity parameter of cloudy media, obtained with approximate formulas and the Mie theory for gamma particle size distribution with effective radius of droplets equal to 4 and 16 μm. The coefficient of variance of the particle size distribution is equal to $1/\sqrt{7}$ (see Table 1.3).

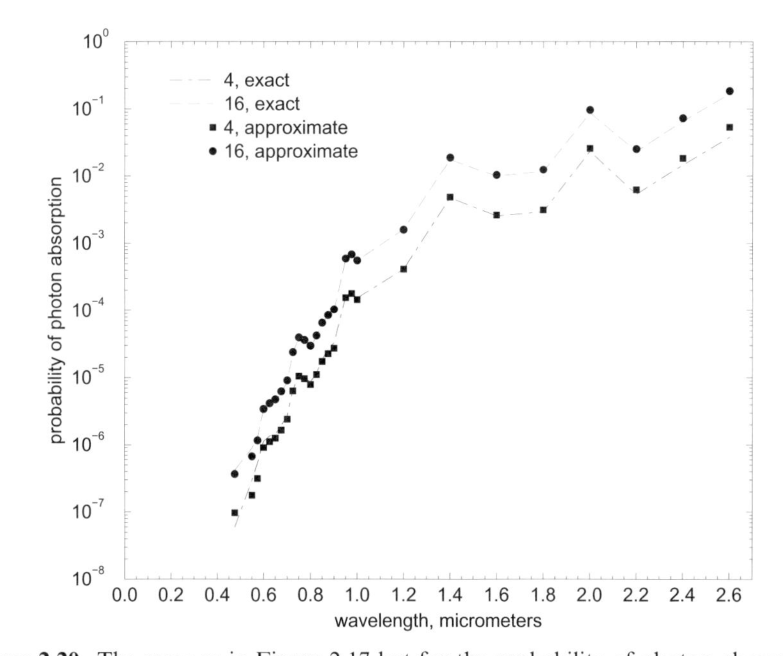

Figure 2.20. The same as in Figure 2.17 but for the probability of photon absorption.

2.4 PARTICLES WITH REFRACTIVE INDEX CLOSE TO THE REFRACTIVE INDEX OF A HOST MEDIUM

2.4.1 Rayleigh–Gans approximation

It is often the case that the refractive index of particles m_p is close to the refractive index of a surrounding medium m_s and

$$|m - 1| \ll 1 \tag{2.98}$$

where $m = m_p/m_s$. Such particles are called optically soft scatterers. Many solid particles in the natural environment have the refractive index around 1.5 in the visible range of the electromagnetic spectrum. Thus, their suspension in water ($m_s \approx 1.3$) will result in refractive indexes close to 1.15. This is the case for soil particles in rivers, lakes, and sea. Organic substances dispersed in oceans have even smaller values of refractive index when compared with soil particles. They are usually in the range 1.02–1.05. The same is true for various bio-liquids, including blood.

Thus, it is important to specifically consider the optical properties of optically soft particles. This question was studied by many authors using both exact and approximate theories of light scattering. A number of important analytical approximations were obtained for such particles (Van de Hulst, 1957; Bryant and Latimer, 1969; Perelman, 1986, 1991, 1994; Chen, 1987, 1992, 1993; Lopatin and Sid'ko, 1988; Zumer, 1988; Bourelly *et al.*, 1989; Sharma and Somerford, 1989, 1990, 1996, 1999; Holoubek, 1991; Klett and Sutherland, 1992; Khlebtsov, 1993; Al Chalabi and Jones, 1994, 1995; Kvien, 1995; Caorsi *et al.*, 1996; Jones *et al.*, 1996).

It should be pointed out that approximate formulae for optically soft particles are very simple and attractive for many applications in specific cases of large ($x \gg 1$) and small ($x \ll 1/(2|m-1|)$) particles. The approximations, which arise under these assumptions, are called the Van de Hulst approximation and the Rayleigh–Gans approximation, respectively. The main limitation of applicability of the Van de Hulst approximation is the size of particles, which should be large. However, the phase shift of a wave propagating inside a particle can have an arbitrary value in the framework of the Van de Hulst approximation. The phase shift in the framework of the Rayleigh–Gans approximation is considered to be small, namely:

$$2|m - 1|x \ll 1 \tag{2.99}$$

which means that sizes of particles are small (under realistic assumptions on the refractive index of particles). Thus, the Rayleigh–Gans approximation is an analogy of the well-known Born approximation of quantum mechanics.

It is clear that a particle in the framework of Rayleigh–Gans approximation does not disturb an incident electromagnetic wave very much. The Rayleigh–Gans particle can be represented by means of an array of noninteracting (contrary to discrete dipole approximation) dipole scatterers. An electromagnetic field inside a particle coincides with the value of the incident field in this case. To obtain the scattering pattern we should add scattering fields from every dipole accounting for the interference of secondary waves.

For the value of the amplitude scattering matrix (1.65) of a spherical particle (see Chapter 1) in this approximation (Van de Hulst, 1957) it follows that:

$$\hat{S}(\theta) = ik^3\alpha \begin{pmatrix} \cos\theta & 0 \\ 0 & 1 \end{pmatrix} R(\theta) \tag{2.100}$$

where $\alpha = (m-1)V/2\pi$ is the polarizability of a particle (see Eq. (2.4) as $m \to 1$). The integral

$$R(\theta) = \frac{1}{V}\int_V \exp(i\Delta(\vec{r}))\,dV \tag{2.101}$$

accounts for the interference effects between scattering fields from different dipoles, $\Delta(\vec{r}) = \vec{r}(\vec{k} - \vec{k}')$ is the phase difference of waves, scattered by dipoles, situated at a distance $|\vec{r}|$ from each other inside a particle, $\vec{k}$ and $\vec{k}'$ are wave vectors of incident and scattered waves, respectively. The values $\Delta(\vec{r})$ and $R(\theta)$ do not depend on the refractive index of particles. The Rayleigh approximation for the matrix $\hat{S}$ of soft small particles follows from equation (2.100) if we neglect interference effects ($\Delta = 0$, $R - 1$) (Van de Hulst, 1957).

It should be pointed out that the integral (2.101) does not depend on the azimuth for randomly oriented particles. Thus, it follows for the phase matrix $\hat{p}(\theta)$ in this case (see equation (2.100)):

$$\hat{p}(\theta) = \psi(\theta) \begin{pmatrix} 1+\cos^2\theta & \cos^2\theta - 1 & 0 & 0 \\ \cos^2\theta - 1 & 1+\cos^2\theta & 0 & 0 \\ 0 & 0 & 2\cos\theta & 0 \\ 0 & 0 & 0 & 2\cos\theta \end{pmatrix} \tag{2.102}$$

where

$$\psi(\theta) = |R(\theta)|^2/F$$

and

$$F = \frac{1}{2}\int_0^\pi (1+\cos^2\theta)|R(\theta)|^2 \sin\theta\,d\theta$$

The phase matrix $\hat{p}(\theta)$ and the asymmetry parameter $g = C/F$, where

$$C = \frac{1}{2}\int_0^\pi (1+\cos^2\theta)\cos\theta\sin\theta|R(\theta)|^2\,d\theta$$

do not depend on the refractive index of particles in this approximation. However, the shape and size of particles are important. For the scattering cross-section C_{sca} (Van de Hulst, 1957) we obtain:

$$C_{sca} = \frac{k^4V^2F|m-1|^2}{2\pi} \tag{2.103}$$

There are analytical results for the values of F and g in the case of spherical particles (Van de Hulst, 1957; Irvine, 1963; Kokhanovsky and Zege, 1997a).

As mentioned before, a particle can be represented as an array of noninteracting dipoles within the framework of the Rayleigh–Gans approximation. Therefore, the absorption cross-section of a particle is the sum of absorption cross-sections of these dipoles:

$$C_{abs} = \int_V \gamma(\vec{r})\, d^3\vec{r} \tag{2.104}$$

Thus, we obtain for uniform particles:

$$C_{abs} = \gamma V \tag{2.105}$$

where $\gamma = \frac{4\pi\chi}{\lambda}$, V is the volume of a particle. It is interesting to note that this result does not depend on the shape of a particle and can be derived from the general equation for the value of the absorption cross-section:

$$C_{abs} = \frac{2k}{\vec{E}_0\vec{E}_0^*}\int_V n(\vec{r})\chi(\vec{r})\vec{E}(\vec{r})\vec{E}^*(\vec{r})\, d^3\vec{r} \tag{2.106}$$

assuming that the electric field inside the particle $\vec{E}(\vec{r})$ coincides with the value of the incident field $\vec{E}_0$ and the refractive index $n \to 1$.

The phase matrix $\hat{p}(\theta)$ (2.102) and the scattering cross-section (2.103) can easily be found in the framework of the Rayleigh–Gans approximation if the following integrals are known:

$$|R(\theta)|^2 = \left|\frac{1}{V}\int_V e^{i\Delta(\vec{r})}\, d^3\vec{r}\right|^2, \qquad F = \frac{1}{2}\int_0^\pi |R(\theta)|^2(1+\cos^2\theta)\sin\theta\, d\theta \tag{2.107}$$

The integration in equations (2.107) should be performed numerically in most cases (see Table 2.9). However, integrals (2.107) can be evaluated analytically assuming the spherical shape of particles. Namely, it follows for the form-factor $\Phi(\theta) \equiv |R(\theta)|^2$ (Van de Hulst, 1957):

$$\Phi(\theta) = \frac{9(\sin u - u\cos u)^2}{u^6}$$

where $u = 2ka\sin(\theta/2)$. It is interesting to note that the form-factor depends on the single parameter u in this case. The simplicity of the expression for the form-factor of a spherical particle allows us to find integrals F (see equation (2.107)) and the asymmetry parameter g analytically. Namely, it follows after simple algebraic calculations (Van de Hulst, 1957; Irvine, 1963; Kokhanovsky and Zege, 1997a):

$$F = \frac{9}{8x^4}\left[\frac{5}{2} + 2x^2 - \frac{\sin 4x}{4x} - \frac{7(1-\cos 4x)}{16x^2} + \left(\frac{1}{2x^2} - 2\right)(\gamma + \lg 4x - Ci(4x))\right]$$

$$g = \frac{4}{Fx^4}\left[\left(\frac{9}{128} - \frac{5x^2}{64}\right)(4\sin 4x + \cos 4x) + \frac{x^6}{2} + \frac{11x^4}{8} - \frac{31x^2}{64} - \frac{9}{128} + x^2\left(x^2 - \frac{3}{8}\right)(Ci(4x) - \gamma - \ln 4x)\right]$$

Table 2.9. Form-factors $\Phi(\theta)$ for particles at random orientation ($k = (2\pi)/\lambda$, λ is the wavelength, θ is the scattering angle) (Kerker, 1969).

Shape	$\Phi(\theta)$	Definition of variables
Spheres	$\dfrac{9(\sin u - u\cos u)^2}{u^6}$	$u = 2ka\sin\dfrac{\theta}{2}$ (a is radius of particle)
Circular cylinders, general case	$\dfrac{2\pi}{vu^2}\displaystyle\int_0^{\pi/2}\dfrac{J^2_{1/2}(v\cos\beta)J^2_1(u\sin\beta)\,d\beta}{\cos\beta}$	$u = 2ka\sin\dfrac{\theta}{2}$, $v = kL\sin\dfrac{\theta}{2}$ (a is radius of cylinder, L is length of cylinder)
Circular cylinders, $a \ll L$ (rods)	$\displaystyle\int_0^{2v}\dfrac{\sin t}{t}dt - \dfrac{\sin^2 v}{v^2}$	$v = kL\sin\dfrac{\theta}{2}$
Circular cylinders, $a \gg L$ (discs)	$\dfrac{2}{u^2}\left(1 - \dfrac{J_1(2u)}{u}\right)$	$u = 2ka\sin\dfrac{\theta}{2}$
Spheroids	$\dfrac{9\pi}{2}\displaystyle\int_0^{\pi/2}\dfrac{J^2_{3/2}(hw(\beta))}{(hw(\beta))^3}\sin\beta\,d\beta$	$h = 2k\sin\dfrac{\theta}{2}$, $w(\beta) = a\sqrt{\cos^2\beta + \xi^2\sin^2\beta}$, $\xi = \dfrac{b}{a}$ (a, b semiaxes of spheroids)
Cubes	$\dfrac{2}{\pi}\displaystyle\int_0^{\pi/2}d\varepsilon\int_0^1 dy\,P(\theta,\varepsilon,y)$, $P(\varepsilon,y) = E^2(v\cos\varepsilon\sqrt{1-y^2}$ $E^2(v\sin\varepsilon\sqrt{1-y^2})E^2(vy)$	$E(s) = \dfrac{\sin s}{s}$, $v = kL\sin\dfrac{\theta}{2}$, L is edge of cube

where $\gamma = 0.577$ is Euler's constant, $Ci(4x) = -\int_{4x}^{\infty}\frac{\cos 4x}{x}dx$. We can obtain that $F = 9/(4x^2)$, $C_{sca} = 2\pi k^2 a^4|m-1|^2$ as $x \to \infty$ and $F = \frac{4}{3}$, $C_{sca} = 2/(3\pi)$ $k^4V^2|m-1|^2$ as $x \to 0$.

The value of C_{abs} (2.104) does not depend on the form-factor $R^2(\theta)$ in this approximation. Thus, it is the same for all shapes of particles. A similar result holds for the degree of polarization $\Pi(\theta) = -p_{12}(\theta)/p_{11}(\theta)$.

Note that the Rayleigh–Gans approximation for stochastically shaped particles was developed by Muinonen (1996). He also considered the case of large and small irregularly shaped particles using geometrical optics and Rayleigh approximations (Muinonen *et al.*, 1996, 1997). The size and shape of a particle was specified by the mean radius and the covariance function of the logarithmic radius (Stoyan and Stoyan, 1994). Shifrin and Mikulinski (1982) showed that the phase function of irregularly shaped Rayleigh–Gans particles can be substituted by the phase function of a single 'averaged' particle. Irregularly shaped optically soft particles were also studied by Al-Chalabi and Jones (1995).

Nonuniform spheres and ellipsoids with a Gaussian profile of the refractive index were studied in the framework of the Rayleigh–Gans approximation by Khlebtsov (1999).

2.4.2 Van de Hulst approximation

2.4.2.1 General equations

Unlike the Rayleigh–Gans approach, the Van de Hulst approximation, which is also called the anomalous diffraction approximation (Van de Hulst, 1957), is applicable for soft particles ($|m-1| \ll 1$) at any value of the phase shift $\rho = 2x|m-1|$ and $x \to \infty$. Thus, the size of particles is much larger than the wavelength of incident radiation. The concept of geometrical optics beams or rays can be applied to the problem in question.

Moreover, it is assumed that the refraction of light by such particles can be ignored due to the condition of the optical softness of a particle ($|m-1 \ll 1$). This is, of course, a drawback of the approximation as far as large scattering angles are considered. However, the assumption of straight rays greatly simplifies the calculation of the phase function at small angles. The extinction efficiency factor can also be easily found for various shapes of particles in this approximation. This is due to the optical theorem, which relates the extinction cross-section with the scattering amplitude in the forward direction.

Thus, it is assumed in the framework of the Van de Hulst approximation that geometrical optics rays pass through a particle without any deflection. However, light beams can undergo a significant phase shift because of the long path length through a particle. This shift depends on the size, the internal structure, and the shape of a particle. It results in alteration of the phase of the original wave on the plane beyond the particle (Van de Hulst, 1957; Bryant and Latimer, 1969; Sharma and Somerford, 1999). Note that a particle within the framework of this approach is replaced by the amplitude-phase screen with the transmission function:

$$\psi(\vec{r}) = 1 - \exp(-i\Delta(\vec{r}))$$

where Δ is the complex phase shift.

Applying Huygens' principle, for the value of the amplitude function at small angles (Van de Hulst, 1957) it follows that:

$$S(\theta) = \frac{k^2}{2\pi} \int_\Sigma \int \psi(\vec{r})\, e^{-i\delta(\vec{r})}\, d^2\vec{r} \tag{2.108}$$

where Σ is the projected area of a particle onto a plane perpendicular to the direction of incident radiation, $\delta(\vec{r})$ is the difference in optical path length of the rays, coming from different infinitesimal discs (Van de Hulst, 1957). Polarization of radiation is ignored in this approximation and the phase matrix reduces to a scalar. This approximation can be used to estimate the extinction cross-section via the optical theorem (Van de Hulst, 1957):

$$C_{ext} = \frac{4\pi}{k^2} \mathrm{Re}[S(0)]$$

The absorption cross-section is determined by the following equation (Van de Hulst, 1957):

$$C_{abs} = \int_{\Sigma}\int (1 - \exp[2\,\mathrm{Im}(\Delta(\vec{r}))])\, d^2\vec{r} \tag{2.109}$$

For the value of the phase function at small angles it follows that:

$$p(\theta) = \frac{4\pi |S(\theta)|^2}{k^2 C_{sca}}$$

where $C_{sca} = C_{ext} - C_{abs}$. Unfortunately, it is not possible to calculate the phase function at large scattering angles within the framework of this approximation. The value of the asymmetry parameter is determined by the following equation:

$$g = \frac{\displaystyle\int_0^{\pi} |S(\theta)|^2 \cos\theta \sin\theta\, d\theta}{\displaystyle\int_0^{\pi} |S(\theta)|^2 \sin\theta\, d\theta}$$

It depends mostly only on the value of the maximal phase shift on the largest dimension of a particle.

Thus, this approximation is important for calculating the values of C_{ext}, C_{abs}, g, and the phase function $p(\theta)$ at small scattering angles for particles of different shapes, including irregularly shaped particles (Aas, 1984; Flatau, 1992; Al-Chalabi and Jones, 1995; Muinonen, 1996; Xu *et al.*, 2003, 2004a).

2.4.2.2 *Spheres*

Let us apply the general equations presented in the preceding section to particles of various shapes. The first example will be the case of isotropic spherical particles. For spherical particles it follows from equation (2.108) that:

$$S(\theta) = \frac{k^2}{2\pi} \iint_{\Sigma} (1 - e^{-i\Delta(\xi,\eta)})\, e^{-ik\xi\theta}\, d\xi\, d\eta$$

where θ is the scattering angle and (ξ, η) are cartesian co-ordinates. Note that we can obtain from this formula as $\Delta \to \infty$:

$$S(\theta) = \frac{k^2}{2\pi} \iint_{\Sigma} e^{-ik\xi\theta}\, d\xi\, d\eta$$

This result is well known (Born and Wolf, 1965) and represents the scattered field behind the aperture with area Σ. It is possible to change from co-ordinates (ξ, η) to polar co-ordinates (r, φ):

$$\xi = r\cos\varphi, \qquad \eta = r\sin\varphi$$

It follows that:

$$\Delta = \rho\sqrt{1 - \left(\frac{r}{a}\right)^2}$$

in this case. Here $\rho = 2ka(m - 1)$ is the phase shift. Thus, we can obtain for the amplitude function the following integral in the polar co-ordinate system:

$$S(\theta) = \frac{k^2}{2\pi}\int_0^a r\,dr\int_0^{2\pi} d\varphi\left(1 - \exp\left(2ik(m-1)\sqrt{a^2 - r^2}\right)\right)\exp(-ikr\theta\cos\varphi)$$

Using the definition of the Bessel function

$$J_0(kr\theta) = \frac{1}{2\pi}\int_0^{2\pi} d\varphi\, e^{-ikr\theta\cos\varphi}\,d\varphi$$

we finally obtain:

$$S(\theta) = k^2\int_0^a (1 - e^{-2ik\sqrt{a^2-r^2}(m-1)})J_0(kr\theta)r\,dr \tag{2.110}$$

The analytical approximation of this integral was obtained by Chen (1993). The light intensity derived from equation (2.110) is an oscillating function at small scattering angles. This is similar to the Fraunhofer diffraction pattern, defined by equation (2.56). However, minima of the curves, obtained from the Fraunhofer theory and equation (2.110), do not coincide. This is due to the presence of the exponent in equation (2.110), which accounts for transmitted light. Such behaviour is the reason for the other name of this approximation, namely the anomalous diffraction approximation.

The result for the extinction cross-section follows from equation (2.110) and the optical theorem (Van de Hulst, 1957):

$$C_{ext} = 2 - 4e^{-\rho\tan\eta}\frac{\cos\eta}{\rho}\sin(\rho - \eta) - 4e^{-\rho\tan\eta}\left(\frac{\cos\eta}{\rho}\right)^2\cos(\rho - 2\eta)$$
$$+ 4\left(\frac{\cos\eta}{\rho}\right)^2\cos 2\eta \tag{2.111}$$

where $\tan\eta = \chi/(n - 1)$. Equation (2.111) can be simplified as $\chi \to 0$ (Van de Hulst, 1957):

$$C_{ext} = \left(2 - \frac{4\sin\rho}{\rho} + \frac{4(1 - \cos\rho)}{\rho^2}\right)\pi a^2 \tag{2.112}$$

This equation shows explicitly the dependence of the extinction cross-section of a spherical particle on the phase shift ρ. It is considered as one of the most important results of approximate light-scattering theory. The accuracy of this equation can be increased (see Figure 2.21) by using the 'edge term' $2x^{-2/3}$ (see equation (2.90)): $C'_{ext} = C_{ext} + 2\pi a^2 x^{-2/3}$ (Ackerman and Stephens, 1987; Kokhanovsky and Zege, 1997). It is of importance only for intermediate values of size parameter x. The edge term disappears as $x \to \infty$.

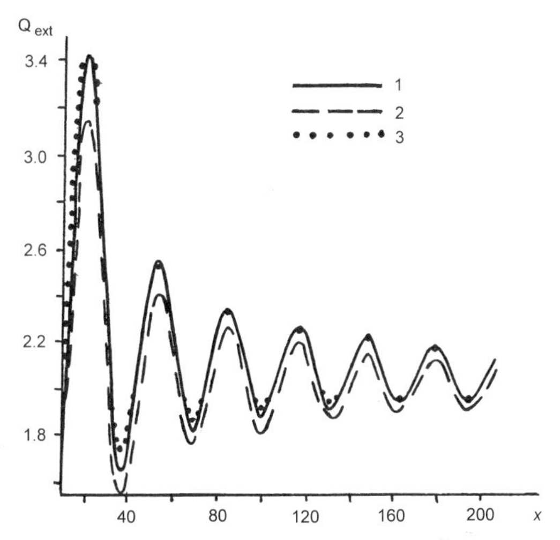

Figure 2.21. The extinction efficiency factor $Q_{ext} = C_{ext}/\pi a^2$ as the function of the size parameter x at the relative refractive index of particles equal to 1.1 according to the Mie theory (1), equation (2.112) (2), and taking account of the edge term $2x^{-2/3}$ (3).

Note that it follows from equation (2.112) for the extinction coefficient $\sigma_{ext} = N\langle C_{ext}\rangle$ that:

$$\sigma_{ext} = N\pi \sum_{j=1}^{\infty} \frac{A_j}{\lambda^{2j}} \langle a^{2+2_j} \rangle$$

where (Lui *et al.*, 1996):

$$A_j = 4(-1)^{j+1}(2j+1)\frac{(4\pi(n-1))^{2j}}{(2j+2)!}$$

and $\langle\ \rangle$ means averaging over the particle size distribution.

The simple equation for the absorption cross-section can also be obtained in the framework of the Van de Hulst approximation (see equation (2.109)). Namely, it follows that:

$$C_{abs} = 2\pi a^2 \int_0^1 (1 - e^{-cr}) r\, dr$$

or

$$C_{abs} = \pi a^2 \left(1 + \frac{2e^{-c}}{c} + \frac{2(e^{-c} - 1)}{c^2}\right) \tag{2.113}$$

where $c = 4\chi x$. This equation can be approximated by the following simple formula (Shifrin and Tonna, 1992): $C_{abs} = \pi a^2(1 - \exp(-\gamma L))$, where $\gamma = 4\pi\chi/\lambda$ and L is the average path of a ray inside a particle. The value of L is determined by the ratio of the volume of a particle to its cross-section: $L = V/\Sigma$. Thus, it follows for spheres that $L = (4a)/3$. We can see that the absorption cross-section does not depend on the real part of the refractive index of particles in this approximation. This is due to the fact that focusing effects can be ignored as $n \to 1$. The value of the absorption efficiency factor $Q_{abs} = C_{abs}/\pi a^2$ is the function of the only parameter c for spheres (see equation (2.113)). This function was tabulated by Van de Hulst (1957).

2.4.2.3 *Ellipsoids*

Formulae for the values of C_{ext} and C_{abs} of spherical particles can be used for ellipsoidal particles with semiaxes a, b, c as well (Lind and Greenberg, 1966; Latimer, 1980; Paramonov *et al.*, 1986; Lopatin and Sid'ko, 1988; Paramonov, 1994), if we change the diameter of spherical particles $2a$ in these equations to the value:

$$h = \frac{2abc}{\sqrt{(bc\cos\alpha)^2 + (ac\cos\beta)^2 + (ab\cos\gamma)^2}}$$

where h is the maximal length of a ray inside a particle, α, β, γ are angles of the incident radiation with the co-ordinate axes X, Y, Z. It follows for spheres that $a = b = c$ and $h = 2a(\cos^2\alpha + \cos^2\beta + \cos^2\gamma = 1)$. On the other hand, we can obtain for spheroids $(a = b)$:

$$h = \frac{2a}{\sqrt{1 + (\xi^2 - 1)\cos^2\gamma}}$$

where $\xi = a/c$ is the shape parameter. Note that it follows from this equation that $h = 2a$ at $\gamma = \pi/2$ and $h = 2c$ at $\gamma = 0$.

We can see that integral light-scattering characteristics of soft large particles depend mostly on the maximal phase shift h, but not on the shape of particles. The value of h for ellipsoids can be represented by the following formula:

$$h = \frac{3V}{2\Sigma}$$

where V is the volume of a particle and Σ is the geometrical cross-section. The geometrical cross-section of an ellipsoidal particle

$$\Sigma = \pi\sqrt{(bc\cos\alpha)^2 + (ac\cos\beta)^2 + (ab\cos\gamma)^2}$$

depends on the semiaxes a, b, c and angles α, β, γ. In many cases particles are randomly oriented and we should average the values of C_{ext} and C_{abs} with regard to orientations of particles. This is a complex procedure. It was shown by Paramonov (1994) that high accuracy can be achieved if we use the value:

$$\langle h \rangle = \frac{3V}{2\langle\Sigma\rangle}$$

in formulae for the extinction and absorption cross-sections instead of the averaging values of C_{abs}, C_{ext}. Here $\langle\Sigma\rangle$ is the average value of the geometrical cross-section.

Note that it follows for randomly oriented convex particles that:

$$\langle\Sigma\rangle = \frac{A}{4}$$

where A is the surface area of particles. Thus, we have:

$$\langle h \rangle = \frac{6V}{A}$$

for randomly oriented particles. Formulae for the values of C_{ext} (2.112), C_{abs} (2.113) of spherical particles with radius $a = \langle h \rangle / 2$ can be used for estimating the extinction and absorption cross-sections of many types of nonspherical particles, including circular cylinders.

The angular distribution of the scattered energy in the case of ellipsoids can be obtained from equation (2.108) (Streekstra *et al.*, 1994; Borovoi *et al.*, 1998).

2.4.2.4 Circular cylinders

General formulae for the values of C_{ext}, C_{abs} of finite circular cylinders with length L and diameter $2a$ were obtained by Cross and Latimer (1970), Volkovitsky *et al.* (1984), Stephens (1984), and Xu (2003). They are rather complex. However, the formulae can be simplified for specific angles of incident radiation ($\alpha = 0, \pi/2$) with the axis of the circular cylinder.

For instance, it follows at $\alpha = 0$ that:

$$C_{ext}(\alpha = 0) = 2(1 - \cos \rho^*)\Sigma$$

$$C_{abs}(\alpha = 0) = (1 - \exp(-c^*))\Sigma$$

where $c^* = 4\chi x^*$, $x^* = (2\pi L)/\lambda$, $\rho^* = 2x^*(m - 1)$, and $\Sigma = \pi a^2$ is the geometrical cross-section of a particle.

It follows in the case of an infinitely long circular cylinder with radius a at an oblique incidence (Chylek and Klett, 1991a) that

$$Q_{ext} = 2 - 2\int_0^1 e^{-(c/2)\sqrt{1-v^2}} \cos\left[\rho\sqrt{1 - v^2}\right] dv$$

$$Q_{abs} = 1 - \int_0^1 e^{-c\sqrt{1-v^2}}\, dv$$

where $c = 4\chi x'$, $\rho = 2x'|m - 1|$, $x' = ka$, $a' = a/\cos\alpha$, α is the angle between the incident beam and normal to the axis of cylinder. We can obtain, for nonabsorbing cylindrical particles at $\alpha = 0$:

$$Q_{ext} = 2 - \frac{4 \sin \rho}{\rho}$$

we have at $\alpha = 30^\circ$ that:

$$Q_{ext} = 2 - \cos \rho' - \frac{\sin \rho'}{\rho}$$

where $\rho' = 4ka|m - 1|/\sqrt{3}$.

2.4.2.5 Hexagonal columns

The model of hexagonal columns is frequently used in the estimation of radiative fluxes in ice clouds (Volkovitsky *et al.*, 1984). Plates and needles can be considered as a special case of columns. General equations for the values of Q_{ext} and Q_{abs} of prismatic columns were obtained by Chylek and Klett (1991b). Their derivation is based on the addition theorem for extinction and absorption cross-sections:

$$C_{ext} = \sum_{i=1}^{N} C_{ext}^{i}, \qquad C_{abs} = \sum_{i=1}^{N} C_{abs}^{i} \tag{2.114}$$

which is valid in the framework of the Van de Hulst approximation. Here C_{ext}^{i}, C_{abs}^{i} are extinction and absorption cross-sections of segments of particles. This theorem is based on the fact that a large optically soft particle can be cut into an arbitrary number of segments with nonoverlapping projected areas by planes parallel to the direction of incident light. Equations (2.114) follow from the additivity of the integration (2.108), (2.109) with respect to nonoverlapping parts of the projected area Σ.

Let us assume that the axis of symmetry of a particle is perpendicular to the direction of the incoming radiation. In this case the position of a hexagonal column in an arbitrary orientation will be determined by the angle α between the incoming radiation and one of the sides of a particle cross-section. It follows at $\alpha = 0°$ (Chylek and Klett, 1991b) that:

$$Q_{ext} = 2 + \frac{4e^{-c}(c\cos\rho - \rho\sin\rho)}{\rho^2 + c^2} - \frac{c\sin\rho\exp(-c/2)}{\rho^2 + c^2}$$

$$Q_{abs} = 1 - e^{-c} + \frac{e^{-2c}}{2}$$

where $\rho = 2ka|m-1|$, $c = 2ka\chi$, a is the side of a hexagon.

We can obtain at the flat incidence ($\alpha = 30°$):

$$Q_{ext} = 2 - \frac{c'}{(\rho')^2 + (c')^2} - e^{-c'}\cos\rho' + \frac{e^{-c'}(c'\cos\rho' - \rho'\sin\rho')}{(\rho')^2 + (c')^2}$$

$$Q_{abs} = 1 - \frac{e^{-2c'}}{2} - \frac{1 - e^{-2c'}}{4c'}$$

where $c' = (\sqrt{3}/2)c$, $\rho' = (\sqrt{3}/2)\rho$.

Analytical equations for columns with arbitrary polygonal bases were obtained by Chylek and Klett (1991b) and Heffels *et al.* (1995). Experimental studies of optical properties of ice clouds were performed by Volkovitsky *et al.* (1984) and Arnott *et al.* (1995). Hess and Wiegner (1994) compiled a special library of optical properties of hexagonal cylinders. Popov and Shefer (1995) considered lidar signals scattered from oriented ice plates (1995). Parametrizations of solar radiative properties of cirrus clouds were obtained by Fu (1996). Sun and Fu derived equations for arbitrarily oriented hexagonal crystals (1999).

2.4.3 Perelman approximation

Another important approximation for soft particles was developed by Perelman (1986, 1991). It can be applied for particles of any size and phase shift as $m \to 1$.

Thus, it is more general than Rayleigh, Rayleigh–Gans, and Van de Hulst approximations in this respect. For instance, we can obtain corresponding equations of the Rayleigh, Rayleigh–Gans and Van de Hulst approximations from the value of Q_{ext} in the Perelman approximation. Only results for uniform spheres (Perelman, 1991) and infinite cylinders (Sharma and Somerford, 1997) have been obtained in the framework of this approximation at the moment. This approximation for the value of the extinction cross-section coincides (Sharma and Somerford, 1996) with the well-known formula obtained by Hart and Montroll (1951).

The Perelman approximation (or S-approximation, as it was called by Perelman, 1986, 1991) is based on the asymptotic properties of the Mie series as $m \to 1$. It is worth pointing out that such a type of approximation as $m \to 1$ can be derived basically for nonuniform and nonspherical particles as well, providing a wider region of applicability than the Rayleigh, Rayleigh–Gans, and Van de Hulst approximate for these more complex types of particles.

Let us derive the approximation formula for the value of the extinction cross-section of spherical particles in the framework of the Perelman approximation. For this we well need the following exact expression for the extinction cross-section (see Appendix 2) :

$$C_{ext} = \frac{2\pi}{k^2} \sum_{n=1}^{\infty} (2n+1) Re(a_n + b_n)$$

where

$$a_n = \frac{\operatorname{tg} \alpha_n (i + \operatorname{tg} \alpha_n)}{\operatorname{tg}^2 \alpha_n + 1}, \qquad b_n = \frac{\operatorname{tg} \beta_n (i + \operatorname{tg} \beta_n)}{\operatorname{tg}^2 \beta_n + 1}$$

$$\operatorname{tg} \alpha_n = -\frac{\psi_n'(y)\psi_n(x) - m\psi_n(y)\psi_n'(x)}{\psi_n'(y)\chi_n(x) - m\psi_n(y)\chi_n'(x)}$$

$$\operatorname{tg} \beta_n = -\frac{m\psi_n'(y)\psi_n(x) - \psi_n(y)\psi_n'(x)}{m\psi_n'(y)\chi_n(x) - \psi_n(y)\chi_n'(x)}$$

and $y = mx$. For the denominators in equations for phase angles α_n, β_n it follows that:

$$\psi_n'(y)\chi_n(x) - m\psi_n(y)\chi_n'(x) \to 1$$
$$m\psi_n'(y)\chi_n(x) - \psi_n(y)\chi_n'(x) \to 1$$

as $m \to 1 (y \to x)$. These asymptotics can be derived from the exact formula for the wronskian (as $m \to 1$):

$$\psi_n'(x)\chi_n(x) - \psi_n(x)\chi_n'(x) = 1$$

It should be pointed out that:

$$\operatorname{tg} \alpha_n \approx -\psi_n'(y)\psi_n(x) + m\psi_n(y)\psi_n'(x) \to 0$$

and

$$\operatorname{tg} \beta_n \approx \psi_n(y)\psi_n'(x) - m\psi_n'(y)\psi_n(x) \to 0$$

as $m \to 1$.

Thus, the values of $\mathrm{tg}^2\,\alpha_n$ and $\mathrm{tg}^2\,\beta_n$ can be ignored in denominators and we can obtain as $m \to 1$:

$$a_n = (i + V_n)V_n, \qquad b_n = (i + W_n)W_n$$

where

$$V_n = m\psi_n(y)\psi_n'(x) - \psi_n'(y)\psi_n(x)$$
$$W_n = \psi_n(y)\psi_n'(x) - \psi_n'(y)\psi_n(x)$$

This approximation for values a_n, b_n is called the soft approximation. Note that more rigorous analysis shows that:

$$a_n = \frac{(i + V_n)V_n}{|m|}, \qquad b_n = \frac{(i + W_n)W_n}{\sqrt{|m|}}$$

as $m \to 1$. We ignore differences $|m| - 1$ and $\sqrt{|m|} - 1$ in the denominators for a_n and b_n in the following calculations.

Let us now consider the case of nonabsorbing particles. Values of V_n and W_n are real numbers in this case and:

$$\mathrm{Re}(a_n) = V_n^2, \qquad \mathrm{Re}(b_n) = W_n^2$$

Thus for the extinction cross-section of a nonabsorbing spherical particle in the framework of the Perelman approximation it follows that:

$$C_{ext} = \frac{2\pi}{k^2}\sum_{n=1}^{\infty}(2n+1)(V_n^2 + W_n^2)$$

It is important that this sum be exactly evaluated by use of the following equations (Perelman, 1991):

$$\sum_{n=1}^{\infty}(2n+1)\psi_n'^2(y)\psi_n^2(x) = \frac{xZ_1}{64y}$$

$$\sum_{n=1}^{\infty}(2n+1)\psi_n(y)\psi_n(x)\psi_n'(y)\psi'(x) = \frac{xZ_2}{64}$$

where

$$Z_1 = (4 + 2\rho)(ci(R) - ci(\rho)) + R^2 - \rho^2 + (R^2 + 4R\rho + \rho^2)(s(R) - s(\rho))$$
$$+ (3R^2 + 4R\rho - \rho^2)c(\rho) + (R^2 - 4R\rho - 3\rho^2)c(\rho)$$
$$Z_2 = 4(ci(R) - ci(\rho)) + (R^2 - \rho^2)(s(R) + s(\rho)) + (3R^2 + \rho^2)c(R) - (R^2 + 3\rho^2)c(\rho)$$

where

$$s(Z) = \frac{\sin Z}{Z}, \qquad ci(Z) = \int_0^Z \frac{1 - \cos t}{t}\,dt, \qquad c(Z) = \frac{1 - \cos Z}{Z^2},$$

Table 2.10. The function $x_0(n)$ (Perelman, 1986).

n	1.02–1.06	1.08	1.1	1.12	1.14	1.16	1.18	1.22
x_0	∞	800	300	130	90	72	49	32

where $x = ka$, $y = mka$, $\rho = 2(m-1)x$, $R = 2(m+1)x$. The answer for non-absorbing particles is (Perelman, 1991; Granovskii and Ston, 1994):

$$C_{ext} = \frac{(m^2-1)^2}{4m^2}\pi a^2 \left\{ \alpha_1 H(R) - \alpha_2 H(\rho) + 2m\left(1 + 4\frac{\cos R - \cos\rho}{R^2 - \rho^2}\right) + \left(\frac{1}{2x^2} - m^2 - 1\right)\int_\rho^R dt \frac{1-\cos t}{t} \right\}$$

where $H(y) = 1 - 2(\sin y)/y + 2(1-\cos y)/y^2$, $R = 2(m+1)x$, $\rho = 2(m-1)x$, $x = ka$, $\alpha_1 = m(\sigma - 1/(\sigma+1))$, $\alpha_2 = m(1 - 1/(\sigma - 1))$, $\sigma = (m^2+1)/2m$. This analytical formula opens fresh opportunities for the solution of inverse problems (e.g. using spectral extinction measurements).

The error of the Perelman approximation for the value of C_{ext} was studied by Perelman (1986). It is smaller than 5% at $n \leq 1.22$ and $x \leq x_0(n)$, where the function $x_0(n)$ is given in Table 2.10.

The same approximation can be used to calculate the absorption cross-section (Perelman, 1994) and the phase function at small scattering angles (Perelman, 1991). This is important for the development of optical particle sizing techniques.

2.5 PARTICLES WITH REFRACTIVE INDEX LARGE COMPARED TO THAT OF A HOST MEDIUM

We have already considered approximations of the scattering theory at large and small values of size parameter x and small values of $\Delta n = n_p - n_s$, where n_p and n_s are refractive indices of particles and a surrounding medium, respectively. Another important approximation in the single scattering theory appears at large values of refractive index n (more precisely, Δn).

The main features of light scattering by perfect reflectors or optically hard particles ($n \to \infty$) will now be considered. For clarity, we choose the special case of spherical scatterers. From general equations of the Mie theory (see Appendix 2) as $n \to \infty$ (Shifrin, 1951) it follows that:

$$a_n = \frac{\psi_n'(x)}{\xi_n'(x)}, \qquad b_n = \frac{\psi_n(x)}{\xi_n(x)} \tag{2.115}$$

where $\xi_n(x) = \psi_n(x) - i\chi_n(x)$. We can see that light-scattering characteristics of such particles are functions of the size parameter x only. It is interesting to note that it follows from equations (2.115) that:

$$\mathrm{Re}(a_n) = |a_n|^2, \qquad \mathrm{Re}(b_n) = |b_n|^2 \tag{2.116}$$

and (see Appendix 2):

$$Q_{ext} = Q_{sca}, \qquad Q_{abs} = 0, \qquad Q_{pr} = (1-g)Q_{sca} \tag{2.117}$$

in the framework of this approximation. There is no absorption of light by such particles. Thus, the single scattering albedo is equal to 1. This is due to the fact that electromagnetic waves cannot penetrate such particles.

Light-scattering characteristics can be obtained from the equations presented in Appendix 2 (see Table A2.1), assuming that a_n and b_n can be derived from equations (2.115).

As in the case of optically soft particles, formulae are simplified for small and large values of n at $n \gg 1$. It follows from expansions in Appendix 3 as $x \to 0$ that:

$$\psi_1(x) = \frac{x^2}{3}, \qquad \xi_1(x) = \frac{x^2}{3} + \frac{i}{x}, \qquad \psi_1'(x) = \frac{2x}{3}, \qquad \xi_1'(x) = \frac{2x}{3} - \frac{i}{x^2}$$

and

$$a_1 = \frac{2ix^3}{3}, \qquad b_1 = -\frac{ix^3}{3} \tag{2.118}$$

Thus, we can obtain (see Appendix 2):

$$Q_{ext} = Q_{sca} = \tfrac{10}{3}x^4, \qquad Q_{pr} = \tfrac{14}{3}x^4, \qquad g = -0.4$$

$$S_{11}(\theta) = -ix^3\left(\frac{1}{2} - \cos\theta\right), \qquad S_{22}(\theta) = ix^3\left(1 - \frac{\cos\theta}{2}\right) \tag{2.119}$$

as $x \to 0$.

The phase function and the degree of polarization for small optically hard spheres can be calculated with the following equations:

$$p(\theta) = \frac{3}{4}\left(1 + \cos^2\theta - \frac{8}{5}\cos\theta\right) \tag{2.120}$$

$$P = \frac{i_1 - i_2}{i_1 + i_2} = \frac{3}{5}\frac{1 - \cos^2\theta}{1 + \cos^2\theta - \frac{8}{5}\cos\theta} \tag{2.121}$$

Phase functions of both Rayleigh and small optically hard ($x \to 0$) spheres are presented in Figure 2.22. We can see that all local scattering characteristics of small optically hard particles ($x \to 0$, $n \to \infty$) differ from the case of Rayleigh scattering ($x \to 0$, $|mx| \to 0$). For instance, the degree of polarization at $\theta = \pi/2$ is not equal to 1. It is much lower. Namely, it follows from equation (2.121) that

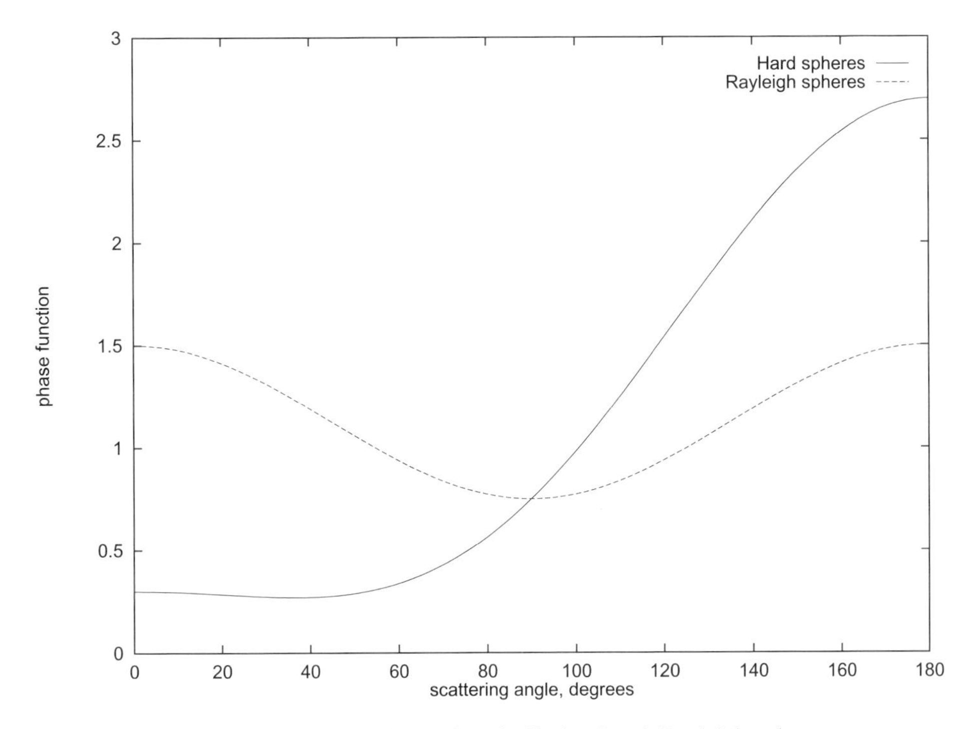

Figure 2.22. Phase functions of optically hard and Rayleigh spheres.

$P(\pi/2) = 0.6$. The phase function and the degree of polarization are not symmetrical functions relative to the angle $\theta = \pi/2$. They have larger values in the backward hemisphere. Thus, the probability of scattering in the backward hemisphere is larger and the asymmetry parameter is negative. Note that it follows:

$$\frac{p(\pi)}{p(0)} = 9 \tag{2.122}$$

for optically hard small spheres.

The physical reason for the difference is that Rayleigh scattering is entirely due to an electric dipole field. But as $n \to \infty$, the surface current gives rise to a magnetic dipole radiation.

Let us now consider the case of large optically hard particles briefly. Thus, it is assumed that the size of particles is much larger than the wavelength of incident radiation and geometrical optics equations can be applied. It follows in this case that $C_{sca} = 2\pi a^2$, $C_{abs} = 0$, $\omega_0 = 1$, $p(\theta) = 1$. Thus, large spheres at $n \gg 1$ scatter light isotropically. This is true for randomly oriented convex nonspherical particles at $n \gg 1$ as well (Van de Hulst, 1957). Such particles do not change the degree of polarization of the incident beam.

2.6 LAYERED PARTICLES

2.6.1 General equations

The particles we considered before were uniform. However, most particles in natural media are nonuniform. It is believed that interstellar dust grains contain a core of silicates and a mantle of organic refractory material (Hage, 1990). Other examples of nonuniform particles are water-coated hailstones, biological cells, foam bubbles and dielectric coated particles, composed of different metals. Even cloud droplets, aerosol particles, and ice crystals have different types of impurities inside, including soot, bubbles, thin films on their surfaces, etc.

All methods demonstrated in the previous sections can be applied to the problem of layered particles as well. We will only consider radially symmetric layered particles in this section. Information on scattering of light by other types of nonuniform particles can be found in the books written by Kerker (1969), Prishivalko *et al.* (1984), Lopatin and Sid'ko (1988), and Babenko *et al.* (2003).

The theory of light scattering by two layered concentric spheres is similar to those of uniform spheres (Aden and Kerker, 1951). The amplitude functions have the same analytical forms as those for uniform spheres. Namely, it follows (see Appendix 2):

$$S_1 = \sum_{n=1}^{\infty} \frac{2n+1}{n(n+1)} \{a_n \tau_n(\cos\theta) + b_n \pi_n(\cos\theta)\} \tag{2.123}$$

$$S_2 = \sum_{n=1}^{\infty} \frac{2n+1}{n(n+1)} \{a_n \pi_n(\cos\theta) + b_n \tau_n(\cos\theta)\} \tag{2.124}$$

where

$$\pi_n(\cos\theta) = \frac{P_n^{(1)}(\cos\theta)}{\sin\theta}, \qquad \tau_n(\cos\theta) = \frac{dP_n^{(1)}(\cos\theta)}{d\theta}$$

and

$$\left.\begin{aligned} a_n &= \frac{\psi_n(y)[\psi_n'(m_2 y) - A_n \chi_n'(m_2 y)] - m_2 \psi_n'(y)[\psi_n(m_2 y) - A_n \chi_n(m_2 y)]}{\xi_n(y)[\psi_n'(m_2 y) - A_n \chi_n'(m_2 y)] - m_2 \xi_n'(y)[\psi_n(m_2 y) - A_n \chi_n(m_2 y)]} \\ b_n &= \frac{m_2 \psi_n(y)[\psi_n'(m_2 y) - B_n \chi_n'(m_2 y)] - \psi_n'(y)[\psi_n(m_2 y) - B_n \chi_n(m_2 y)]}{m_2 \xi_n(y)[\psi_n'(m_2 y) - B_n \chi_n'(m_2 y)] - \xi_n'(y)[\psi_n(m_2 y) - B_n \chi_n(m_2 y)]} \end{aligned}\right\} \tag{2.125}$$

It follows for the values of A_n, B_n:

$$A_n = \frac{m_2 \psi_n(m_2 x)\psi_n'(m_1 x) - m_1 \psi_n'(m_2 x)\psi_n(m_1 x)}{m_2 \chi_n(m_2 x)\psi_n'(m_1 x) - m_1 \chi_n'(m_2 x)\psi_n(m_1 x)}$$

$$B_n = \frac{m_2 \psi_n(m_1 x)\psi_n'(m_2 x) - m_1 \psi_n(m_2 x)\psi_n'(m_1 x)}{m_2 \chi_n'(m_2 x)\psi_n(m_1 x) - m_1 \psi_n'(m_1 x)\chi_n(m_2 x)}$$

Here m_1 and m_2 are refractive indexes of the core and the mantle relative to that of the external medium, $x = ka$, $y = kb$, a is the radius of a core, b is the radius of the particle. Equations (2.125) for the amplitude coefficients a_n, b_n coincide with the solution of the Mie problem at the following special limits: $m_1 \to m_2$, $a \to b$, $a \to 0$ or $m_2 \to 1$, as it should be (see Appendix 2).

2.6.2 Rayleigh approximation

It follows from equations (2.123)–(2.125) as $y \to 0$ that:

$$S_1(\theta) = \tfrac{3}{2} a_1 \cos\theta$$
$$S_2(\theta) = \tfrac{3}{2} a_1$$

where

$$a_1 = \frac{2i}{3} y^3 F(m), \quad F(m) = \frac{(m_2^2 - 1)(m_1^2 + 2m_2^2) + q^3(2m_2^2 + 1)(m_1^2 - m_2^2)}{(m_2^2 + 2)(m_1^2 + 2m_2^2) + q^3(2m_2^2 - 2)(m_1^2 - m_2^2)}, \quad q = \frac{a}{b}$$

The values of $S_1(\theta)$ and $S_2(\theta)$ of Rayleigh particles can be expressed through the polarizability α (Van de Hulst, 1957):

$$S_1(\theta) = ik^3 \alpha \cos\theta, \qquad S_2(\theta) = ik^3 \alpha$$

Thus, for the polarizability of a two-layered sphere (Van de Hulst, 1957) we obtain:

$$\alpha = b^3 F(m)$$

The value of α can be used to calculate the absorption and scattering cross-sections of two-layered Rayleigh particles:

$$C_{abs} = -4\pi k \,\mathrm{Im}(\alpha)$$
$$C_{sca} = \tfrac{8}{3} \pi k^4 |\alpha|^2$$

It should be noted that $\alpha = 0$, $C_{abs} = 0$, $C_{sca} = 0$ at $F(m) = 0$. Thus, the particle does not scatter or absorb radiation. It does not act as an optical inhomogenity in this case. This property can be used in technological applications. If the denominator of $F(m)$ is close to zero, the polarizability α of a two-layered particle is large and there is a maximum in spectral dependencies of cross-sections $C_{abs}(\lambda)$ and $C_{sca}(\lambda)$.

The case of a coated ellipsoid small compared with the wavelength was studied by Bohren and Huffman (1983). For instance, it was shown that polarizabilities α_j of coated ellipsoids can be calculated from the following equation:

$$\alpha_j = \frac{V\{(\varepsilon_2 - \varepsilon_m)[\varepsilon_2 + (\varepsilon_1 - \varepsilon_2)(L_j^{(1)} - q^3 L_j^{(2)})] + q^3 \varepsilon_2(\varepsilon_1 - \varepsilon_2)\}}{[\varepsilon_2 + (\varepsilon_1 - \varepsilon_2)(L_j^{(1)} - q^3 L_j^{(2)})][\varepsilon_m + (\varepsilon_2 - \varepsilon_m)L_j^{(2)}] + q^3 L_j^{(2)} \varepsilon_2(\varepsilon_1 - \varepsilon_2)}$$

where V is the volume of the ellipsoid, ε_1, ε_2, ε_m are dielectric permittivities of the core, shell, and a host medium, respectively, $q = \sqrt{(a_1 b_1 c_1)/(a_2 b_2 c_2)}$, a_1, b_1, c_1(a_2, b_2, c_2) are semiaxes of the core (shell), and:

$$L_1^{(k)} = \frac{a_k b_k c_k}{2} \int_0^\infty \frac{ds}{(a_k^2 + s) f_k(s)}$$

$$L_2^{(k)} = \frac{a_k b_k c_k}{2} \int_0^\infty \frac{ds}{(b_k^2 + s) f_k(s)}$$

$$L_3^{(k)} = \frac{a_k b_k c_k}{2} \int_0^\infty \frac{ds}{(c_k^2 + s) f_k(s)}$$

where

$$f_k(s) = \sqrt{(a_k^2 + s)(b_k^2 + s)(c_k^2 + s)}$$

It is easy to show that the values of α and $|\alpha|^2$ in formulae for the amplitude functions, absorption, and scattering cross-sections can be found from the following equations in the case of randomly oriented particles:

$$\alpha = \frac{\alpha_1 + \alpha_2 + \alpha_3}{3}$$

$$|\alpha|^2 = \frac{|\alpha_1|^2 + |\alpha_2|^2 + |\alpha_3|^2}{3}$$

Clearly, it follows that $\alpha \equiv \alpha_1$ if the applied electric field is parallel to the axis a_1 of an ellipsoid. Multilayered ellipsoids were studied by Farafonov (2000).

2.6.3 Rayleigh–Gans approximation

The amplitude scattering matrix $\hat{S}(\theta)$ of a coated sphere in the framework of the Rayleigh–Gans approximation has the following form:

$$\hat{S}(\theta) = ik^3 \alpha \begin{pmatrix} \cos\theta & 0 \\ 0 & 1 \end{pmatrix} R(\theta)$$

where the value of polarizability α is as in the previous section and:

$$R(\theta) = R_b(\theta) + \frac{m_1 + m_2}{m_2 - 1} q^3 R_a(\theta)$$

Here m_1 and m_2 are relative refractive indexes of a core and a mantle, respectively. Values of $R_a(\theta)$ and $R_b(\theta)$ are defined by equation (2.101) for a mantle and a core, respectively. It follows in the case of two-layered spherical particles that:

$$R_a(\theta) = 3\frac{\sin u - u\cos u}{u^3}, \qquad R_b(\theta) = 3\frac{\sin u' - u'\cos u'}{u'^3} \tag{2.126}$$

where $u = 2ka\sin(\theta/2)$, $u' = 2kb\sin(\theta/2)$.

Thus, it follows for the form-factor in the case under study that:

$$\Phi(\theta) = \left| R_b(\theta) + \left(\frac{m_1 - m_2}{m_2 - 1} \right) q^3 R_a(\theta) \right|^2$$

This equation can be used to find the values of g, C_{sca}, $\hat{p}(\theta)$ of two-layered spheres according to formulae in Section 2.4.1. The absorption cross-section in this approximation is represented by the following formula:

$$C_{abs} = \gamma_1 V_1 + \gamma_2 (V - V_1)$$

where $\gamma_i = (4\pi\chi_i)/\lambda$, V_1 and V_2 are volumes of a core and a particle mantle, respectively, χ_1 and χ_2 are imaginary parts of their refractive indexes. These equations can be easily generalized for the case of an N-layered particle with the radius b:

$$\Phi(\theta) = \left| R_N(\theta) + \sum_{i=1}^{N-1} \frac{m_i - m_{i+1}}{m_N - 1} \frac{V_i}{V_n} R_i(\theta) \right|^2, \qquad C_{abs} = \sum_{i=1}^{N} \gamma_i V_i$$

where V_i and $m_i = n_i - \chi_i$ are the volume and the refractive index of an i-layer, respectively, $\gamma_i = (4\pi\chi_i)/\lambda$. Form-factors $R_i(\theta)$ and $R_N(\theta)$ are determined by equations (2.126) at $u = 2ka_i \sin(\theta/2)$ and $u = 2kb \sin(\theta/2)$, respectively. Here a_i is the radius of an i-layer. These equations are important for calculating the local optical characteristics of nonuniform spherically symmetrical particles that have a permanently varying refractive index $m(r)$ as well. In this case we should divide a particle into N thin layers, where $N \to \infty$.

2.6.4 Van de Hulst approximation

2.6.4.1 Coated ellipsoids

As shown above, the Van de Hulst approximation (or the anomalous diffraction approximation) is of importance in problems of light scattering by nonspherical optically soft particles. Here we will consider the case of an N-layered ellipsoid in the framework of this approximation. The case of an N-layered sphere will follow from the result obtained at $a_i = b_i = c_i$, where a_i, b_i, and c_i are semiaxes of an ellipsoid.

An ellipsoid with semiaxes $c < b < a$ is a smooth particle of the most general type. By varying the ratio of the axes of this particle we can obtain rod-shaped, disc-shaped, spheroidal, and spherical particles. Moreover, an ellipsoid is the most universal model of the shape of living microparticles (e.g. bacteria and spores, yeast and microalgae, plant and animal cells). The light extinction and absorption cross-sections of a homogeneous ellipsoid were investigated within the framework of the anomalous diffraction approximation by Lopatin and Sid'ko (1988).

Let us assume that characteristic sizes of a particle are large compared with the wavelength of the incident light λ, while the optical constants are close to those of a surrounding medium, which makes it possible to use the Van de Hulst approximation. The absorption C_{abs}, extinction C_{ext} cross-sections and the forward scattering amplitude function $S(0)$ are described by the following equations in the framework of this approximation (Lopatin and Sid'ko, 1988):

$$C_{abs} = \int_0^{\Sigma} [1 - \exp(2\,\mathrm{Im}(l))]\, d\sigma \tag{2.127}$$

$$C_{ext} = \frac{4\pi}{k^2}\mathrm{Re}(S(0)), \qquad S(0) = \frac{k^2}{2\pi}\int_0^{\Sigma} (1 - \exp(-il))\, d\sigma \tag{2.128}$$

where $k = 2\pi/\lambda$, l is the phase shift of the ray, Σ and $d\sigma$ are the projection area and elementary area of the particle in the plane perpendicular to the incident radiation, respectively.

Equations (2.127), (2.128) are applicable to large optically soft particles with arbitrary shape and structure. The primary element for using these equations is the establishment of the relation $l(\sigma)$ for a particle of the specific type. It follows for a homogeneous ellipsoid that:

$$l(\sigma) = kh(m - 1)\sqrt{1 - \frac{\sigma}{\Sigma}} \tag{2.129}$$

where $h = 2\pi abc/\Sigma$ is the maximum ray length within a particle and $m = n - i\chi$ is the refractive index of the scatterer. The function $l(\sigma)$ becomes a discontinuous function for a layered ellipsoid. It follows in this case that:

$$l(\sigma) = k\sum_{j=1}^{N} h_j(m_j - m_{j+1})p_j^{1/2}\Theta(p_j) \tag{2.130}$$

where $h_i = 3V_j/2\Sigma_j$, $p_j = 1 - (\sigma/\Sigma_j)$, $\Theta(p_j)$ is a step function (equal to zero for $p_j < 0$ and equal to unity in the remaining cases), $V_j = \frac{4}{3}\pi a_j b_j c_j$ is the volume of the jth ellipsoid, Σ_j is the area of its projection onto the plane perpendicular to the direction of incident radiation, N is the number of layers, $m_j = n_j - ix_j$ is the refractive index of the jth layers, $\Sigma_N = \Sigma$, and $m_{N+1} = 1$. The quantity Σ_j depends on the orientation of the jth ellipsoid and is determined by the following equation:

$$\Sigma_j = \pi\sqrt{(a_j c_j \varphi_j)^2 + (b_j c_j \mu_j)^2 + (a_j b_j \eta_j)^2}$$

where φ_j, μ_j, and η_j are the direction cosines of the incident radiation (relative to the co-ordinate system related to the axis of the jth ellipsoid).

It was assumed in deriving equation (2.130) that the intersection points of the axes of the internal and external ellipsoids are identical, while their arbitrary spatial orientation is assumed. Equation (2.130) reduces to equation (2.129) at $N = 1$. Thus, integrals (2.128) could be evaluated analytically (Lopatin and Sid'ko, 1988). For a two-layered ellipsoid ($N = 2$) we have from equation (2.130):

$$l(\sigma) = kh_1(m_1 - m_2)\sqrt{1 - \frac{\sigma}{\Sigma_1}}\Theta\left(1 - \frac{\sigma}{S_1}\right) + kh_2(m_2 - 1)\sqrt{1 - \frac{\sigma}{\Sigma}} \tag{2.131}$$

which leads to the following expressions for the values of C_{abs}, C_{ext}, and $S(0)$:

$$C_{abs} = 2\Sigma \int_0^1 \{1 - \exp[2\,\mathrm{Im}(l(u))]\} u\,du \tag{2.132}$$

$$C_{ext} = \frac{4\pi}{k^2}\mathrm{Re}(S(0)), \qquad S(0) = \frac{k^2\Sigma}{\pi}\int_0^1 [1 - \exp(-il(u))] u\,du \tag{2.133}$$

where a new integration variable:

$$u = \sqrt{1 - \frac{\sigma}{S}}$$

is introduced. It follows from equation (2.131) that:

$$l(u) = \begin{cases} l_1 = kh_2(m_2 - 1)u, & 0 \le u \le \alpha \\ l_2 = l_1 + kh^*(m_1 - m_2)\sqrt{u^2 - \alpha^2}, & \alpha < u \le 1 \end{cases} \tag{2.134}$$

where $\alpha = \sqrt{1 - \nu^2}$, $\nu = \sqrt{\Sigma_1/\Sigma}$, and $h^* = h_1/\nu$.

Equations (2.132)–(2.134) have been analysed in detail by Lopatin and Sid'ko (1988), based on the two-layer sphere model of a particle (then $\Sigma_1 = \pi a_1^2$, $\Sigma = \pi a_2^2$, $h_2 = h^* = 2a_2$; a_1 and a_2 are the inner and outer radii of the sphere).

Note that the relation $l(\sigma)$ is invariant with respect to a change in shape from ellipsoidal to spherical: only the parameters h_j and Σ_j change (see equations (2.131) and (2.134)). Thus, an optically soft large layered ellipsoidal particle can be replaced by an equivalent layered sphere. This equivalence is, of course, not complete, and is only observed with respect to the quantities (2.127), (2.128). For these equivalent particles it holds that:

$$h_j^s(m_j^s - m_{j+1}^s) = h_j^{el}(m_j^{el} - m_{j+1}^{el}), \qquad \Sigma_j^s = \Sigma_j^{el} \tag{2.135}$$

where the indices *s*, *el* refer to a spherical layer or an ellipsoidal layer, respectively. For a homogeneous sphere and an ellipsoid these equivalence conditions can be recast as:

$$h^s(m^s - 1) = h^{el}(m^{el} - 1), \qquad \Sigma^s = \Sigma^{el}$$

Thus, quantities (2.127), (2.128) for optically soft large spheres and ellipsoids are identical when two conditions hold:

(1) equal areas Σ of the projections of the sphere and ellipsoid onto the plane normal to the direction of incidence of the radiation;
(2) equality of the maximal phase shifts $L = h(m - 1)$ of a sphere and ellipsoid.

For multilayered particles, as follows from equation (2.135), these are necessary yet insufficient conditions. They must hold not only for the entire particle itself but also for each internal ellipsoid.

We can numerically evaluate single integrals (2.132), (2.133) utilizing equation (2.134). They are much simpler than equations (2.127), (2.128).

For small and large phase shifts $L = k\sum_{j=1}^{N} h_j(m_j - m_{j+1})$ further simplifications are possible. Let us demonstrate this. As $\text{Im}(L) \to 0$ we have from equations (2.127) and (2.130) that:

$$C_{abs} = 2k\sum_{j=1}^{N} h_j(\chi_j - \chi_{j+1})\int_0^{\Sigma} p_j^{1/2}\Theta(p_j)\,d\sigma$$

or, after integration:

$$C_{abs} = \frac{4k}{3}\sum_{j=1}^{N} h_j\Sigma_j(\chi_j - \chi_{j+1})$$

This equation can be rewritten as follows ($V_j = (2h_j\Sigma_j)/3$):

$$C_{abs} = 2k\sum_{j=1}^{N} V_j(\chi_j - \chi_{j+1}), \qquad \chi_{N+1} = 0$$

Let us assume now that all layers (with the exception of a core) do not absorb. Then we can obtain that:

$$\text{Im}(l) = -h_1\chi_1 p_1^{1/2}\Theta(p_1)$$

and the absorption cross-section of an ellipsoid is identical to the absorption cross-section of the core:

$$C_{abs} = \Sigma_1\left(1 + \frac{e^{-2\gamma}}{\gamma} + \frac{e^{-2\gamma} - 1}{2\gamma^2}\right)$$

where $\gamma = k\chi_1 h_1$.

Let us consider expressions (2.128), (2.133) for the extinction cross-section C_{ext} in more detail. It follows from equations (2.128), (2.133) that the value of $C_{ext} \to 2\Sigma$ as $\text{Re}(L) \to \infty$. On the other hand, expanding the exponent in the expression for C_{ext} (2.133) we obtain:

$$C_{ext} = \int_0^{\Sigma} l^2(\sigma)\,d\sigma \tag{2.136}$$

where terms of fourth and higher orders are ignored. Clearly, this expression is valid only at small values of the phase shift.

Integral (2.136) can be evaluated analytically. Indeed, it follows from equation (2.130), ignoring the absorption of light inside the particle, that:

$$l^2(\sigma) = k^2\sum_{j=1}^{N} h_j^2(m_j - m_{j+1})^2 p_j\Theta(p_j)$$
$$+ k^2\sum_{i\neq j}^{N} h_j h_i(m_i - m_{j+1})(m_i - m_{i+1})p_j^{1/2}p_i^{1/2}\Theta(p_i)\Theta(p_j)$$

Substitution of this formula into equation (2.136) yields:

$$C_{ext} = k^2\sum_{j=1}^{N} h_j^2(m_j - m_{j+1})^2\Sigma_j/2 + 2k^2\sum_{i>j} F_{ij}h_i h_j(m_j - m_{j+1})(m_i - m_{i+1}) \tag{2.137}$$

where

$$F_{ij} = \int_0^{\Sigma} p_i^{1/2} p_j^{1/2} \Theta(p_j)\Theta(p_i)\, d\sigma$$

or after integration:

$$F_{ij} = \frac{1}{4}\left\{\Sigma_i + \Sigma_j - \frac{(\Sigma_i - \Sigma_j)^2}{2\sqrt{\Sigma_i \Sigma_j}} \ln\left|\frac{\sqrt{\Sigma_i} + \sqrt{\Sigma_j}}{\sqrt{\Sigma_i} - \sqrt{\Sigma_j}}\right|\right\}$$

It results from equation (2.137) at $N = 1$ that:

$$C_{ext} = k^2 h_1^2 (m_1^2 - 1)\Sigma/2 \tag{2.138}$$

Equation (2.138) determines the extinction cross-section of a large optically soft homogeneous ellipsoid at small phase shifts (the Rayleigh–Gans approximation).

We derive from equation (2.137) for a two-layered ellipsoid ($N = 2$):

$$C_{ext} = k^2 h_1^2 (m_1 - m_2)^2 \Sigma_1/2 + k^2 h_2^2 (m_2 - 1)^2 \Sigma/2 + 2k^2 F_{12} h_1 h_2 (m_1 - m_2)(m_2 - 1) \tag{2.139}$$

where

$$F_{12} = \frac{\Sigma}{4}\left\{1 - \nu^2 + \frac{(1 - \nu^2)^2}{2\nu} \ln\left|\frac{1 + \nu}{1 - \nu}\right|\right\}$$

The well-known result (Lopatin and Sid'ko, 1988) follows from equation (2.139) in a two-layered sphere limit ($\nu = a_1/a_2$, $h_1 = 2a_1$, $h_2 = 2a_2$, $\Sigma_1 = \pi a_1^2$, $\Sigma_2 = \pi a_2^2$):

$$C_{ext} = 4k^2 a_1^2 (m_1 - m_2)^2 \Sigma_1/2 + 4k^2 a_2^2 (m_2 - 1)^2 \Sigma/2 + 2k^2 F_{12} a_1 a_2 (m_1 - m_2)(m_2 - 1)$$

which confirms our derivations.

Note that general expression (2.137) for an N-layered particle can be written in the form of equation (2.138) for a uniform particle:

$$C_{ext} = k^2 h_N^2 (\bar{m} - 1)^2 \Sigma/2$$

where the effective refractive index $\bar{m}$ is found from the following relationship:

$$(\bar{m} - 1)^2 = (m_N - 1)^2 + \sum_{j=1}^{N-1} \frac{h_j^2 \Sigma_j}{h_N^2 \Sigma}(m_j - m_{j+1})^2 + 4\sum_{i>j}^{N} F_{ij} \frac{h_i h_j}{h_N^2 \Sigma}(m_j - m_{j+1})(m_i - m_{i+1})$$

This takes into account the nonuniformity of a particle through introduction of the 'effective' refractive index of a particle. It simplifies calculations of light extinction by ensembles of layered ellipsoidal particles.

2.6.4.2 Layered spheres

Let us consider the case of layered spheres in more detail. Equations (2.108), (2.109) and the expression for the extinction efficiency factor for spherical two-layered particles became:

$$\left.\begin{aligned} S(\theta) &= \tfrac{1}{2}y^2 \int_0^{\pi/2} (1 - e^{-i\delta}) J_0(y\theta \cos\tau) \sin 2\tau \, d\tau \\ Q_{abs} &= \int_0^{\pi/2} (1 - e^{-\mathrm{Re}(i\delta)}) \sin 2\tau \, d\tau \\ Q_{ext} &= \frac{4}{y^2} \mathrm{Re}[S(0)] \end{aligned}\right\} \tag{2.140}$$

where θ is the scattering angle, $y = 2\pi b/\lambda$, b is the radius of a particle, λ is the wavelength of the incident radiation, $\tau = \pi/2 - \varphi$, φ is the angle of incidence of a ray on the sphere, and δ is the phase delay at a given incidence angle ϕ. To derive equations (2.140), an incident plane wave was represented as a large number of rays that pass through the sphere without changing direction. It follows that the Van de Hulst approximation is most effective for large scatterers ($b \gg \lambda$) owing to the use of ray representations. In addition, the refractive index of the particle material must be close to that of the surrounding medium ('soft' scatterers), so that refraction and reflection at the interface may be neglected. Note that this approximation for $S(\theta)$ takes into account the change of phase of the ray field within the scatterers. This is an advantage of this approximation compared with the more usual version of ray optics in which phase relationships are ignored.

For coated spherical particles, the value of b in equation (2.140) is the external radius of the scatterer, and the phase difference δ can be derived from a consideration of the paths of the rays in the particle (see equation (2.134)):

$$\delta = \begin{cases} \delta_1 = 2yu(m_2 - 1), & 0 \le u \le \alpha \\ \delta_2 = \delta_1 + 2y(m_1 - m_2)\sqrt{u^2 - \alpha^2}, & \alpha \le u \le 1 \end{cases} \tag{2.141}$$

where $\alpha = \sqrt{1 - \nu^2}$, $\nu = a/b$, $u = \sin\tau$, a is the internal radius of the particle, and $m_j = n_j - i\chi_j$ is the complex relative refractive index of the core ($j = 1$) or the coating ($j = 2$). Note that particles with non-concentric cores or with more than two layers can be treated in a quite analogous fashion. Substituting equation (2.141) into (2.140), we obtain after simple calculations:

$$\left.\begin{aligned} Q_{ext} &= 2 + \frac{4A}{(d^2 + f^2)^2} + \frac{4(d\cos\alpha f - \sin\alpha f)\alpha\, e^{-\alpha d}}{d^2 + f^2} - 4J \\ Q_{abs} &= 1 + \frac{\alpha\, e^{-2\alpha d}}{d} + \frac{e^{-2\alpha d} - 1}{2d^2} - 2M \\ S(0) &= y^2\{\tfrac{1}{2} + t^{-2}(1 - (1 + it\alpha)\, e^{-it\alpha}) - H\} \end{aligned}\right\} \tag{2.142}$$

where $t = 2y(m_2 - 1)$, $d = -\mathrm{Im}(t)$, $f = \mathrm{Re}(t)$:

$$J = \int_\alpha^1 u e^{\xi} \cos\psi \, du, \qquad M = \int_\alpha^1 u e^{2\xi} \, du, \qquad H = \int_\alpha^1 e^{-i\delta_2} u \, du,$$

$$\xi = \mathrm{Im}(\delta_2), \qquad \psi = \mathrm{Re}(\delta_2) \qquad A = e^{-\alpha d}(d^2 - f^2)\cos\alpha f - 2df\, e^{-\alpha d} \sin\alpha f + f^2 - d^2$$

By substituting equation (2.141) into (2.140) we can also find the small-angle part of the scattering phase function $p(\theta)$:

$$p(\theta) = \frac{4|S(\theta)|^2}{y^2 Q_{sca}}, \qquad Q_{sca} = Q_{ext} - Q_{abs} \tag{2.143}$$

Let us analyse equations (2.142). When $\alpha = b$ and $m_1 = m_2$ they transform to the familiar formulae of the Van de Hulst approximation for homogeneous spherical particles (Van de Hulst, 1957). In this approximation, the absorption efficiency factor Q_{abs} (2.142) is independent of the refractive indices of the core and coating n_1 and n_2. This is a general property of soft particles ($n_1 \to n_2 \to 1$). For a non-absorbing coating ($\chi_2 = 0$), we obtain:

$$C_{abs} = 2\pi a^2 K(\omega) \tag{2.144}$$

where $K(\omega) = \frac{1}{2} + e^{-\omega}/\omega + (e^{-\omega} - 1)/\omega^2$ is the Van de Hulst function, and $\omega = 4\chi_1 x$, $x = 2\pi a/\lambda$. If the core is weakly absorbing ($\omega \to 0$), then equation (2.144) is simplified to:

$$C_{abs} = \gamma_1 V \tag{2.145}$$

where $\gamma_1 = 4\pi\chi_1/\lambda$ is the absorption factor and V is the volume of the sphere. The numerical results obtained from equations (2.144) and (2.145) differ by less than 1% if the attenuation of light over the diameter of the particle ω is less than $\frac{8}{3} \times 10^{-2}$. From the above discussion the important conclusion follows. Namely, we find that absorption cross-sections of coated particles with soft nonabsorbing coatings are close to those of their cores. The physical cause of this property is obvious: the coating does not affect C_{abs} because it does not itself absorb and, in addition, owing to its softness ($n_2 \to 1$) it produces little focusing or defocusing of the radiation incident on the core. Significantly, under these circumstances the absorption cross-section of the core can be approximated by a wide range of formulae from the theory of scattering by uniform particles (e.g. the Rayleigh approximation, Rayleigh–Gans approximation, ray optics approximation), or calculations may be carried out with the Mie theory. For example, in the framework of the ray optics approximation we can obtain for large cores (see equation (2.49)):

$$C_{abs} = \pi a^2 \left[1 - \frac{2n^2}{c^2}(e^{-bc}(1 + bc) - e^{-c}(1 + c)) - W(n)(1 - e^{-bc})^2\right] \tag{2.146}$$

where $b = \sqrt{1 - n^{-2}}$.

The effect of the total internal reflection of light from the core must be taken into account in equation (2.146) at $n_1 < n_2$ (Kokhanovsky, 1988). Calculations of $Q_{abs} = C_{abs}/\pi b^2$ for a coated particle with a nonabsorbing coating, made using equations (2.144) (Zege and Kokhanovsky, 1989), (2.146), and the exact theory, indicate that the ray optics approximation (2.146) is preferable. This is to be expected, because equation (2.144) is the limit of equation (2.146) as $n \to 1$. The more n differs from unity, the greater the advantage of equation (2.146).

Concluding our analysis of equation (2.143) for Q_{abs}, we consider its asymptotic form for weak ($4\chi_2 y \ll 1$, $4\chi_1 \chi \ll 1$) and strong ($4\chi_2 y(1-\nu) \gg 1$) absorption. In the last case we obtain $Q_{abs} = 1$ from equation (2.142) (i.e. all radiation incident on the particle is absorbed by the shell). This asymptotic case becomes increasingly accurate as the coating becomes softer ($n_2 - 1 \ll 1$). With increasing n_2 we must allow for the reflection of light from the surface of a particle. It follows that $Q_{abs} = 1 - W(n_2)$ in this case. For small absorption, we obtain from equation (2.142):

$$C_{abs} = \frac{4\pi a^3}{3}\gamma_1 + \frac{4\pi(b^3 - a^3)}{3}\gamma_2 \tag{2.147}$$

where $\gamma_j = 4\pi\chi_j/\lambda$ is the absorption coefficient of the core ($j = 1$) or the coating ($j = 2$).

Let us now consider the extinction efficiency Q_{ext}. We should expect that, as in the case of uniform particles, the principal characteristic governing Q_{ext} is the phase difference L between a ray passing through the particle without deviation (the central ray) and a ray travelling the same path outside the sphere $L = 2y(n_2 - 1 + \nu(n_1 - n_2))$. As $L \to \infty(2y|n_2 - 1| \to \infty$, $2x|n_1 - n_2| \to \infty)$, it follows from equation (2.142) that $Q_{ext} \to 2$, in accordance with the ray optics approximation. Analysis of calculations made by the rigorous theory indicates that with an error of less than 10%, $Q_{ext} = 2$ for $L > 50$. For absorbing particles this asymptotic case is reached even more rapidly. As $L \to 0(2y|n_2 - 1| \to 0$, $2x|n_1 - n_2| \to 0)$, equation (2.142) for Q_{ext} is reduced to the Rayleigh–Gans approximation. It follows for nonabsorbing particles ($\chi_1 = \chi_2 = 0$) (see equation (2.139)) that:

$$Q_{ext} = 2y^2(n^* - 1)^2 \tag{2.148}$$

where

$$(n^* - 1)^2 = (n_2 - 1)^2 + (n_1 - n_2)^2\nu^4 + (n_2 - 1)(n_1 - n_2)(\nu + \nu^3 - \tfrac{1}{2}(1 - \nu^2)^2 \ln[(1 + \nu)/(1 - \nu)])$$

It follows, on general physical grounds, that the extinction curve $Q_{ext}(y)$ should experience several interference oscillations between the asymptotic limit in the region of large L and the parabolic dependence (2.148).

In principle, the positions of the extrema on the extinction curve can be found from the fact that the derivative $\partial Q_{ext}/\partial y$ will be equal to zero. But a simpler and clearer method involves considering the conditions of destructive interference for the central ray and a ray that does not intersect the particle: $L = (2l + 1)\pi$, where $l = 0$, 1, 2, ... It follows from this equation that the maxima on the curve $Q_{ext}(\beta)$ arise at $y_e = (2l + 1)\pi/2v$, where $v = n_2 - 1 + \nu(n_1 - n_2)$ and that the distance between neighbouring maxima is $\Delta y = 2y_0$. This simple estimate is confirmed by calculations made using the rigorous theory for coated particles. For example, even for rather 'hard' particles ($n_1 = 1.34$, $n_2 = 1.2$), for $\nu = 0.9$ we obtain $y_0 = 4.8$ ($\Delta y = 9.6$) from the approximate formula and $y_0 = 6$ ($\Delta y = 8$) from the rigorous theory (see Appendix 2). As $n_2 \to 1$, $n_1 \to n_2$, the parameter υ tends to zero, and the quantities y_0 and Δy tend to infinity. Thus, for soft particles, as we should expect, the first maximum on the extinction curve is attained only at large values of the diffraction parameter y, and the distances between the maxima of Δy are large. We can apply the same reasoning to multilayered particles. The equation given above for y_0 is valid for nonabsorbing and weakly absorbing particles. With increasing absorption, the interference maxima are suppressed starting at large orders l owing to a decrease in the intensity of the central ray compared with the ray that bypasses the sphere.

To conclude this section, we will briefly discuss the accuracy of the Van de Hulst approximation for coated particles by comparing it with the rigorous theory (Zege and Kokhanovsky, 1989). Exact calculations were carried out with parameter y varying from 1 to 200 at $\Delta y = 1$ increments for various values of m_1, m_2, and ν. The maximum error of the Van de Hulst approximation for the value of Q_{ext} is located near the first maximum of the extinction curve. It could be reduced considerably by adding the edge term $2y^{-2/3}$. We found that the accuracy of the Van dc Hulst approximation for Q_{ext} increases with increasing χ_1 and χ_2 and with decreasing ν.

A similar relationship is found for the absorption efficiency Q_{abs}. The maximum error of the approximation for the absorption efficiency factor is located in the vicinity of the first maximum on the extinction curve $Q_{ext}(y)$. As expected, the accuracy of the Van de Hulst approximation decreases with increasing values of $|n_1 - n_2|$ and $|n_2 - 1|$. For example, the maximum relative error in calculating Q_{abs} at $1 \leq y \leq 200$, $\nu = 0.9$, $\chi_1 = 10^{-2}$, $\chi_2 = 10^{-4}$ varies as follows: 2% ($n_1 = 1.01$; $n_2 = 1.02$), 4% ($n_1 = 1.01$; $n_2 = 1.1$), 11% ($n_1 = 1.05$; $n_2 = 1.1$, 17% ($n_1 = 1.1$; $n_2 = 1.05$). It follows that, even for $n_1 - 1 \approx 0.1$, the accuracy of the approximation for the absorption efficiency Q_{abs} is low and that we must resort to the ray optics approximation, which considers a particle as a kind of spherical lens which accounts for the focusing effect. But it is evident from the example given above that the Van de Hulst approximation is valid for particles with dimensions approximately equal to (or even smaller than) the wavelength, where the ray representations on which this approximation is based become invalid. The same applies to Q_{ext}.

Thus, in the case of soft-coated particles, the extinction efficiency factor Q_{ext} and the absorption efficiency factor Q_{abs} can be calculated with the Van de Hulst approximation (2.142). Equation (2.142) for Q_{ext} gives the same locations of maxima on the plot of $Q_{ext}(y)$ as the exact theory does.

The absorption cross-section of a coated particle with soft nonabsorbing coating is close to that of a uniform core. This conclusion is general and because the absorption cross-sections of spheroidal and cylindrical particles in the Van de Hulst approximation are known, for example, they may also be used for coated particles with soft, nonabsorbing coatings and with spheroidal and cylindrical cores. Particles with nonspherical, soft, nonabsorbing coatings or multilayered particles (in which all layers are soft and nonabsorbing) can also be analysed with this theory. This allows us to go beyond the exact solutions of electromagnetic scattering problems derived up to date.

2.6.5 Geometrical optics approximation

To derive geometrical optics formulae for the two-layered spheres we can follow the procedure outlined in Section 2.3.1 (i.e. by considering the history of rays incident on the surface of a particle using Fresnel equations and Snell's law (Xu *et al.*, 2004b)). However, let us take another route here and consider high-frequency asymptotics of the exact solution (Kerker, 1969) for a two-layered spherical particle. Both ways lead to the same result, of course. The solution of the electromagnetic scattering problem for a two-layered spherical particle is similar to that for uniform spheres (see Appendix 2). Expressions for the values of Mie coefficients a_l and b_l are given by equations (2.125). The coefficients a_l can be rewritten in terms of phase angles Ω_l as follows:

$$a_l = \frac{\tan\Omega_l}{\tan\Omega_l - i} = \tfrac{1}{2}(1 - e^{-i\Omega_l}) \tag{2.149}$$

where

$$\tan\Omega_l = \frac{\tan\delta_1 - \tan\delta_2\tan\gamma}{1 - \tan\alpha\tan\delta_2} \tag{2.150}$$

and

$$\left.\begin{aligned}
\tan\delta_1 &= (\psi_l(y)\psi_l'(m_2y) - M\psi_l'(y)\psi_l(m_2y))/t_l \\
\tan\gamma &= (\psi_l(y)\chi_l'(m_2y) - M\psi_l'(y)\chi_l(m_2y))/t_l \\
\tan\alpha &= (\chi_l(y)\chi_l'(m_2y) - M\chi_l'(y)\chi_l(m_2y))/t_l \\
\tan\delta_2 &= (\psi_l(m_2x)\psi_l'(m_1x) - N\psi_l'(m_2x)\psi_l(m_1x))/s_l \\
t_l &= \chi_l(y)\psi_l'(m_2y) - M\chi_l'(y)\psi_l(m_2y) \\
s_l &= \chi_l(m_2x)\psi_l'(m_1x) - N\chi_l'(y)\psi_l(m_2y)
\end{aligned}\right\} \tag{2.151}$$

Here $y = 2\pi b/\lambda$, $x = 2\pi a/\lambda$, m_j is the complex refractive index of the core ($j = 1$) and the shell ($j = 2$), a and b are the radii of the core and shell, $\psi_l(z) = \sqrt{(\pi z/2)}J_{l+1/2}(z)$, $\chi_l(z) = -\sqrt{(\pi z/2)}Y_{l+1/2}(z)$, $J_{l+1/2}(z)$ and $Y_{l+1/2}(z)$ are Bessel functions of the first and second kinds respectively, $M = m_2$, $N = m_1/m_2$, and λ is the wavelength of incident light. The coefficient b_l is determined in a similar way to a_l in equation (2.149), but M and N in equations (2.151) must be replaced by M^{-1} and N^{-1}. Consequently, we shall consider only the asymptotic behaviour of a_l.

The parameters Ω_l, δ_1 and δ_2 are phase angles for two-layered and uniform (with radii b and a) spheres, respectively. The advantage of using phase angles Ω_l is that they are real-valued for nonabsorbing particles. The new phase angles α and γ (see equations (2.151)) depend only upon the characteristics of the shell.

Let us consider the asymptotic behaviour of the coefficients a_l in equation (2.149) as $\lambda \to 0$. In this situation, the arguments of the Riccati–Bessel functions $\psi_l(z)$ and $\chi_l(z)$, that appear in the phase-angle equations (2.151), are large, and can be replaced by Debye asymptotic values (restricting derivation at first to the case of nonabsorbing scatterers) (see Appendix 3):

$$\left.\begin{aligned} \sqrt{\sin\tau}\,\psi_l(z) &= -\frac{\chi_l'(z)}{\sqrt{\sin\tau}} = \cos\left(zf - \frac{\pi}{4}\right) \\ -\sqrt{\sin\tau}\,\chi_l(z) &= -\frac{\psi_l'(z)}{\sqrt{\sin\tau}} = \sin\left(zf - \frac{\pi}{4}\right) \end{aligned}\right\} \tag{2.152}$$

where $\cos\tau = (l + \frac{1}{2})/z$, $f = \sin\tau - \tau\cos\tau$. Approximations (2.152) are applicable when $z \gg 1$ and $l + \frac{1}{2} + z^{1/2} \leq z$. Substituting equation (2.152) into (2.151), we obtain after elementary but cumbersome derivations that:

$$\left.\begin{aligned} &\operatorname{tg}\delta_1 = \frac{\sin\beta - r\cos\beta^+}{\cos\beta^- + r\sin\beta^+}, \quad \operatorname{tg}\delta_2 = \frac{\sin\delta^- - R\cos\delta^+}{\cos\delta^- + R\sin\delta^+} \\ &\operatorname{tg}\alpha = \frac{\sin\beta^- + r\cos\beta^+}{\cos\beta^- + r\sin\beta^+}, \quad \operatorname{tg}\gamma = \frac{-\cos\beta^- + r\sin\beta^+}{\cos\beta^- + r\sin\beta^+} \\ &\beta^{\pm} = y(\sin\tau \pm m_2\sin\tau' - (\tau \pm \tau')\cos\tau) \\ &\delta^{\pm} = x(m_2\sin\tau_2 \pm m_1\sin\tau_1 - (\tau_2 \pm \tau_1)m_2\cos\tau_2), \quad R = \frac{N\sin\tau_2 - \sin\tau_1}{N\sin\tau_2 + \sin\tau_1} \\ &y\cos\tau = n_2 y\cos\tau' = n_2 x\cos\tau_2 = n_1 x\cos\tau_1 = l + \tfrac{1}{2}, \quad r = \frac{M\sin\tau - \sin\tau'}{M\sin\tau + \sin\tau'} \end{aligned}\right\} \tag{2.153}$$

Once the phase angles δ_1, δ_2, α, and γ are calculated, we can readily obtain the scattering coefficients a_l and b_l, amplitude matrix $\hat{S}(\theta)$ and other light-scattering characteristics (Van de Hulst, 1957).

Note that τ and τ' (or τ_2 and τ_1) can be interpreted as the angles of incidence and refraction of rays by the shell and core, respectively, relative to the tangent to the interface. Then equations $\cos\tau = n_2\cos\tau'$ and $n_2\cos\tau_2 = n_1\cos\tau_1$, given above, represent Snell's law for the shell and core, respectively, and r and R are the Fresnel reflection coefficients of the first and second interfaces. By the localization principle (Van de Hulst, 1957), the conditions $l + \frac{1}{2} + y^{1/2} \leq y$, $l + \frac{1}{2} + x^{1/2} \leq x$ mean that rays incident on the shell or core at grazing angles ($\tau \sim \tau' \sim \tau_1 \sim \tau_2 \sim 0$) are not considered. Thus, we ignore the edge effects, which are significant in the region just outside the region of applicability of geometrical optics. We can obtain from equation (2.153):

$$\tau_2 = \arccos\left(\frac{\cos\tau}{n_2\nu}\right), \qquad \nu = \frac{x}{y} \tag{2.154}$$

We may therefore conclude that when $n_2\nu < \cos\tau$, the rays are not incident on the core. In this situation it is logical to make the substitutions $\delta_2 = 0$ and $\Omega_l = \delta_1$ in equation (2.150). If $n_2\nu \geq 1$, then the rays will be focused by a shell on the core at all angles of incidence (i.e. there will be a real-valued angle τ_2 for every τ). The same conclusion also follows from direct analysis of ray paths in the particle (Kokhanovsky, 1989).

To test equations (2.153) we can consider certain limiting cases. If we assume that $m_1 = m_2$, then $R = 0$, $\delta_2 = 0$, and $\Omega_l = \delta_1$. When $m_2 = 1$, we obtain $r = 0$, $\delta_1 = \alpha = 0$, $\gamma = \pi/2$, and $\Omega_l = \delta_2$. The former case represents a transfer from a two-layered to a uniform particle with radius equal to that of the shell, while in the second case the radius of the resultant particle is that of the core.

Equations (2.153) were derived for nonabsorbing particles ($\chi_j = 0$). But they are amenable to analytic continuation in the complex plane if $\chi_j \ll n_j$ (Van de Hulst, 1957). Under such circumstances we must treat m_1 and m_2 in equations (2.153) as complex numbers.

The use of asymptotic equations (2.153) instead of equations (2.151) for tangents of phase angles δ_1, δ_2, α, and γ greatly simplifies calculation of scattering characteristics for two-layered particles. They are also of importance for understanding the physics of light scattering by a layered particle. Naturally, the approximation error decreases with increasing size of two-layered scatterers (i.e. in the region where numerical calculations with exact formulas (2.149) become difficult).

The conversion from multipole expansions (e.g. the extinction efficiency factor $Q_{ext} = (2/y^2) \sum_{l=1}^{\infty} (2l+1) \operatorname{Re}(a_l + b_l))$ to the corresponding geometrical optics formulae is important on general grounds.

For large x, the exact scattering series contain many terms that are comparable in magnitude, and can therefore be replaced with little loss of accuracy by integrals:

$$\Sigma \to \int dl \tag{2.155}$$

We readily obtain from equations (2.153):

$$2l + 1 = 2y \cos \tau, \qquad dl = -y \sin \tau \, d\tau \tag{2.156}$$

Substituting (2.155) and (2.156) into series in terms of partial waves (see Appendix 2), we obtain:

$$\left.\begin{aligned} Q_{ext} &= 2 \operatorname{Re} \int_0^{\pi/2} (a(\tau) + b(\tau)) \sin 2\tau \, d\tau \\ Q_{sca} &= 2 \int_0^{\pi/2} (|a(\tau)|^2 + |b(\tau)|^2) \sin 2\tau \, d\tau \\ g &= \frac{2}{Q_{sca}} \operatorname{Re} \int_0^{\pi/2} (a(\tau)a^*(t) + b(\tau)b^*(t)) \sin 2\tau \, d\tau \end{aligned}\right\} \tag{2.157}$$

where $t = \arccos(\cos \tau + 1/y)$, and $a(\tau)$ and $b(\tau)$ are the coefficients of the scattering series (2.149) written in terms of asymptotic formulae (2.153). Note that integrals (2.157) with respect to the angle of incidence τ are equivalent to summation of all rays incident on the particle.

We know from numerical calculations that in the limit of large scatterers, $Q_{ext} \to 2$. Let us demonstrate this assertion by substituting equation (2.149) into (2.157) (for simplicity we consider only the case of nonabsorbing particles):

$$\left.\begin{aligned} Q_{ext} &= 2 \operatorname{Re} \int_0^{\pi/2} (1 - \tfrac{1}{2}(e^{-2i\Omega(\tau)} + e^{-2i\Omega'(\tau)})) \sin 2\tau \, d\tau \\ &= 2 - \int_0^{\pi/2} (\cos 2\Omega(\tau) + \cos 2\Omega'(\tau)) \sin 2\tau \, d\tau \end{aligned}\right\} \tag{2.158}$$

where Ω and Ω' are phase angles for a_l and b_l, respectively. For large particles, when $x \gg 1$, $(n_1 - n_2)x \gg 1$, $(n_2 - 1)y \gg 1$, parameters $\beta^{\pm}$ and $\delta^{\pm}$ are large, and consequently, so is Ω. As a result, $\cos 2\Omega$ and $2\Omega'$ take positive and negative values with equal frequency on the range of integration from 0 to $\pi/2$. Thus, from equation (2.158) it follows that $Q_{ext} \to 2$ as $\Omega, \Omega' \to \infty$. For absorbing scatterers, the cosines in equation (2.158) are multiplied by an additional factor (the attenuation of light within the core).

As $n = n_2 \to 1$, we readily obtain from equation (2.158) the familiar Van de Hulst equation for a homogeneous soft particle with a large diffraction parameter. For $n_1 \to n_2 \to 1 (n_1 = n_2)$, the anomalous diffraction approximation for two-layered particles follows from equation (2.158). Let us demonstrate this now.

Suppose that $n_1 \to n_2 \to 1$. Then, it follows from equation (2.153) that $r \to R \to 0$, $\alpha \to \delta_1 \to \beta^-$, $\delta_2 \to \delta^-$, $\operatorname{tg}\gamma \to -1$, $\Omega \to \delta^- + \beta^-$. Therefore, the phase angle Ω of a soft two-layered particle is equal to the sum of the phase angles of the core and shell. In addition it follows in this case that $M = N = 1$ and $a(\tau) = b(\tau)$. Substituting equations (2.149), (2.150), (2.153) into (2.158) and making the changes described above, we easily obtain the anomalous diffraction formulae for optically soft large two-layered scatterers, which were analysed in the previous section.

Obviously, for optically soft N-layered particles:

$$\Omega = \sum_{k=1}^{N} \delta_k$$

where some of the δ_k vanish, depending upon the angle of incidence τ (see equation (2.154)). The physical meaning of values δ_k is clear from the previous discussion. Namely, δ_1 is the phase angle for a core and δ_k at $k \geqslant 2$ are phase angles of shells.

The accuracy of equations (2.157) was studied in detail by Kokhanovsky (1990).

3

Radiative transfer

3.1 RADIATIVE TRANSFER EQUATION

Generally speaking, we need to solve the wave equation for a system of N particles ($N \to \infty$) for studies of optical properties of multiply light-scattering media. This is a difficult problem, as can be easily understood from previous chapters, where the complexity of light scattering by a single particle was presented. However, under some approximations it is possible to derive a simpler radiative transfer equation (RTE) from a general wave equation for N particles in an electromagnetic field (Ishimaru, 1978; Apresyan and Kravtsov, 1983; Papanicolaou and Burridge, 1975; Mishchenko, 2002):

$$\sigma_{ext}^{-1}(\vec{n}\,\mathbf{grad})I_t(\vec{r},\vec{n}) = -I_t(\vec{r},\vec{n}) + \frac{\omega_0}{4\pi}\int_{4\pi} p(\vec{n},\vec{n}')I_t(\vec{r},\vec{n}')\,d\Omega' + B_0(\vec{r},\vec{n}) \tag{3.1}$$

where $\vec{r} = x\vec{l}_x + y\vec{l}_y + z\vec{l}_z$ is the radius-vector of the observation point, the vector $\vec{n} = l\vec{e}_x + m\vec{e}_y + n\vec{e}_z$ determines the direction of beam with intensity I_t, $B_0(\vec{r},\vec{n})$ is the internal source function. Note that equation (3.1) does not account for the possible frequency charge due to a scattering process. Also we assume that both light source and disperse medium characteristics do not change with time.

Equation (1.40) follows from equation (3.1) at $B_0(\vec{r},\vec{n}) = 0$ using the equality:

$$(\vec{n}\,\mathbf{grad})I_t(\vec{r},\vec{n}) \equiv \frac{dI_t(\vec{r},\vec{n})}{dL}$$

For a plane-parallel light-scattering layer, illuminated at every point on the top by a unidirectional beam of light, the RTE can be written in the following form (Sobolev, 1972):

$$\cos\vartheta\frac{dI(\tau,\vartheta,\phi)}{d\tau} = -I(\tau,\vartheta,\phi) + B(\tau,\vartheta,\phi) \tag{3.2}$$

where

$$B(\tau, \vartheta, \phi) = \frac{\omega_0}{4\pi} \int_0^{2\pi} d\phi' \int_0^{\pi} I(\tau, \vartheta', \phi') p(\theta') \sin \vartheta' \, d\vartheta' + \frac{\omega_0 I_0}{4} p(\theta) \, e^{-\tau/(\cos \vartheta_0)} \quad (3.3)$$

is the source function and the value of B_0 is ignored. Here $\tau = \sigma_{ext} L$ is the optical thickness of a layer, L is the geometrical thickness, ϑ_0 is the incidence angle, ϑ is the observation angle, ϕ is the azimuth of observed radiation:

$$\cos \theta' = \cos \vartheta \cos \vartheta' + \sin \vartheta \sin \vartheta' \cos(\phi - \phi')$$

$$\cos \theta = \cos \vartheta \cos \vartheta_0 + \sin \vartheta \sin \vartheta_0 \cos \phi \quad (3.4)$$

ω_0 is the single scattering albedo, $p(\theta)$ is the phase function, $I(\tau, \vartheta, \phi)$ is the diffused intensity at the optical thickness τ in the direction (ϑ, ϕ), πI_0 is the net flux per unit area normal to the incident light beam. We assume that the azimuth of the incident radiation ϕ_0 is equal to 0. The total intensity I_t (see equation (3.1)) is:

$$I_t = I(\tau, \vartheta, \phi) + \pi I_0 \exp\left(-\frac{\tau}{\xi} \right) \delta(\vec{\Omega} - \vec{\Omega}_0) \quad (3.5)$$

where $\delta(\vec{\Omega} - \vec{\Omega}_0)$ is the delta function, $\vec{\Omega}_0$ is the solid angle in the direction of incidence $\xi = \cos \vartheta_0$. Equation (3.2) is simpler than equation (3.1) because it only holds for diffused light. We can account for direct light with equation (3.5).

Boundary conditions for equation (3.2) can be presented in the following form (Sobolev, 1972):

$$I(0, \vartheta, \varphi) = 0 \quad \text{at } \vartheta < \frac{\pi}{2}, \qquad I(\tau_0, \vartheta, \varphi) = 0 \quad \text{at } \vartheta > \frac{\pi}{2} \quad (3.6)$$

These conditions underline the fact that diffused radiation does not enter a scattering layer either from the top ($\tau = 0$) or from the bottom ($\tau = \tau_0$). We can solve the differential equation (3.2) and obtain:

$$I(\tau, \eta, \phi) = \frac{e^{-\tau/\eta}}{\eta} \int_0^{\tau} B(\tau', \eta, \phi) \, e^{\tau'/\eta} \, d\tau' \quad \text{at } \eta > 0 \quad (3.7)$$

and

$$I(\tau, \eta, \phi) = \frac{e^{-\tau/\eta}}{\eta} \int_{\tau_0}^{\tau} B(\tau', \eta, \phi) \, e^{\tau'/\eta} \, d\tau' \quad \text{at } \eta < 0 \quad (3.8)$$

where $\eta = \cos \vartheta$. Also, we define: $\mu = |\eta|$, $\mu_0 \equiv \xi$. It follows from equation (3.4) in terms of these angles that $\cos \theta = (-1)^l \mu \mu_0 + \sqrt{(1 - \mu^2)(1 - \mu_0^2)} \cos \phi$, where $l = 1$ for reflected and $l = 2$ for transmitted light, respectively. If the function $B(\tau', \eta, \phi)$ is known, equations (3.7) and (3.8) can be used to calculate the intensity of the light field at any optical thickness τ inside a layer. Equation (3.7) represents the downward radiation and equation (3.8) represents the upward radiation. The first boundary condition in equation (3.6) can be obtained from equation (3.7) at $\tau = 0$. The second boundary condition follows from equation (3.8) at $\tau = \tau_0$.

Very often we need to know the light field intensity escaping from the top ($I_\uparrow(0, \eta, \xi, \varphi)$) and the bottom ($I_\downarrow(\tau_0, \eta, \xi, \varphi)$) of a scattering layer. They can be obtained from general equations (3.7), (3.8):

$$I_\uparrow(0, \eta, \xi, \phi) = -\frac{1}{\eta}\int_0^{\tau_0} B(\tau', \eta, \xi, \phi)\, e^{\tau'/\eta}\, d\tau', \quad \eta < 0 \tag{3.9}$$

$$I_\downarrow(\tau_0, \eta, \xi, \phi) = \frac{e^{-\tau_0/\eta}}{\eta}\int_0^{\tau_0} B(\tau', \eta, \xi, \varphi)\, e^{\tau'/\eta}\, d\tau', \quad \eta > 0 \tag{3.10}$$

The function $B(\tau', \eta, \xi, \phi)$ is not known a priori in most of cases. The integral equation for this function can be obtained from equations (3.3), (3.7), (3.8):

$$\begin{aligned} B(\tau, \eta, \xi, \phi) = &\frac{\omega_0}{4\pi}\int_0^{2\pi} d\phi' \left\{ \int_0^1 p(\theta')\, d\eta' \int_0^{\tau} B(\tau', \eta', \xi, \phi')\, e^{(\tau'-\tau)/\eta'} \frac{d\tau'}{\eta'} \right. \\ &\left. - \int_{-1}^0 p(\theta')\, d\eta' \int_{\tau}^{\tau_0} B(\tau', \eta', \xi, \phi')\, e^{(\tau'-\tau)/\eta'} \frac{d\tau'}{\eta'} \right\} + \frac{I_0\omega_0}{4} p(\theta)\, e^{-\tau/\xi} \end{aligned} \tag{3.11}$$

Equation (3.11) cannot be solved analytically. Different numerical and approximate methods are used to solve equations (3.2), (3.7)–(3.11) (Lenoble, 1985). Benchmark results were published by Garcia and Siewert (1985).

It is evident that the solution of equation (3.2) for the diffused intensity should depend on optical thickness τ_0, single scattering albedo ω_0, phase function $p(\theta)$, depth τ, and geometrical parameters (angles ϑ, ϑ_0, φ). Though this solution is not known in the general case, it can be derived approximately at small or large values of the optical thickness τ_0, the single scattering albedo ω_0, and the asymmetry parameter g. We will consider these asymptotics now. Note that the accuracy of different multiple scattering approximations was studied in detail by King and Harshvardhan (1986).

3.2 THIN LAYERS

For a thin layer ($\tau \to 0$) we can ignore photons, scattered more than once (the integral in equation (3.11) is assumed to be equal to 0) and obtain the following analytical solution for the source function (see equation (3.3)):

$$B(\tau, \vartheta, \vartheta_0, \phi) = \frac{I_0\omega_0 p(\theta)}{4}\, e^{-\tau/\xi} \tag{3.12}$$

From equations (3.9), (3.10), (3.12), under the assumption that values of ω_0 and $p(\theta)$ do not depend on depth τ (a homogeneous layer), it follows (Sobolev, 1956) that:

$$I_\uparrow^d = \frac{\omega_0 I_0 \xi}{4(\mu+\xi)}\left\{1 - \exp\left[-\left(\frac{1}{\mu}+\frac{1}{\xi}\right)\tau_0\right]\right\} p(\theta) \tag{3.13}$$

$$I_\downarrow^d = \frac{\omega_0 I_0 \xi}{4(\mu-\xi)}\left\{\exp\left[-\frac{\tau_0}{\mu}\right] - \exp\left[\frac{\tau_0}{\xi}\right]\right\} p(\theta) \quad \text{at } \mu \neq \xi \tag{3.14}$$

and

$$I_{\downarrow}^{d} = \frac{\omega_0 I_0 \tau_0}{4\mu} \exp\left[-\frac{\tau_0}{\mu}\right] p(\theta) \quad \text{at } \mu = \xi \tag{3.15}$$

where $I_{\downarrow}^{d}$ ($I_{\uparrow}^{d}$) denotes the downward (upward) diffused light intensity and $\mu = |\eta|$. Let us introduce the reflection function:

$$R = \frac{I_{\uparrow}^{d}}{\xi I_0}$$

and the transmission function:

$$T = \frac{I_{\downarrow}^{d}}{\xi I_0}$$

It follows from equations (3.13), (3.14) that:

$$R = \frac{\omega_0 p(\theta)}{4(\mu + \xi)} \{1 - e^{-(1/\mu + 1/\xi)\tau_0}\} \tag{3.16}$$

$$T = \frac{\omega_0 p(\theta)}{4(\mu - \xi)} \{e^{-\tau_0/\mu} - e^{-\tau_0/\xi}\} \tag{3.17}$$

where $\theta = \arccos(\mu\xi + \sqrt{(1-\mu^2)(1-\xi^2)}\cos\phi)$ for transmitted light and $\theta = \arccos(-\mu\xi + \sqrt{(1-\eta^2)(1-\mu^2)}\cos\phi)$ for reflected light. These two equations are extremely important and have already been used in many applications. As $\tau_0 \to \infty$ it follows from equations (3.16), (3.17) that:

$$R = \frac{\omega_0 p(\theta)}{4(\mu + \xi)}, \qquad T = 0$$

The first equation represents the contribution of singly scattered light to the reflection function of a semi-infinite medium. We can see that this contribution is larger for weakly absorbing media ($\omega_0 \approx 1$) and large incidence and observation angles ($\mu \approx \xi \approx 0$ or $\vartheta \approx \vartheta_0 \approx \pi/2$). It depends on the phase function $p(\theta)$ of a scattering medium as well. It follows that:

$$R = \frac{\omega_0}{4(\mu + \xi)}$$

for the isotropic scattering ($p(\theta) = 1$) and:

$$R = \frac{3\omega_0\left(1 + \mu^2\xi^2 + (1-\mu^2)(1-\xi^2)\cos^2\phi - 2\mu\xi\sqrt{(1-\mu^2)(1-\xi^2)}\cos\phi\right)}{16(\mu + \xi)}$$

for the Rayleigh phase function $p(\theta) = \frac{3}{4}(1 + \cos^2\theta)$.

As $\tau_0 \to 0$ we can obtain from equations (3.16), (3.17):

$$R = \frac{\omega_0 \tau_0}{4\mu\xi} p\left(-\mu\xi + \sqrt{(1-\mu^2)(1-\xi^2)}\cos\phi\right)$$

$$T = \frac{\omega_0 \tau_0}{4\mu\xi} p\left(\mu\xi + \sqrt{(1-\mu^2)(1-\xi^2)}\cos\phi\right)$$

It follows for the isotropic scattering ($p(\theta) = 1$) that:

$$R = T = \frac{\omega_0 \tau_0}{4\mu\xi}$$

We can see that functions:

$$R^* = 4\mu\xi R$$

and

$$T^* = 4\mu\xi T$$

do not depend on incidence and observation angles for isotropic scattering as $\tau_0 \to 0$. Thus, their usage has some advantages as pointed out by Chandrasekhar (1950). It follows for these functions that:

$$R^* = \frac{4\mu I_\uparrow^d}{I_0}, \qquad T^* = \frac{4\mu I_\downarrow^d}{I_0}$$

We can take account of the reflection of the underlying Lambertian surface with albedo A_s by using the following equations (Sobolev, 1972):

$$\hat{R} = R + \frac{A_s t(\eta) t(\xi)}{1 - r_s A_s} \tag{3.18}$$

$$\hat{T} = T + \frac{A_s r(\eta) t(\xi)}{1 - r_s A_s} \tag{3.19}$$

where $\hat{R}$ and $\hat{T}$ are reflection and transmission functions of a layer with an underlying Lambertian surface, R and T are the same functions for a layer without an underlying surface and:

$$t(\xi) = \frac{1}{\pi} \int_0^{2\pi} d\phi \int_0^1 T(\eta, \xi, \tau_0) \eta \, d\eta + e^{-\tau_0/\xi} \tag{3.20}$$

$$r(\xi) = \frac{1}{\pi} \int_0^{2\pi} d\phi \int_0^1 R(\mu, \xi, \tau_0) \mu \, d\mu \tag{3.21}$$

$$r_s = 2 \int_0^1 r(\xi) \xi \, d\xi \tag{3.22}$$

The values $t(\xi)$, $r(\xi)$, and r are called the total transmittance, the plane albedo, and the spherical albedo, respectively. The integral in equation (3.20) gives the diffused transmittance.

3.3 THICK LAYERS

3.3.1 General solutions

Let us now derive the approximate solution of the radiative transfer equation for optically thick layers. To this end we should consider the RTE for intensity $I(\tau, \eta)$ averaged over the azimuth, which follows from equation (3.2):

$$\eta \frac{I(\tau, \eta)}{d\tau} = -I(\tau, \eta) + B(\tau, \eta) + B_0^*(\tau_0, \xi)$$

where

$$B(\tau,\eta) = \frac{\omega_0}{2}\int_{-1}^{1} p(\eta,\eta')I(\tau,\eta')\,d\eta'$$

$$B_0^*(\tau_0,\xi) = \frac{\omega_0 I_0}{4}\overline{p(\theta)}\,e^{-\tau_0/\xi}$$

$$p(\eta,\eta') = \frac{1}{2\pi}\int_0^{2\pi} p(\theta)\,d\phi$$

Here $\overline{p(\theta)}$ is the phase function averaged over the azimuthal angle. At large thickness τ_0 we can ignore the term $B_0^*(\tau_0,\eta)$ and use the following representation (Sobolev, 1972):

$$I(\tau,\eta) = i(\eta)\,e^{-k\tau}$$

where k is the diffusion exponent. Thus, it follows from the RTE that:

$$i(\eta) = \frac{\omega_0}{2(1-k\eta)}\int_{-1}^{1} p(\eta,\eta')i(\eta')\,d\eta'$$

The function $i(\eta)$ describes the angular distribution of light intensity in deep layers of semi-infinite disperse media. Multiplying the RTE by values of $i(\eta)$ and $i(-\eta)$ and integrating over η from -1 to 1 we obtain the following equations (Sobolev, 1984):

$$\frac{d}{d\tau}\int_{-1}^{1} I(\tau,\eta)i(\eta)\eta\,d\eta = -k\int_{-1}^{1} I(\tau,\eta)i(\eta)\eta\,d\eta + \int_{-1}^{1} B_0^*(\tau,\eta)i(\eta)\,d\eta$$

$$\frac{d}{d\tau}\int_{-1}^{1} I(\tau,\eta)i(-\eta)\eta\,d\eta = k\int_{-1}^{1} I(\tau,\eta)i(-\eta)\eta\,d\eta + \int_{-1}^{1} B_0^*(\tau,\eta)i(-\eta)\,d\eta$$

Solutions of these linear differential equations for integrals

$$K_1(\tau) = \int_{-1}^{1} I(\tau,\eta)i(\eta)\eta\,d\eta$$

$$K_2(\tau) = \int_{-1}^{1} I(\tau,\eta)i(-\eta)\eta\,d\eta$$

subject to the conditions that $I(0,\eta) = 0$ for $\eta > 0$ and $I(\tau_0,\eta) = 0$ for $\eta < 0$ are:

$$K_1(\tau) = -e^{-k\tau}\int_{-1}^{1} I(0,-\eta)i(-\eta)\eta\,d\eta + C(\tau)$$

$$K_2(\tau) = e^{-k(\tau_0-\tau)}\int_{-1}^{1} I(\tau_0,\eta)i(-\eta)\eta\,d\eta - D(\tau)$$

where

$$C(\tau) = \int_0^\tau e^{-k(\tau-t)}\, dt \int_{-1}^1 B_0^*(t,\eta) i(\eta)\, d\eta$$

$$D(\tau) = \int_\tau^{\tau_0} e^{k(\tau-t)}\, dt \int_{-1}^1 B_0^*(t,\eta) i(-\eta)\, d\eta$$

Thus, for reflection

$$R(\eta, \xi, \tau_0) = \frac{I(0, -\eta, \xi)}{I_0 \xi}$$

and transmission

$$T(\eta, \xi, \tau_0) = \frac{I(\tau_0, \eta, \xi)}{I_0 \xi}$$

functions, it follows that:

$$2\int_0^1 R(\eta,\xi,\tau_0) i(\eta)\eta\, d\eta + 2e^{-k\tau_0} \int_0^1 T(\eta,\xi,\tau_0) i(-\eta)\eta\, d\eta = i(-\xi)(1 - e^{-k(1+1/\xi)\tau_0})$$

$$2\int_0^1 T(\eta,\xi,\tau_0) i(\eta)\eta\, d\eta + 2e^{-k\tau_0} \int_0^1 R(\eta,\xi,\tau_0) i(-\eta)\eta\, d\eta = i(\xi)(e^{-k\tau_0} - e^{-\tau_0/\xi})$$

where we accounted for the following equalities:

$$C(\tau_0) = \tfrac{1}{2} I_0 i(\xi)\xi(\exp(-k\tau_0) - \exp(-\tau_0/\xi))$$
$$D(0) = \tfrac{1}{2} I_0 i(-\xi)\xi(1 - \exp(-\tau_0(k + 1/\xi)))$$

It is possible to assume from physical grounds and symmetry relations for values of $R(\eta, \xi, \tau)$, $T(\eta, \xi, \tau)$ that (Sobolev, 1984):

$$R_\infty(\eta, \xi) - R(\eta, \xi, \tau_0) = f(\tau_0) K(\eta) K(\xi)$$

and

$$T(\eta, \xi, \tau_0) = g(\tau_0) K(\eta) K(\xi)$$

where $R_\infty(\eta, \xi) \equiv R(\eta, \xi, \tau_0 = \infty)$, $K(\eta)$ is the intensity of escaped radiation in Mulne's problem (i.e. for radiation sources at infinite optical depth). As $\tau_0 \to \infty$ we get $g = f = 0$. Substituting these equations into integral relations for $R(\eta, \xi, \tau_0)$, $T(\eta, \xi, \tau_0)$ and ignoring small terms containing $e^{-\tau_0/\xi}$ it is possible to derive the following equations:

$$f(\tau_0) = g(\tau_0) l\, e^{-k\tau_0}, \qquad g(\tau_0) = \frac{m\, e^{-k\tau_0}}{1 - l^2\, e^{-2k\tau_0}}$$

where

$$m = 2\int_{-1}^1 i^2(\eta)\eta\, d\eta, \qquad l = 2\int_0^1 K(\eta) i(-\eta)\eta\, d\eta$$

Here the following well-known formulae, which follow from invariance principles (Ambarzumian, 1961) were used:

$$i(-\xi) = 2\int_0^1 R_\infty(\eta,\xi)i(\eta)\eta\,d\eta$$

$$i(\xi) = 2\int_0^1 R_\infty(\eta,\xi)i(-\eta)\eta\,d\eta + mK(\xi)$$

Note that functions $i(\eta)$ and $K(\eta)$ are normalized by the following conditions:

$$\frac{\omega_0}{2}\int_0^1 i(\eta)\eta\,d\eta = 1, \qquad 2\int_0^1 K(\eta)i(\eta)\eta\,d\eta = 1$$

Thus, it follows for thick layers ($\tau \to \infty$) (Sobolev, 1972) that:

$$\left.\begin{aligned} R(\tau_0,\mu,\mu_0,\phi) &= R_\infty(\mu,\mu_0,\phi) - T(\tau_0,\mu,\mu_0)l\,e^{-k\tau_0}, \\ T(\tau_0,\mu,\mu_0) &= tK(\mu)K(\mu_0), \qquad \omega_0 < 1, \\ R(\tau_0,\mu,\mu_0,\phi) &= R^0_\infty(\mu,\mu_0,\phi) - T(\tau_0,\mu,\mu_0), \\ T(\tau_0,\mu,\mu_0) &= \frac{K(\mu)K(\mu_0)}{0.75(1-g)\tau + \Delta}, \qquad \omega_0 = 1 \end{aligned}\right\} \tag{3.23}$$

where $\mu = |\eta|$, $\mu_0 = |\xi|$, $R_\infty(\mu,\mu_0,\phi)$ is the reflection function of a semi-infinite medium with the same local optical characteristics $(\omega_0, p(\theta))$ as a finite layer, $R^0_\infty(\mu,\mu_0,\phi)$ is the same function at $\omega_0 = 1$, $t = (m\,e^{-k\tau_0})/(1 - l^2\,e^{-2k\tau_0})$, $l = 2\int_0^1 K(\mu)i(-\mu)\,d\mu$, $m = 2\int_{-1}^1 i^2(\mu)\mu\,d\mu$, $\Delta = 3\int_0^1 K(\mu)\mu^2\,d\mu$. Functions $i(\mu)$ and $R_\infty(\mu,\mu_0,\phi)$ can be derived from the following integral equations (Ambarzumian, 1961; Sobolev, 1972; Van de Hulst, 1980):

$$i(\mu) = \frac{\omega_0}{2(1-k\mu)}\int_{-1}^1 p(\mu,\mu')i(\mu')\,d\mu' \tag{3.24}$$

$$\begin{aligned} R_\infty(\mu_0,\phi_0,\mu,\phi) &= \frac{\omega_0}{4(\mu+\mu_0)}p(\theta) \\ &\quad + \frac{\mu_0\omega_0}{4\pi(\mu_0+\mu)}\int_0^1\int_0^{2\pi} p(\mu,\phi,\mu',\phi')R_\infty(\mu',\phi',\mu_0,\phi_0)\,d\mu'\,d\phi' \\ &\quad + \frac{\mu\omega_0}{4\pi(\mu_0+\mu)}\int_0^1\int_0^{2\pi} p(\mu_0,\phi_0,\mu',\phi')R_\infty(\mu',\phi',\mu,\phi)\,d\mu'\,d\phi' \\ &\quad + \frac{\omega_0\mu\mu_0}{4\pi^2(\mu_0+\mu)}\int_0^1 d\phi'\int_0^{2\pi} R_\infty(\mu',\phi',\mu,\phi)\,d\mu' \\ &\quad \times \int_0^{2\pi} d\phi''\int_0^1 p(-\mu',\phi',\mu'',\phi'')R_\infty(\mu'',\phi'',\mu_0,\phi_0)\,d\mu'' \end{aligned} \tag{3.25}$$

where θ is the scattering angle. The details of the solution of equation (3.25) are given by Mishchenko *et al.* (1999).

Note that the function $i(\mu)$ is the solution of equation (3.24) at the minimal characteristic number k and (Sobolev, 1972) that:

$$K(\mu) = \frac{i(\mu)}{m} - \frac{2}{m}\int_0^1 R_\infty(\mu, \mu_0) i(-\mu_0)\mu_0 \, d\mu_0$$

where

$$R_\infty(\mu, \mu_0) = \frac{1}{2\pi}\int_0^{2\pi} R_\infty(\mu, \mu_0, \phi) \, d\phi$$

Asymptotic relations (3.23) have already been used in many applications. It is difficult to apply them directly for calculation of reflection and transmission functions of particulate media, because they include two unknown functions $R_\infty(\mu, \mu_0, \phi)$, $K(\mu)$ and three unknown constants k, l, m. However, it is important that these functions do not depend on the optical thickness τ_0. They depend only on the geometry, single scattering albedo ω_0 and phase function, and can be calculated, for example, with codes developed by Konovalov (1975) and Nakajima and King (1992).

Equations (3.23) can be used to generate simple approximations in some particular cases. Here we consider three of them: isotropic scattering ($p(\theta) = 1$), nonabsorbing media ($\omega_0 = 1$) and weakly absorbing media ($\omega_0 \to 1$).

3.3.2 Isotropic scattering

It follows in the case of isotropic scattering ($p(\theta) = 1$) from equation (3.24):

$$i(\mu) = \frac{\omega_0}{1 - k\mu} \tag{3.26}$$

where the normalization condition

$$\frac{1}{2}\int_{-1}^{1} i(\mu) \, d\mu = 1 \tag{3.27}$$

was used. Note that we have: $i(\mu) = 1$ at $\omega_0 = 1$. To find the value of k we need to solve the following equation:

$$\frac{1}{2k}\ln\left[\frac{1+k}{1-k}\right] = \frac{1}{\omega_0} \tag{3.28}$$

or $k = \omega_0 \operatorname{arctanh}(k)$, which can be derived substituting equation (3.26) into equation (3.27). The function $k(\omega_0)$ obtained from equation (3.28) is presented in Table 3.1.

It follows for the value of $R_\infty(\mu, \mu_0)$ from equation (3.25) at $p(\theta) = 1$:

$$R_\infty(\mu, \mu_0) = \frac{\omega_0}{4(\mu + \mu_0)}\left\{1 + 2\mu_0\int_0^1 R_\infty(\mu', \mu) \, d\mu' + 2\mu\int_0^1 R_\infty(\mu', \mu_0) \, d\mu' \right.$$
$$\left. + 4\mu\mu_0\int_0^1 R_\infty(\mu', \mu_0) \, d\mu' \int_0^1 R_\infty(\mu'', \mu) \, d\mu''\right\}$$

Table 3.1. Dependence of the diffuse exponent k on the single scattering albedo ω_0 for isotropic scattering.

ω_0	k
0	1
0.2	0.9999
0.3	0.9974
0.4	0.9856
0.5	0.9575
0.6	0.9073
0.7	0.8286
0.8	0.7104
0.9	0.5254
0.92	0.4740
0.94	0.4140
0.96	0.3408
0.98	0.2430
0.99	0.1725
1	0

or

$$R_\infty(\mu, \mu_0) = \frac{\omega_0 H(\mu) H(\mu_0)}{4(\mu + \mu_0)} \tag{3.29}$$

where the H-function is defined as follows:

$$H(\mu) = 1 + 2\mu \int_0^1 R_\infty(\mu, \mu_0)\, d\mu_0 \tag{3.30}$$

This function is of great importance in the radiative transfer theory (Chandrasekhar, 1960).

It follows from substitution of equation (3.29) into equation (3.30) that:

$$H(\mu) = 1 + \frac{\omega_0}{2} H(\mu)\mu \int_0^1 \frac{H(\mu_0)}{\mu + \mu_0}\, d\mu_0 \tag{3.31}$$

The implicit solution for this equation was obtained by Fock (1944):

$$H(\mu) = \exp\left(-\frac{\mu}{\pi} \int_0^\infty \ln\left(1 - \omega_0 \frac{\arctan x}{x} \right) \frac{dx}{1 + \mu^2 x^2} \right) \tag{3.32}$$

and very accurate approximation (errors less than 1%) by Hapke (1993):

$$H(\mu) = \left\{ 1 - [1 - \varsigma]\mu \left[r + \left(1 - \frac{r}{2} - r\mu \right) \ln \frac{1 + \mu}{\mu} \right] \right\}^{-1}$$

where $\varsigma = \sqrt{1 - \omega_0}$, r is the spherical albedo. It follows for the moments $H_j(\mu) = \int_0^1 H(\mu)\mu^j \, d\mu$ in the framework of the same approximation that:

$$H_j = \frac{2}{(1+j)(1+\varsigma)} + \frac{jr}{(1+j)(2+j)(1+\varsigma)}$$

If the value of r is not known, we can use an even more simple equation for the value of $H(\mu)$ (Hapke, 1993):

$$H(\mu) = \frac{1 + 2\mu}{1 + 2\mu\varsigma}$$

It differs by less than 4% from exact values everywhere. This formula can be used with equation (3.31) to find the exact solution by means of iterations.

We derive from equation (3.32) that $H(0) = 1$. It follows for the reflection function at $\mu = 1$, $\mu_0 = 0$ that $R_\infty(0, 1) = \omega_0 H(1)/4$ (see equation (3.29)) for the isotropic case. The results of calculation of the function $H(\mu)$ for isotropic scattering and $\omega_0 = 0.1, 0.5, 1.0$ using the exact equation (3.32) are presented in Table 3.2.

The data at $\omega_0 = 1$ can be approximated by the following simple equation: $H(\mu) = 1 + 2\mu$. Substituting this formula into equation (3.29) we obtain at $\omega_0 = 1$:

$$R_\infty^0(\mu, \mu_0) = R_\infty^{MS} + R_\infty^{SS} \tag{3.33}$$

Table 3.2. Function $H(\mu)$ at $\omega_0 = 0.1, 0.5, 1.0$ for the isotropic scattering.

μ	$\omega_0 = 0.1$	$\omega_0 = 0.5$	$\omega_0 = 1.0$
0	1.0000	1.0000	1.0000
0.05	1.0078	1.0444	1.1368
0.10	1.0124	1.0724	1.2474
0.15	1.0158	1.0447	1.3508
0.20	1.0186	1.1135	1.4503
0.25	1.0210	1.1297	1.5473
0.30	1.0230	1.1439	1.6425
0.35	1.0248	1.1566	1.7364
0.40	1.0263	1.1680	1.8293
0.45	1.0277	1.1783	1.9213
0.50	1.0289	1.1878	2.0128
0.55	1.0300	1.1964	2.1037
0.60	1.0311	1.2044	2.1941
0.65	1.0320	1.2117	2.2842
0.70	1.0328	1.2186	2.3740
0.75	1.0336	1.2250	2.4635
0.80	1.0344	1.2309	2.5527
0.85	1.0350	1.2365	2.6417
0.90	1.0357	1.2417	2.7306
0.95	1.0363	1.2466	2.8193
1.00	1.0368	1.2513	2.9078

where $R_\infty^{MS} = \frac{1}{2} + (\mu\mu_0)/(\mu+\mu_0)$ accounts for the multiple light scattering contribution and $R_\infty^{SS} = 1/(4(\mu+\mu_0))$ accounts for the single scattering contribution.

It follows for the isotropic scattering (Sobolev, 1972) that:

$$K(\mu) = \frac{\omega_0}{2} a_0 \frac{H(\mu)}{1-k\mu}, \qquad m = \frac{4}{k}\left(\frac{1}{1-k^2} - \frac{1}{\omega_0}\right),$$

$$l = \frac{m a_0^2 \omega_0}{2k}, \qquad a_0 = \frac{2k}{m}\int_0^1 \frac{H(\mu)}{1-k^2\mu^2}\mu\, d\mu$$

and (see equations (3.23)):

$$R(\tau_0, \mu, \xi) = \frac{\omega_0 H(\mu) H(\mu_0)}{4(\mu+\mu_0)} - \frac{T(\mu,\mu_0) m a_0^2 \omega_0 \, e^{-k\tau_0}}{2k}$$

$$T(\tau_0, \mu, \mu_0) = t\left(\frac{\omega_0 a_0}{2}\right)^2 \frac{H(\mu)H(\mu_0)}{(1-k\mu)(1-k\mu_0)}, \qquad t = \frac{m\, e^{-k\tau_0}}{1 - l^2 e^{-2k\tau_0}}$$

These formulae are simplified at $\omega_0 = 1$ (Sobolev, 1956):

$$\left.\begin{aligned} R(\mu,\mu_0) &= \frac{H(\mu)H(\mu_0)}{4(\mu+\mu_0)}\left\{1 - \frac{\mu+\mu_0}{\tau_0+\delta}\right\} \\ T(\mu,\mu_0) &= \frac{H(\mu)H(\mu_0)}{4(\tau_0+\delta)}, \qquad \delta = \sqrt{3}\int_0^1 H(\mu)\mu^2\, d\mu \approx 1.42\cdots \end{aligned}\right\} \tag{3.34}$$

Reflection and transmission functions in this case do not depend on the azimuth.

3.3.3 Nonabsorbing optically thick media

Let us investigate the case of nonabsorbing media in more detail. It follows from equations (3.23) at $\omega_0 = 1$ that:

$$R(\tau_0, \mu, \mu_0, \phi) = R_\infty^0(\mu, \mu_0, \phi) - T(\tau_0, \mu, \mu_0),\ T(\tau_0, \mu, \mu_0) = \frac{K_0(\mu)K_0(\mu_0)}{0.75(1-g)\tau_0 + \Delta} \tag{3.35}$$

where

$$\Delta = 3\int_0^1 K_0(\mu)\mu^2\, d\mu$$

and $K_0(\mu)$ denotes the function $K(\mu)$ at $\omega_0 = 1$. Equations (3.34) follow from equation (3.35) at $p(\theta) = 1$, $g = 0$. Note that values of $K_0(\mu)$, $R_\infty^0(\mu, \mu_0, \phi)$, depend on the phase function of a scattering medium only. They do not depend on the optical thickness. Moreover, even the dependence of these functions on the phase function is rather weak. It was shown (Zege *et al.*, 1991) that with the error less than 5%:

$$K_0(\mu) = \tfrac{3}{7}(1+2\mu)$$

at $\mu > 0.2$. Thus, it follows that $\Delta = \frac{15}{14}$ in this approximation.

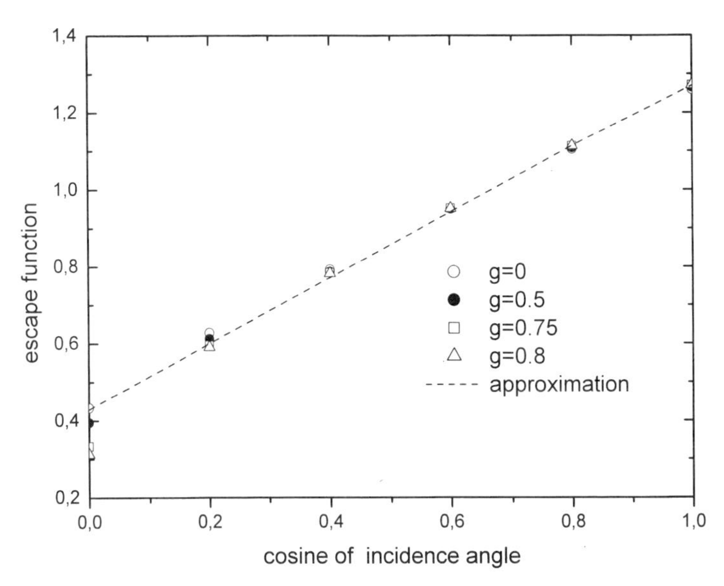

Figure 3.1. The escape function $K_0(\mu_0)$ at different values of the asymmetry parameter g and according to the linear dependence: $K_0(\mu_0) = \frac{3}{7}(1 + 2\mu_0)$.

The data for the function $K_0(\mu_0)$, obtained with this simple approximate equation and the exact radiative transfer code for the Heney–Greenstein phase function:

$$p(\theta) = \frac{1 - g^2}{(1 + g^2 - 2g\cos\theta)^{3/2}}$$

which has the following expansion on the Legendre polynomials $P_i(\cos\theta)$:

$$p(\theta) = \sum_{i=0}^{\infty}(2i + 1)g^i P_i(\cos\theta)$$

are presented in Figure 3.1 for different values of the asymmetry parameter g. We can see that indeed the escape function $K_0(\mu_0)$ only weakly depends on the asymmetry of the phase function g at $\mu > 0.2$. The function R_∞^0 for water clouds of different microstructure is presented in Figure 3.2. It is almost insensitive to the size of particles.

It is important that this simple expression for the escape function is consistent with the exact normalization condition:

$$2\int_0^1 K_0(\mu)\mu\, d\mu = 1 \tag{3.36}$$

It follows for the second moment in the same approximation that:

$$2\int_0^1 K_0(\mu)\mu^2\, d\mu = C \tag{3.37}$$

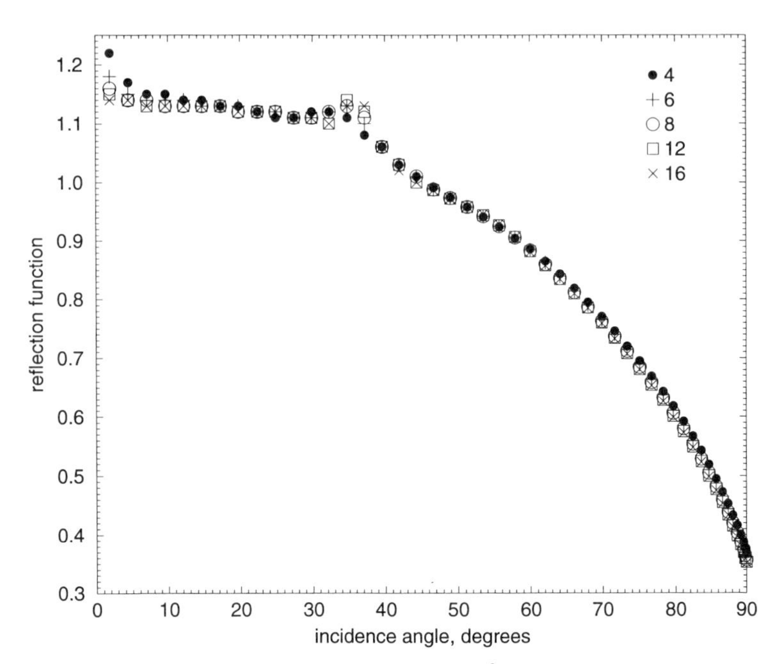

Figure 3.2. The dependence of the reflection function R_∞^0 of semi-infinite nonabsorbing water clouds of different microstructure ($a_{ef} = 4, 6, 8, 12, 16\,\mu\text{m}$) on the incidence angle at the nadir observation. The phase function of water clouds was calculated using the Mie theory assuming gamma particle size distribution with half-width parameter $\mu = 6$ at the wavelength $0.55\,\mu\text{m}$.

where $C = \frac{5}{7}$.

To apply equations (3.35) for light-scattering problems we need to have the function $R_\infty^0(\mu, \mu_0, \phi)$. Unfortunately, there are no simple expressions for this function. The reflection function $R_\infty^0(\mu, \mu_0, \phi)$ depends on the phase function and angles ϑ, ϑ_0, ϕ. We will assume that this function can be represented as a combination of two terms: the first does not depend on the phase function and the other term strongly depends on the phase function. It is reasonable to assume that the first term is proportional to:

$$R_\infty^{MS} = \frac{1}{2} + \frac{\mu\mu_0}{\mu + \mu_0}$$

as for isotropic scattering. The second term is proportional to the contribution of single scattering:

$$R_\infty^{SS} = \frac{p(\theta)}{4(\mu + \mu_0)}$$

Thus, it follows that:

$$R_\infty^0(\mu, \mu_0, \phi) = \alpha\left(\frac{1}{2} + \frac{\mu\mu_0}{\mu + \mu_0}\right) + \beta\frac{p(\theta)}{4(\mu + \mu_0)} \tag{3.38}$$

where α and β are unknown constants.

This expression can be used for determination of constants α and β for specific scattering media from exact calculations of the function $R^0_\infty(\mu, \mu_0, \phi)$ or from experiments.

It follows for the isotropic scattering that $\alpha = \beta = 1$ (see equation (3.33)). We can obtain for liquid water clouds (Kokhanovsky *et al.*, 1998):

$$\alpha = 1, \qquad \beta = 8 - 4.5\exp(-5(\pi - \theta)) - 5\exp(-5(\theta^* - \theta)) \tag{3.39}$$

where θ is the scattering angle, θ^* is the rainbow angle in radians ($\theta^* = 2.4$ in the visible).

It is difficult to obtain a simple equation for the value $R^0_\infty(\mu, \mu_0, \phi)$ valid at any phase function. The dependence of this function on the phase function is more important than it is for the function $K_0(\mu)$. Equation (3.25) at $\omega_0 = 1$ can be used to find this function numerically. Nonlinear intergral equation (3.25) is also a good starting point for the derivation of approximate results for the function $R^0_\infty(\mu, \mu_0, \phi)$.

Another approach to this problem is based on the fact that the reflection function

$$R^0_\infty(\mu, \mu_0) = \frac{1}{2\pi}\int_0^{2\pi} R(\mu, \mu_0, \phi)\, d\phi$$

averaged over the azimuth only slightly depends on the phase function. Thus, we can use the well-known result for isotropic scattering in this case (see equation (3.29)):

$$R^0_\infty(\mu, \mu_0) = \frac{H(\mu)H(\mu_0)}{4(\mu + \mu_0)} \tag{3.40}$$

where $H(\mu) \approx a + b\mu$, a and b are constants. Thus, it follows that:

$$R^0_\infty(\mu, \mu_0) = \frac{A + B\mu\mu_0 + C(\mu + \mu_0)}{\mu + \xi} \tag{3.41}$$

where constants $A = a^2$, $B = b^2$, $C = ab$ can be found from exact calculations of the function:

$$R^0_\infty(\mu, \mu_0) = \frac{1}{2\pi}\int_0^{2\pi} R^0_\infty(\mu, \mu_0, \phi)\, d\phi \tag{3.42}$$

for specific media. It follows that $A = \frac{1}{2}$, $B = 2$, $C = 0$ with error less than 10% at $\mu, \mu_0 > 0.2$ (Zege *et al.*, 1991) for clouds. We have at $\mu = 1$ or $\mu_0 = 1$: $R^0_\infty(\mu, \mu_0, \phi) \equiv R^0_\infty(\mu, \mu_0)$ (i.e. the reflection function does depend on the azimuth).

Another approach to find constants A, B, C calls for the use of exact integral relationships (Sobolev, 1972):

$$2\int_0^1 R^0_\infty(\mu, \mu_0)\mu_0\, d\mu_0 = 1$$

$$\frac{3}{2}\int R^0_\infty(\mu, \mu_0)(\mu + \mu_0)\mu_0\, d\mu_0 = K_0(\mu)$$

$$\lim_{\mu, \mu_0 \to 0} R^0_\infty(\mu, \mu_0)(\mu + \mu_0) = \tfrac{1}{4}p(0, 0)$$

where

$$p(0,0) = \frac{1}{2\pi}\int_0^{2\pi} p(0,0,\phi)\,d\phi$$

3.3.4 Weakly absorbing optically thick media

3.3.4.1 Reflection and transmission functions

General equations (3.23) can be simplified not only for isotropic scattering and nonabsorbing media ($\omega_0 \to 1$) but for weakly absorbing media as well. It was shown (Van de Hulst, 1980; Minin, 1988; Zege *et al.*, 1991) that as $\omega_0 \to 1$:

$$k = \sqrt{3(1-\omega_0)(1-\omega_0 g)} + O(1-\omega_0) \tag{3.43}$$

$$l = 1 - 2q_0 k + 2q_0^2 k^2 + O(k^3) \tag{3.44}$$

$$m = \frac{8k}{3(1-g)} + O(k^3) \tag{3.45}$$

$$K(\mu) = K_0(\mu)(1 - q_0 k) + O(k^2) \tag{3.46}$$

$$i(\mu) = 1 + \frac{k\mu}{1-g},$$

$$R_\infty(\mu,\mu_0,\phi) = R_\infty^0(\mu,\mu_0,\phi) - \frac{4k}{3(1-g)}K_0(\mu)K_0(\xi) + O(k^2) \tag{3.47}$$

Here ω_0 is the single scattering albedo, g is the asymmetry parameter and

$$q_0 = \frac{2}{1-g}\int_0^1 K_0(\mu)\mu^2\,d\mu$$

We can find next terms of these expansions in the book by Minin (1988). Results of numerical computations of these values at any value of ω_0 were published by Yanovitskij (1997). Analytical results for functions $R_\infty^0(\mu,\mu_0,\phi)$, $K_0(\mu)$ were presented in the previous section.

Unfortunately, these equations can be used only at $\beta = 1 - \omega_0 \leq 0.005$. But for many applications (e.g. snow, cloud, and foam optics) it is important to have simple formulae which can be applied at $\beta > 0.005$ as well.

To derive these equations we note that it follows from equation (3.46) that:

$$K(\mu)K(\mu_0) \approx K_0(\mu)K_0(\mu_0) \tag{3.48}$$

Thus equations (3.23) can be written in the following forms:

$$R(\tau_0,\mu,\mu_0,\phi) = R_\infty(\mu,\mu_0,\phi) - T(\tau_0,\mu,\mu_0)\,l\exp(-k\tau_0) \tag{3.49}$$

$$T(\tau_0,\mu,\mu_0) = tK_0(\mu)K_0(\mu_0) \tag{3.50}$$

where

$$t = \frac{m\,e^{-k\tau_0}}{1 - l^2\,e^{-2k\tau_0}} \tag{3.51}$$

Equation (3.51) determines the global transmittance (see equations (3.36), (3.50)):

$$t = 4\int_0^1 \mu\, d\mu \int_0^1 \mu_0\, d\mu_0 T(\tau_0, \mu, \mu_0) \tag{3.52}$$

The spherical albedo:

$$r = \frac{2}{\pi}\int_0^{2\pi} d\phi \int_0^1 \mu\, d\mu \int_0^1 \mu_0\, d\mu_0\, R(\tau, \mu, \mu_0, \phi) \tag{3.53}$$

can be derived from the following formula (see equations (3.49), (3.36), (3.50), (3.53)):

$$r = r_\infty - tl\exp(-k\tau_0) \tag{3.54}$$

Approximate formulae for r_∞, l and m in equations (3.49)–(3.51), (3.54) can be found by comparison of equations (3.51), (3.54) with well-known approximate equations derived by Rozenberg (1962, 1967):

$$r = \frac{\sinh(x)}{\sinh(x+y)}, \qquad t = \frac{\sinh(y)}{\sinh(x+y)} \tag{3.55}$$

where (Zege *et al.*, 1991) $x = k\tau_0$, $y = 4kq$, $q = 1/(3(1-g))$. Thus, it follows that $l = e^{-y}$, $m = 1 - l^2$ or $l = r_\infty$, $m = 1 - r_\infty^2$.

For the function $R_\infty(\mu, \mu_0, \phi)$ we can approximately obtain (Zege *et al*, 1991):

$$R_\infty(\mu, \mu_0, \phi) = R_\infty^0(\mu, \mu_0, \phi)\exp(-yD(\mu, \mu_0, \phi)) \tag{3.56}$$

where $D(\mu, \mu_0, \phi) = K_0(\mu)K_0(\mu_0)[R_\infty^0(\mu, \mu_0, \phi)]^{-1}$. This equation coincides with exact result (3.47) as $\omega_0 \to 0$ but can be used even at $\omega_0 = 0.95$, which is not the case for equation (3.47).

Thus, equations (3.49), (3.50) can be written in the following simpler form:

$$R(\mu, \mu_0, \phi) = R_\infty^0(\mu, \mu_0, \phi)\exp(-yD(\mu, \mu_0, \phi)) - T(\mu, \mu_0)\, e^{-y-x} \tag{3.57}$$

$$T(\mu, \mu_0) = \frac{\sinh y}{\sinh(x+y)} K_0(\mu)K_0(\mu_0) \tag{3.58}$$

These important equations were proposed for the first time by Rozenberg (1962). Approximate formulae for functions $R_\infty^0(\mu, \mu_0, \phi)$ and $K_0(\mu)$ were presented in the previous section. Note that equations (3.57), (3.58) can be used for analytical calculations of integral radiative characteristics as well (see Table 3.3).

Examination of Table 3.3 shows that radiative characteristics depend on two local optical characteristics (y, z), cosines of two angles (μ, μ_0), and two functions $R_\infty^0(\mu, \mu_0, \phi)$, $K_0(\mu)$, which should be precalculated for the case of semi-infinite nonabsorbing media. These functions only slightly depend on the microstructure of the media in question as can be seen from results of calculation of functions $K_0(\mu)$ and $R_\infty^0(\mu, 1)$ for the Heney–Greenstain phase function presented in Tables 3.4 and 3.5. The value of C (see equation (3.37)) is also almost independent of the phase function. This justifies our choice of this constant as $C = \frac{5}{7} \approx 0.7143$.

Table 3.3. Radiative characteristics of weakly absorbing optically thick layers

$$\left(y = 4\sqrt{\frac{1-\omega_0}{3(1-g)}}, \qquad z = \tfrac{3}{4}(1-g)\tau_0\right).$$

Radiative characteristic	Definition	Formula
Reflection function $R(\mu, \mu_0, \phi)$		$R^0_\infty(\mu, \mu_0, \phi)\exp\left(-y\frac{K_0(\mu)K_0(\mu_0)}{R^0_\infty(\mu, \mu_0, \phi)}\right)$ $-tK_0(\mu_0)K_0(\mu)\exp(-(1+z)y)$
Plane albedo $r(\mu_0)$	$\frac{1}{\pi}\int_0^{2\pi} d\phi \int_0^1 R(\mu, \mu_0, \phi)\mu\, d\mu$	$\exp(-yK_0(\mu_0)) - tK_0(\mu_0)\exp(-(1+z)y)$
Spherical albedo r	$2\int_0^1 R(\mu_0)\mu_0\, d\mu_0$	$\frac{\sinh(yz)}{\sinh((1+z)y)}$
Transmission function $T(\mu, \mu_0)$		$tK_0(\mu_0)K_0(\mu)$
Diffused transmittance $t(\mu_0)$	$2\int_0^1 T(\mu, \mu_0)\mu\, d\mu$	$tK_0(\mu_0)$
Global transmittance t	$2\int_0^1 t(\mu_0)\mu_0\, d\mu_0$	$\frac{\sinh(y)}{\sinh((1+z)y)}$
Absorptance A	$1 - r - t$	$1 - \frac{\sinh(y) + \sinh(yz)}{\sinh((1+z)y)}$
Reflection function of a semi-infinite nonabsorbing medium $R^0_\infty(\mu, \mu_0)$ averaged over the azimuth	$\frac{1}{2\pi}\int_0^{2\pi} R^0_\infty(\mu, \mu_0, \phi)\, d\phi$	$\frac{1+4\mu\mu_0}{2(\mu+\mu_0)}$
Escape function $K_0(\mu_0)$		$\frac{3}{7}(1+2\mu_0)$

Table 3.4. The function $K_0(\mu)$ for the Heney–Greenstein phase function at different values of the asymmetry parameter g (Danielson *et al.*, 1969). The value of C (see equation (3.37)) is presented in the last row of the table.

μ	$g=0$	$g=0.25$	$g=0.50$	$g=0.75$	$g=0.875$
0	0.443	0.424	0.395	0.333	0.272
0.1	0.540	0.534	0.516	0.484	0.463
0.3	0.711	0.708	0.701	0.692	0.688
0.5	0.872	0.870	0.869	0.869	0.869
0.7	1.028	1.028	1.030	1.033	1.035
0.9	1.182	1.184	1.188	1.192	1.194
1	1.259	1.261	1.265	1.270	1.271
C	0.7104	0.7109	0.7120	0.7134	0.7140

Table 3.5. The value of $R^0_\infty(\mu, 1)$ at different values of the cosine of the observation angle μ and the asymmetry parameter g at illumination at nadir (Danielson *et al.*, 1969).

μ	$g=0$	$g=0.25$	$g=0.50$	$g=0.75$	$g=0.875$
0	0.727	0.683	0.577	0.439	0.344
0.1	0.824	0.789	0.711	0.619	0.572
0.3	0.918	0.898	0.861	0.821	0.802
0.5	0.976	0.968	0.958	0.949	0.944
0.7	1.015	1.018	1.026	1.035	1.039
0.9	1.044	1.056	1.075	1.095	1.106
1	1.057	1.072	1.095	1.119	1.131

3.3.4.2 *Polarization of reflected and transmitted light*

Accounting for the light polarization in equations (3.57), (3.58) is simple. The main formulae have the same form. However, we should use reflection $\hat{R}(\mu, \mu_0, \varphi)$ and transmission $\hat{T}(\mu, \mu_0)$ matrices (Domke, 1978a, b) in equations (3.57), (3.58) instead of reflection and transmission functions (Kokhanovsky, 2001):

$$\hat{R}(\mu, \mu_0, \phi) = \hat{R}^0_\infty(\mu, \mu_0, \phi) \exp(-y\hat{D}(\mu, \mu_0, \phi)) - \hat{T}(\mu, \mu_0) \exp(-x - y) \tag{3.59}$$

$$\hat{T}(\mu, \mu_0) = \frac{\sinh y}{\sinh(x+y)} \vec{K}_0(\mu) \vec{K}_0^T(\mu_0) \tag{3.60}$$

where $\hat{D}(\mu, \mu_0, \phi) = \hat{R}^{0^{-1}}_\infty(\mu, \mu_0, \phi)\vec{K}_0(\mu)\vec{K}_0^T(\mu_0)$ and $\hat{R}^0_\infty(\mu, \mu_0, \phi)$ is the reflection matrix of a semi-infinite nonabsorbing layer having the same phase matrix as the absorbing layer of finite thickness under study. The vector $\vec{K}_0(\mu)$ describes polarization and intensity of light in the Milne problem for nonabsorbing semi-infinite media (Wauben, 1992). The components $K_{01}(\mu)$ and $K_{02}(\mu)$ of this vector were calculated by Chandrasekhar (1950) for Rayeigh particles ($g=0$) and by Wauben (1992) for spherical particles with refractive index $n = 1.44$ and gamma particle size

Table 3.6. The dependence K_{01} on cosine of the observation angle μ at $g = 0$ (Chandrasekhar, 1950) and $g = 0.718$ (Wauben, 1992).

g/μ	0	0.2	0.4	0.6	0.8	1.0
0	0.414	0.614	0.784	0.948	1.109	1.269
0.718	0.305	0.594	0.778	0.949	1.114	1.277

Table 3.7. The same as Table 3.6 but for K_{02}.

g/μ	0	0.2	0.4	0.6	0.8	1.0
0	0.04853	0.03774	0.02378	0.05141	0.00757	0
0.718	0.003623	0.003582	0.002393	0.001359	0.000582	0

Table 3.8. The dependence of functions $R^0_{\infty 11}(1, \mu_0)$, $R^0_{\infty 21}(1, \mu_0)$, and $P(1, \mu_0)$ at nadir observation on the cosine of the incidence angle μ_0 for the model of particles in clouds on Venus (Wauben, 1992).

μ_0	0	0.2	0.4	0.6	0.8	1.0
$R^0_{\infty 11}(1, \mu_0)$	0.3929	0.7069	0.8489	0.9541	1.061	1.231
$R^0_{\infty 21}(1, \mu_0)$	0.01426	0.01786	0.01578	0.01368	0.01244	0
$P(1, \mu_0)$	−0.036	−0.025	−0.019	−0.014	−0.012	0

distribution:

$$f(a) = Ar^{\mu} \exp\left(-\mu \frac{r}{r_0} \right)$$

where $A = \text{const}$, $\mu = 11.3$, $r_0 = 0.83\,\mu\text{m}$. The wavelength of λ was equal to $0.55\,\mu\text{m}$. Note that the model of spheres with $\mu = 11.3$, $r_0 = 0.83\,\mu\text{m}$, and $n = 1.44$ is usually used to characterize particles in clouds on Venus (Hansen and Travis, 1974). It follows for the effective size a_{ef}, the effective variance Δ_{ef}, and the asymmetry parameter g in this case: $a_{ef} = 1.05\,\mu\text{m}$, $\Delta_{ef} = 0.07$ and $g = 0.718$. The results of numerical calculations are presented in Tables 3.6, 3.7.

We can see that component K_{01} only weakly depends on the phase function (see Table 3.4 as well). Component K_{02} is small in comparison with K_{01} and increases with observation angle ϑ. It is equal to 0 at nadir observation ($\vartheta = 0°$). The degree of polarization of transmitted light $p = -T_{21}/T_{11}$ does not depend on the optical thickness. It is equal to $-K_{02}/K_{01}$. It changes from zero value at $\mu = 1$ to -0.117 at $\mu = 0$ for Rayleigh particles and from zero value at $\mu = 1$ to -0.012 at $\mu = 0$ for particles with $a_{ef} = 1.05\,\mu\text{m}$ (see Tables 3.6, 3.7). Thus, the degree of polarization of transmitted light is extremely low as might be expected. This is because polarization of light is mostly due to single light scattering of incident solar light. However, the probability of the existence of singly scattered photons at large optical thickness is low.

Functions $R^0_{\infty 11}(1, \mu_0)$, $R^0_{\infty 21}(1, \mu_0)$, and the degree of polarization $P(1, \mu_0) = -R^0_{\infty 21}(1, \mu_0)/R^0(1, \mu_0)$ for particles in clouds on Venus are presented in Table 3.8.

One can see that the degree of polarization of light reflected from a semi-infinite layer with such particles at normal observation is small.

Note that formulae (3.59), (3.60) can be simplified for nonabsorbing media ($y = 0$):

$$\hat{R}(\mu, \mu_0, \varphi) = \hat{R}^0_\infty(\mu, \mu_0, \varphi) - \hat{T}(\mu, \mu_0) \tag{3.61}$$

$$\hat{T} = t\vec{K}_0(\mu)\vec{K}_0^T(\mu_0) \tag{3.62}$$

where

$$t = \frac{1}{1 + \frac{3}{4}\tau_0(1 - g)} \tag{3.63}$$

is the global transmittance.

Let us apply equation (3.61) to a particular problem, namely, to the derivation of a simple relationship between the spherical albedo $r = 1 - t$ and the degree of polarization of reflected light $P(\mu_0)$ at nadir viewing geometry ($\mu = 1$). We will assume that a scattering layer is illuminated on the top by a wide unidirectional unpolarized light beam. The value of $P(\mu_0)$ is given simply by $-R_{21}(\mu_0, 1)/R_{11}(\mu_0, 1)$ in this case. Thus, it follows from equation (3.61) that:

$$P(\mu_0) = \frac{P_\infty(\mu_0)}{1 - u(\mu_0)(1 - r)} \tag{3.64}$$

where

$$u(\mu_0) = \frac{K_{01}(1)K_{01}(\mu_0)}{R^0_{\infty 11}(\mu_0, 1)}, \qquad P_\infty(\mu_0) = -\frac{R^0_{\infty 21}(\mu_0, 1)}{R^0_{\infty 11}(\mu_0, 1)}$$

and we account for the equality (Wauben, 1992): $K_{02}(1) = 0$.

Our calculations show that the value of $u(\mu_0)$ is close to 1 for most observation angles, which implies inverse proportionality (see equation (3.64)) between the brightness of a turbid medium and the degree of polarization of reflected light. This inverse proportionality between spherical albedo r and degree of polarization $P(\mu_0)$ was discovered experimentally by Umow (1905). It takes place both for nonabsorbing and absorbing (Vitkin and Studinski, 2001) media.

The accuracy of equation (3.64) is studied in Figure 3.3, where data obtained from this simple formula and results of numerical solution of the vector radiative transfer equation for cloudy media by the doubling method (Van de Hulst, 1980) are presented. We have found that function $u(\mu_0)$ varies from 1 to 1.36 with the incidence angle for water clouds. However, this variation is of a secondary importance in calculations of the degree of polarization (see equation (3.64)). Thus, we have ignored the variation of the function $u(\mu_0)$ with incidence angle and used the constant value $u(\mu_0) = 1.18$ in all calculations. The value of $P_\infty(\mu_0)$ was obtained from exact radiative transfer calculations for a semi-infinite layer. The gamma droplet size distribution was given by $f(a) = Aa^\mu \exp(-1.5a)$ (model C1) where A is the normalization constant ($\int_0^\infty f(a)\,da = 1$). The optical thickness of a layer was equal to 8, 15, and 30. Calculations of the phase matrix were performed with the Mie theory at wavelength 443 nm, where light absorption by water droplets can be ignored. The obtained value of the asymmetry parameter was equal to 0.8541 in the case under investigation.

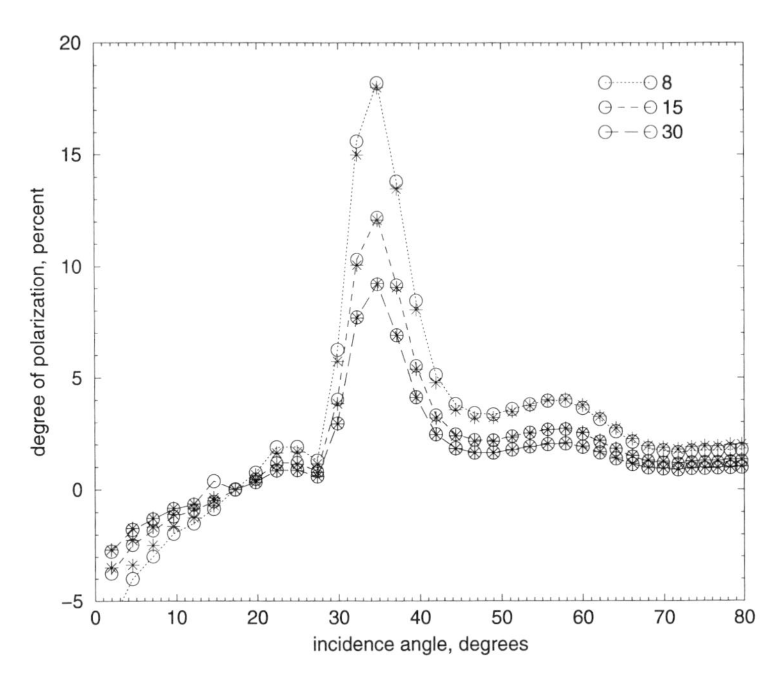

Figure 3.3. The dependence of the degree of polarization of light reflected from a water cloud with gamma particle size distribution (C1 model, $a_{ef} = 6\,\mu m$, $\mu = 6$) on the incidence angle at wavelength 443 nm, nadir observation and $\tau = 8$, 15, 30 according to numerical radiative transfer calculations (broken lines) and equation (3.64) (symbols) (from Kokhanovsky, 2001).

It follows from Figure 3.3 that equation (3.64) describes transformation of the polarization curves due to the change in optical thickness with a high accuracy at $\tau \geq 8$. As a matter of fact this equation can be applied to media with optical thickness as low as 5 (see Figure 3.4).

Equation (3.64) is easily generalized to account for absorption of light in a medium. Namely, it follows that:

$$P(\mu_0) = \frac{P_\infty^*(\mu_0)}{1 - u^*(\mu_0) t \exp(-x - y)} \tag{3.65}$$

where the global transmittance t is presented in Table 3.1 and:

$$u^*(\mu_0) = \frac{K_{01}(1) K_{01}(\mu_0)}{R_{\infty 11}^*(\mu_0, 1)}$$

Values of $P_\infty^*(\mu_0)$ and $R_\infty^*(\mu_0, 1)$ represent the degree of polarization and reflection function of a semi-infinite weakly absorbing medium at nadir viewing geometry. Note that equation (3.65) can be written in the following form:

$$P(\mu_0) = c(\mu_0, \tau) P_\infty^*(\mu_0) \tag{3.66}$$

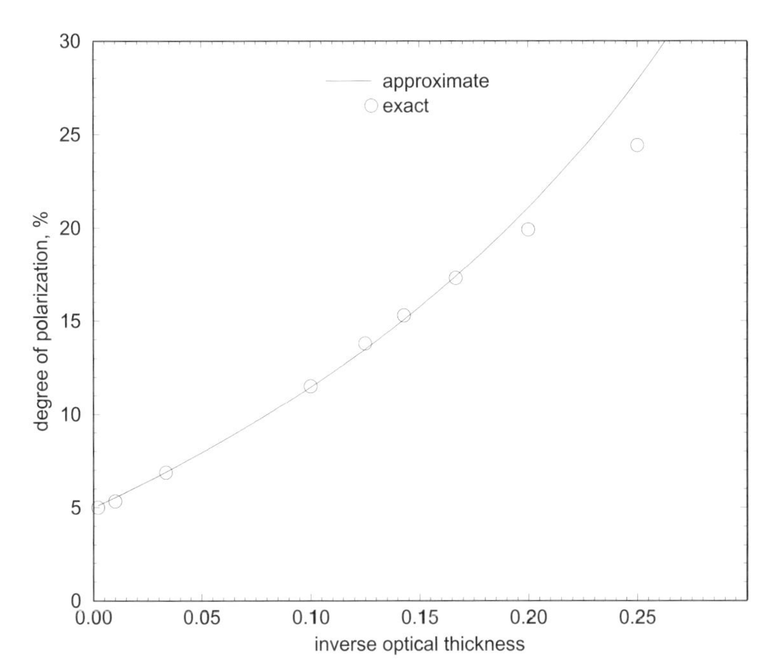

Figure 3.4. The dependence of the degree of polarization of light reflected from a water cloud with gamma particle size distribution (C1 model, $a_{ef} = 6\,\mu\text{m}$, $\mu = 6$) at wavelength 443 nm on inverse optical thickness at nadir observation and incidence angle equal to 37 degrees according to approximate equation (3.64) (solid line) and numerical radiative transfer calculations (symbols) (from Kokhanovsky, 2001).

where

$$c(\mu_0, \tau) = \frac{1}{1 - u^*(\mu_0) t \exp(-x - y)} \tag{3.67}$$

can be interpreted as the enhancement factor, which is solely due to the finite depth of a turbid layer. It follows for semi-infinite layers that transmittance t is equal to 0 and therefore $c = 1$, as it should be. Also, it follows from equation (3.66) that zeroes of polarization curves for semi-infinite and optically thick finite layers coincide, which is supported by numerical calculations using the radiative transfer code (see Figure 3.3). Generally, multiple scattering diminishes polarization of singly scattered light.

The dependencies of the degree of polarization at fixed incident angles as functions of the probability of photon absorption $\beta = 1 - \omega_0$, calculated using equation (3.65) and the vector radiative transfer code, are presented in Figure 3.5 for the same phase matrix as in Figures 3.3 and 3.4. We can see that the accuracy of equation (3.65) is better than 2% at $\beta < 0.1$ at these geometries.

The formulae presented here open new ways for the parametrization of global optical characteristics of disperse media. They also could be useful in the solution of inverse problems, particularly those involving spectropolarimetric measurements.

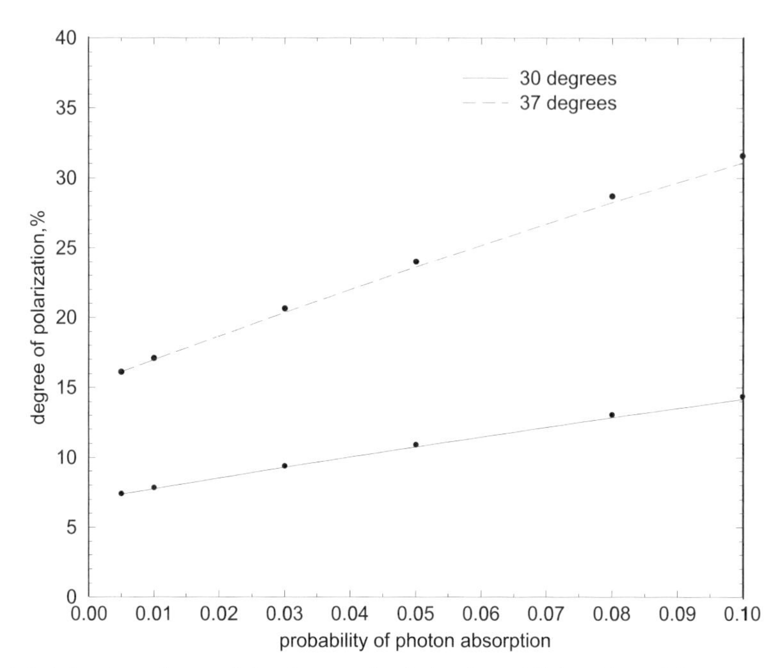

Figure 3.5. The dependence of the degree of polarization of light reflected from a water cloud with the gamma particle size distribution (C1 model, $a_{ef} = 6\,\mu\text{m}$, $\mu = 6$) on the probability of photon absorption at nadir observation and incidence angle equal to 30 and 37 degrees according to numerical radiative transfer calculations and equation (3.65) (lines) at optical thickness 7 (from Kokhanovsky, 2001).

3.4 ISOTROPICALLY LIGHT-SCATTERING LAYERS

Let us now consider the case of isotropic scattering in more detail. We have already touched on this problem in the previous sections for thin and thick layers. Now we will consider radiative transfer in a layer of any optical thickness. The RTE can be written in the following form at $p(\theta) = 1$:

$$\mu \frac{dI(\tau, \mu, \mu_0)}{d\tau} = -I(\tau, \mu, \mu_0) + \frac{\omega_0}{2} \int_0^1 I(\tau, \mu', \mu_0)\, d\mu' + \frac{\omega_0 I_0}{4} \exp\left(-\frac{\tau}{\mu_0}\right) \tag{3.68}$$

The RTE (3.68) is much simpler than equation (3.2), because it does not depend on the azimuth ϕ and can be solved in terms of special functions of three arguments $X(\omega_0, \tau, \mu)$ and $Y(\omega_0, \tau, \mu)$. These functions were introduced to radiative transfer theory by Ambarzumian (1943) and investigated in detail by Chandrasekhar (1950). They are called Chandrasekhar or Ambarzumian–Chandrasekhar functions. The results for reflection and transmission functions $R(\tau, \mu, \mu_0)$, $T(\tau, \mu, \mu_0)$ of isotropic layers in terms of these functions are the following (Chandrasekhar, 1950; Van de Hulst, 1980):

$$R(\tau_0,\mu,\mu_0)=\frac{\omega_0}{4}\frac{X(\omega_0,\tau_0,\mu)X(\omega_0,\tau_0,\mu_0)-Y(\omega_0,\tau_0,\mu)Y(\omega_0,\tau_0,\mu_0)}{\mu+\mu_0} \tag{3.69}$$

$$T(\tau_0,\mu,\mu_0)=\frac{\omega_0}{4}\frac{X(\omega_0,\tau_0,\mu)Y(\omega_0,\tau_0,\mu_0)-X(\omega_0,\tau_0,\mu_0)Y(\omega_0,\tau_0,\mu)}{\mu_0-\mu} \tag{3.70}$$

It follows at $\mu=\mu_0$ that:

$$R(\tau_0,\mu)=\frac{[X^2(\omega_0,\tau_0,\mu)-Y^2(\omega_0,\tau_0,\mu)]\omega_0}{8\mu} \tag{3.71}$$

$$T(\tau_0,\mu)=\frac{\omega_0}{4}X^2(\omega,\tau_0,\mu)\frac{d}{d\mu}\left(\frac{Y(\omega_0,\tau_0,\mu)}{X(\omega_0,\tau_0,\mu)}\right) \tag{3.72}$$

A physical derivation of these formulae from Abmarzumian's principles of invariance was presented by Van de Hulst (1980). Special functions X and Y were calculated in many papers (Van de Hulst, 1980). They have the following asymptotic properties (Chandrasekhar, 1950):

$$\lim_{\tau_0\to 0} X(\omega_0,\tau_0,\mu)=1 \tag{3.73}$$

$$\lim_{\tau_0\to\infty} X(\omega_0,\tau_0,\mu)=H(\omega_0,\mu) \tag{3.74}$$

$$\lim_{\tau_0\to 0} Y(\omega_0,\tau_0,\mu)=e^{-\tau_0/\mu} \tag{3.75}$$

$$\lim_{\tau_0\to\infty} Y(\omega_0,\tau_0,\mu)=0 \tag{3.76}$$

where $H(\omega_0,\mu)$ is the H-function (see equation (3.32)). From equations (3.69)–(3.76) we can derive equations (3.16), (3.17) (at $p(\theta)=1$), (3.29) for thin and semi-infinite layers.

Note that equations (3.69), (3.70) can be used for approximate calculations of the radiative characteristics (averaged on the azimuth ϕ) of scattering layers with any phase function. This conclusion is evident from the following approximate representation of the phase function (Joseph *et al.*, 1976; Wiscombe, 1977; Nakajima and Tanaka, 1988; Ito, 1993):

$$p(\theta)=\gamma+(1-\gamma)\,\delta(\theta) \tag{3.77}$$

where $\delta(\theta)$ is the delta function and $\gamma=\text{const}$. Note that expansion (3.77) is not unique (Zege *et al.*, 1993). Substituting (3.77) into equation (3.2) we have:

$$\mu\frac{dI}{d\tau}=-(1-(1-\gamma)\omega_0)I+\gamma\omega_0\int I\frac{d\Omega}{4\pi}+\frac{\omega_0 I_0}{4}p(\theta)\,e^{-\tau/\mu_0} \tag{3.78}$$

This equation coincides with the exact equation (3.68) for isotropic scattering if we neglect last terms in both equations, which is possible at large values of τ and introduce the following similarity relations:

$$\omega_0^* = \omega_0 \frac{\gamma}{1-(1-\gamma)\omega_0}, \qquad \tau^* = (1-(1-\gamma)\omega_0)\tau \tag{3.79}$$

It follows that:

$$(1-\omega_0^*)\tau^* = (1-\omega_0)\tau \tag{3.80}$$

Thus, we can approximately replace anisotropically light-scattering layers with single scattering albedo ω_0 and optical thickness τ by isotropically scattering ones with single scattering albedo ω_0^* and optical thickness τ^* (see equations (3.69)–(3.72), (3.79), (3.80)). This is an important result. It follows that $\tau^* < \tau$ and $\omega_0^* < \omega_0$. The value of γ can be found from the equality of asymmetry parameters $g = \frac{1}{2}\int_0^\pi p(\theta)\sin\theta\cos\theta\,d\theta$ of the real world and approximate (see equation (3.77)) phase functions. This equality gives $\gamma = 1-g$, where g is the asymmetry parameter. The importance of parameter g can be understood from equations (3.57), (3.58) as well. It follows at $\gamma = 1-g$ from equations (3.79):

$$\omega_0^* = \omega_0 \frac{1-g}{1-g\omega_0}, \qquad \tau^* = \tau(1-g\omega_0) \tag{3.81}$$

Also we obtain: $y = \sqrt{1-\omega_0^*}, z = \tau^*$ (see Table 3.3). This choice of similarity relations is not unique. As stated by Van de Hulst (1980), 'the similarity relations resemble the problem of mapping the spherical earth on a plane map: there are many possibilities, each of which leaves a certain distortion'. He proposed using the following similarity relations:

$$1-\omega_0^* = (1-\omega_0)\frac{k^*}{k}, \qquad \tau^* = \tau\frac{k}{k^*} \tag{3.82}$$

where k^* is the minimal root of equation (3.28) and k is the minimal root of the characteristic equation for anisotropic scattering:

$$1-\omega_0 = \cfrac{k^2}{3(1-\omega_0 g) - \cfrac{4k^2}{5-\omega_0 x_2 - \cfrac{9k^2}{7-\omega_0 x_3} - \cdots}} \tag{3.83}$$

where

$$x_l = \frac{2l+1}{2}\int_0^\pi p(\theta)P_l(\cos\theta)\sin\theta\,d\theta \tag{3.84}$$

$P_l(\cos\theta)$ is the Legendre polynomial.

It follows from equation (3.83) for weakly absorbing media ($\omega_0 \to 1$) that:

$$k = \sqrt{3(1-\omega_0 g)(1-\omega_0)} \tag{3.85}$$

Equations (3.81) coincide with equations (3.82) ($k^* \equiv k(g=0)$) if we take account of equation (3.85).

3.5 HIGHLY ANISOTROPICALY LIGHT-SCATTERING LAYERS

3.5.1 Small-angle approximation

Let us now consider the case of a highly anisotropic light-scattering layer (the asymmetry parameter $g \approx 1$) illuminated along the normal. Examples of such media are oceanic waters, bioliquids, and tissues. In this case most scattered photons are within the small-angle scattering region and there is the possibility of simplifying the RTE. Thus, we assume that $\cos\theta = 1$ in equation (3.2) and obtain (Wang and Guth, 1951; Dolin, 1964; Borovoi, 1982; Alexandrov *et al.*, 1993):

$$\frac{dI(\tau,\mu)}{d\tau} = -I(\tau,\mu) + \frac{\omega_0}{2}\int_0^1 d\mu'\, I(\tau,\mu')p(\mu,\mu') \tag{3.86}$$

where

$$p(\mu,\mu') = \frac{1}{2\pi}\int_{-1}^{2\pi} p(\mu,\mu',\phi)\,d\phi \tag{3.87}$$

We used the fact that the intensity of scattered light fields for layers with randomly oriented particles does not depend on the azimuth at illumination of a layer along the normal. Note that the value of $I(\tau,\mu)$ is the total intensity (not the diffused intensity as in equation (3.2)) and it includes direct light. The phase function $p(\mu,\mu',\varphi)$ in equation (3.87) can be represented in the following form (Minin, 1988):

$$p(\mu,\mu',\phi) = p(\mu,\mu') + 2\sum_{m=1}^{\infty}\cos m(\phi-\phi')\sum_{i=m}^{\infty} c_i^m P_i^m(\mu)P_i^m(\mu') \tag{3.88}$$

where $P_i(\mu)$ and $P_i^m(\theta)$ are Legendre and associated Legendre polynomials, respectively, and:

$$p(\mu,\mu') = \sum_{i=0}^{\infty} x_i P_i(\mu)P_i(\mu'), \qquad x_i = \frac{2i+1}{2}\int p(\theta)P_i(\theta)\sin\theta\,d\theta, \qquad c_i^m = x_i\frac{(i-m)!}{(i+m)!} \tag{3.89}$$

We will seek the solution of equation (3.86) in the following form:

$$I(\tau,\mu) = \sum_{i=0}^{\infty} b_i(\tau)P_i(\mu) \tag{3.90}$$

Substituting equations (3.88), (3.90) into equations (3.86), (3.87) we get:

$$\frac{db_i(\tau)}{d\tau} = -c_i b_i(\tau) \tag{3.91}$$

where

$$c_i = 1 - \omega\frac{x_i}{2i+1} \tag{3.92}$$

where the orthogonality of Legendre polynomials (see Appendix 3) was used. Thus, it follows (Zege *et al.*, 1991):

$$b_i(\tau) = A_i\exp(-c_i\tau) \tag{3.93}$$

where $A_i = \text{const}$. It is possible to obtain the intensity of transmitted light with the following formula:

$$I(\tau, \mu) = \sum_{i=0}^{\infty} A_i e^{-c_i \tau} P_i(\mu) \tag{3.94}$$

Values of A_i can be found from initial conditions. We will assume that:

$$I(0, \mu) = I_0 \, \delta(1 - \mu) \tag{3.95}$$

where $\delta(1 - \mu)$ is the delta function, I_0 is the density of incident light flux. This function can be represented in the following form:

$$\delta(1 - \mu) = \frac{1}{4\pi} \sum_{i=0}^{\infty} (2i + 1) P_i(\mu) \tag{3.96}$$

Thus, it follows from equations (3.94)–(3.96) that:

$$A_i = \frac{2i + 1}{4\pi} \tag{3.97}$$

and

$$I(\tau, \mu) = B \sum_{i=0}^{\infty} \frac{2i + 1}{2} e^{-c_i \tau} P_i(\mu) \tag{3.98}$$

where $B = I_0/2\pi$.

This is a solution to the problem under consideration. Equation (3.98) describes the angular distribution of transmitted light at normal incidence. This important formula can be rewritten in integral form. Indeed, the phase function $p(\theta)$ has a sharp peak in the forward-scattering direction ($\theta = 0$) and the main contribution to the integral (3.89) arises at small scattering angles. Thus, it follows from equation (3.89) that:

$$x_i = \frac{2i + 1}{2} \int_0^{\infty} p(\theta) J_0\left(\theta\left(i + \tfrac{1}{2}\right)\right) \theta \, d\theta \tag{3.99}$$

where the asymptotic relationship

$$\lim_{\theta \to 0} P_i(\cos\theta) = J_0\left(\theta\left(i + \tfrac{1}{2}\right)\right) \tag{3.100}$$

was used.

From equations (3.98)–(3.100) and the sum formula

$$\sum_{i=0}^{\infty} f\left(i + \tfrac{1}{2}\right) \approx \int_0^{\infty} f(\sigma) \, d\sigma \tag{3.101}$$

it follows that:

$$I(\tau, \vartheta) = \frac{I_0}{2\pi} \int_0^{\infty} d\sigma \, J_0(\sigma\vartheta) \exp(-\tau(1 - \omega_0 P(\sigma))) \tag{3.102}$$

where (see equations (3.92), (3.99)):

$$P(\sigma) = \frac{1}{2} \int_0^{\infty} p(\theta) J_0(\theta\sigma) \theta \, d\theta \tag{3.103}$$

Equation (3.102) is easier to handle than equation (3.98) (Kokhanovsky, 1997). For

Table 3.9. Phase function $p(\theta)$ and their Fourier–Bessel transforms $P(\sigma)$ (Υ is the normalization constant, x is the size parameter) (from Zege *et al.*, 1991).

$p(\theta)$	$P(\sigma)$
$\dfrac{2\Upsilon\exp(-\Upsilon\theta)}{\theta}$	$\dfrac{\Upsilon}{\sqrt{\Upsilon^2+\sigma^2}}$
$2\Upsilon^2\exp(-\Upsilon\theta)$	$\dfrac{\Upsilon^3}{(\Upsilon^2+\sigma^2)^{3/2}}$
$\dfrac{2}{\Upsilon^2}\exp\left(-\dfrac{\theta^2}{2\Upsilon^2}\right)$	$\exp\left(-\dfrac{\Upsilon^2\sigma^2}{2}\right)$
$\dfrac{4J_1^2(\theta x)}{\theta^2}$	$\begin{cases}\dfrac{2}{\pi}\left\{\arccos\left(\dfrac{\sigma}{2x}\right)-\dfrac{\sigma}{2x}\sqrt{1-\left(\dfrac{\sigma}{2x}\right)^2}\right\}, & \sigma\le 2x\\ 0, & \sigma>2x\end{cases}$

instance, equation (3.103) can be analytically integrated for special types of phase functions $p(\theta)$ (see Table 3.9).

Note that for the diffused intensity $I_d(\tau,\vartheta)$ it follows from equation (3.102) that:

$$I_d(\tau,\vartheta)=\frac{I_0}{2\pi}\int_0^{\infty}\left[e^{-\tau(1-\omega_0P(\sigma))}-e^{-\tau}\right]J_0(\sigma\vartheta)\sigma\,d\sigma \tag{3.104}$$

where we extracted the peak of light intensity exactly in the forward direction:

$$\frac{I_0}{2\pi}e^{-\tau}\int_0^{\infty}J_0(\sigma\vartheta)\sigma\,d\sigma=e^{-\tau}\,\delta(\vartheta) \tag{3.105}$$

Equation (3.104) is important for solution of both direct and inverse problems of light-scattering media optics. It is valid at $\tau\le 5$. Results of calculation (see equation (3.104)) of the normalized diffused intensity:

$$\mathrm{i}=\frac{I_d(\tau,\vartheta)}{I_d(\tau,0)}$$

as the function of parameter $b=2x\vartheta$ at $\omega_0=0.5$ and $P(\sigma)$ given by the last row in Table 3.9 is presented in Figure 3.6. We can see that the phenomenon of multiple light scattering is responsible for broadening the angular spectrum of the light transmitted by a scattering layer at small angles. The same effect is also observed in the single-scattering regime if the size of particles decreases.

The approximation, already considered in this section, is called the small-angle approximation (SAA). Moments of different radiative characteristics in this approximation were studied by Lutomirski *et al.* (1995). The SAA can be used not just for studies of the transmission of light, uniformly illuminating a layer at every point on

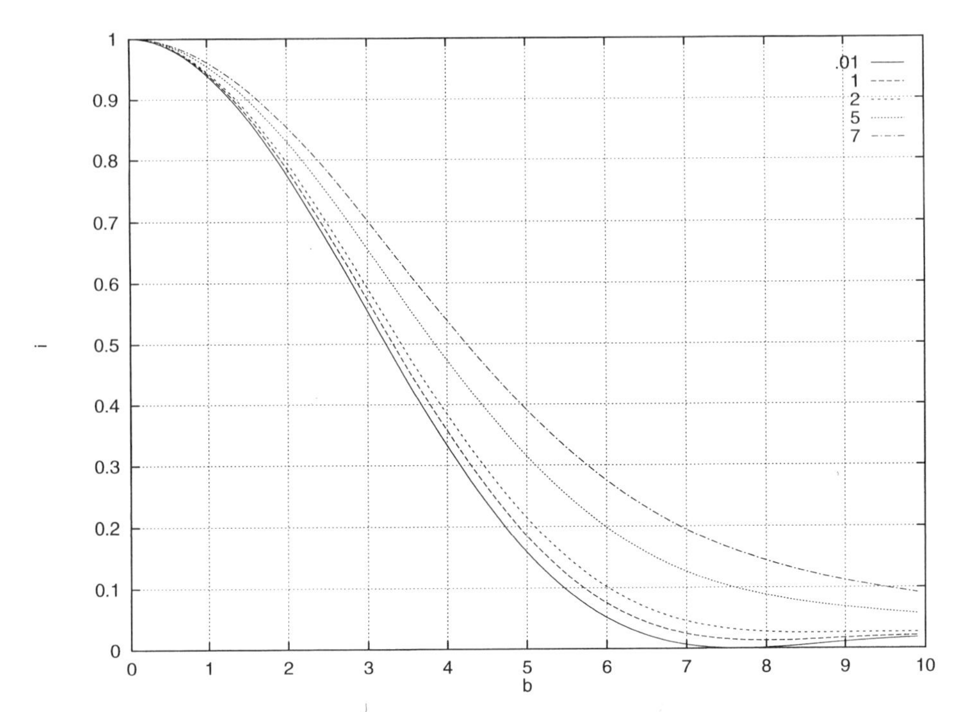

Figure 3.6. The dependence of the normalized intensity i on parameter $b = 2x\vartheta$ at optical thickness $\tau = 0.01$, 1, 2, 5, 7 for monodispersed spheres in the framework of the Fraunhofer approximation.

the top, as presented here, but also for more complex cases, including a vertically inhomogeneous layer illuminating by narrow beam with light distribution $I_0(\vec{r}, \vec{s})$. In this case we should use the RTE (3.1) and not equation (3.2).

The result for angular distribution of the transmitted intensity in this more general case is (Ishimaru, 1978; Zege *et al.*, 1991):

$$I(\vec{r}_\perp, \vec{n}_\perp, z) = \frac{1}{(2\pi)^4} \int_{-\infty}^{\infty} \int_{-\infty}^{\infty} d\vec{\gamma} \int_{-\infty}^{\infty} \int_{-\infty}^{\infty} I(\vec{\gamma}, \vec{\sigma}, z)\, e^{i(\vec{\gamma}\vec{r}_\perp + \vec{\sigma}\vec{n}_\perp)}\, d\vec{\sigma} \tag{3.106}$$

where

$$I(\vec{\gamma}, \vec{\sigma}, z) = I_0(\vec{\gamma}, \vec{\sigma} + \vec{\gamma}z) \exp\left(-\int_0^z \Psi(z - \xi, \vec{\sigma} + \vec{\gamma}\xi)\, d\xi\right) \tag{3.107}$$

$$\Psi(z, \alpha) = \sigma_{ext}(z) - \sigma_{sca}(z) P(\alpha, z) \tag{3.108}$$

$$P(\alpha, z) = \frac{1}{2} \int_0^{\infty} p(\theta, z) J_0(\alpha\theta)\theta\, d\theta \tag{3.109}$$

$$I_0(\vec{\gamma}, \vec{\sigma}, z) = \int_{-\infty}^{\infty} \int_{-\infty}^{\infty} d\vec{r}_\perp \int_{-\infty}^{\infty} \int_{-\infty}^{\infty} I_0(\vec{r}_\perp, \vec{n}_\perp)\, e^{-i(\vec{\gamma}\vec{r}_\perp + \vec{\sigma}\vec{n}_\perp)}\, d\vec{n}_\perp \tag{3.110}$$

Here $\vec{r}_\perp$ and $\vec{n}_\perp$ are projections of vectors $\vec{r}$, $\vec{n}$ in equation (3.1) on the plane, perpendicular to the light beam, z is the geometrical depth.

These equations are more complex than equation (3.102) but they allow us to calculate light distributions $I(\vec{r}_\perp, \vec{n}_\perp, z)$ of transmitted light for narrow (e.g. laser) incident beams. For instance, for layers illuminated by a unit point monodirectional source of light $I_0(\vec{r}, \vec{n}_\perp) = \delta(\vec{r})\,\delta(\vec{n}_\perp)$, we obtain (see equation (3.107)):

$$I(\vec{\gamma}, \vec{\sigma}, z) = \exp\left\{-\int_0^z \Psi(z - \xi, \vec{\sigma} + \vec{\gamma}\xi)\, d\xi\right\} \tag{3.111}$$

Note that the value (see equation (3.106)):

$$I(\vec{\gamma}, \vec{\sigma} = 0, z) = \int_{-\infty}^{\infty}\int_{-\infty}^{\infty} e^{-i\vec{\gamma}\vec{r}_\perp}\, d\vec{r}_\perp \int_{-\infty}^{\infty}\int_{-\infty}^{\infty} I(\vec{r}_\perp, \vec{n}_\perp, z)\, d\vec{n}_\perp$$

defines the optical transfer function $S(\vec{\gamma})$, which determines the loss of contrast in the image during its transfer through a scattering layer (Zege *et al.*, 1991). Thus, this important function can be calculated from equation (3.111) at $\vec{\sigma} = 0$:

$$S(\vec{\gamma}) = \exp\left(-\int_0^z \Psi(z - \xi, \vec{\gamma}\xi)\, d\xi\right) \tag{3.112}$$

The value

$$I(\vec{\gamma} = 0, \vec{\sigma}, z) = \int_{-\infty}^{\infty}\int_{-\infty}^{\infty} I(z, \vec{n}_\perp)\, e^{-i\vec{\sigma}\vec{n}_\perp}\, d\vec{n}_\perp \tag{3.113}$$

where $I(z, \vec{n}_\perp) = \int_{-\infty}^{\infty}\int_{-\infty}^{\infty} I(\vec{r}_\perp, \vec{n}_\perp, z)\, d\vec{r}_\perp$ is the Fourier transform of the intensity. It coincides with the transverse spatial coherence function (Ishimaru, 1978; Apresyan and Kravtsov, 1983) in the framework of the SAA. Thus, this approach allows us to investigate the propagation of transverse optical coherence in random multiple light-scattering media (Borovoi *et al.*, 1986; Zege and Kokhanovsky, 1995; John *et al.*, 1996; Wax and Thomas, 1998) and study the correlation of light fields in a scattering layer. Note that the coherent input beam loses most of its coherence almost exactly at the input plane of a disperse medium at the distance approximately equal to σ_{ext}^{-1}.

It follows from equation (3.113) that:

$$I(z, \vec{n}_\perp) = \frac{1}{4\pi^2}\int_{-\infty}^{\infty}\int_{-\infty}^{\infty} I(\vec{\gamma} = 0, \vec{\sigma}, z)\, e^{i\vec{\sigma}\vec{n}_\perp}\, d\vec{\sigma} \tag{3.114}$$

Equation (3.102) can be derived from equation (3.114).

The formulae presented here can be used not only for calculations of transmitted light intensity, they can also be applied to calculations of reflected light in the simple model 'small-angle multiple light scattering + single light scattering backward'. This approach was proposed by De Wolf (1971) and has already been used in many applications (Barun, 1995; Borovoi, 1995; Zege *et al.*, 1995; Borovoi *et al.*, 1998; Katsev *et al.*, 1998).

If particles both scatter photons mainly in the small angle region and are characterized by strong absorption, the angular distribution of reflected light is determined mostly by single scattering and we obtain (Gordon, 1973) for the reflection function of semi-infinite

$$R_\infty(\mu, \mu_0, \psi) = \frac{A\omega_0 p(\theta)}{4(\mu + \mu_0)} \tag{3.115}$$

and finite

$$R(\mu, \mu_0, \phi) = R_\infty(\mu, \mu_0, \phi)\left(1 - e^{-\tilde{\tau}_0(1/\mu + 1/\mu_0)}\right) \tag{3.116}$$

layers. Here the term $(\omega_0 p(\theta))/(4(\mu + \mu_0))$ coincides with the contribution of singly scattered light to value $R_\infty(\mu, \mu_0, \phi)$ and $\tilde{\tau}_0 = \tau_0(1 - \Phi)$. The value of $A = 1/(1 - \omega_0 \Phi)$ approximately accounts for multiply scattered light ($\Phi = \frac{1}{2}\int_0^{\pi/2} p(\theta) \sin\theta \, d\theta$). This approximation holds for μ, $\mu_0 > 0.5$, $1 - \omega_0 \geq 0.15(1 - g)$ (Zege *et al.*, 1991).

Note that the accuracy of the small-angle approximation is low at $\tau \geq 7$ (Zege *et al.*, 1995; Jaffe, 1995). In this case the transmitted intensity is not a sharp peaked function of the angle of observation and the integral in equation (3.1) can be replaced by the differential term. This approximation is called the small-angle diffusion approximation (see below). It is based on the Fokker–Planck equation for diffused intensity and was considered in details by Zege *et al.* (1991).

Another important approximation for thick layers (the diffusion approximation) was initially developed in neutron physics (Case and Zweifel, 1967; Apresyan and Kravzov, 1983). It was applied for optical problems as well (Furutsu, 1980a, b, 1997; Zege *et al.*, 1991; Aronson, 1995). This approximation is important for solution of the time-dependent RTE which is used for pulse propagation problems (Minin, 1988; Zege *et al.*, 1991). In this case the term $dI/(c\,dt)$ (t is time and c is light speed) should be added to equation (3.1). The possibility and limits of such an extension were discussed by Lagendijk and Van Tiggelen (1996). The most powerful tool for the solution of these problems is the Laplace transform technique (Minin, 1988). Laplace transforms and moments of the light intensity for a time-dependent problem can be expressed by the solution of the time-independent RTE (Minin, 1988).

3.5.2 Small-angle patterns of optically dense disperse media

Some of the theoretical results presented in the previous section have been checked against special experiments by Belov *et al.* (1984) and Borovoi *et al.* (1986). The experimental set-up used by Kokhanovsky *et al.* (2001) is presented in Figure 3.7.

A He–Ne laser beam with wavelength $\lambda = 0.6328\,\mu\text{m}$ passes through a beam expander and a sample with monodispersions of polystyrene spheres. The diffraction pattern is measured using a detector with thirty-one semicircular elements of different widths. Polystyrene spheres with certified mean diameters d_0 equal to $100\,\mu\text{m} \pm 2.0\,\mu\text{m}$ and $9.685 \pm 0.064\,\mu\text{m}$ were used in the experiment. The coefficient

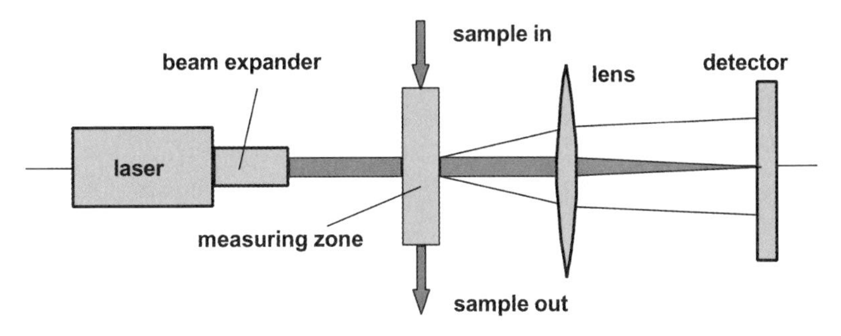

Figure 3.7. The experimental set-up.

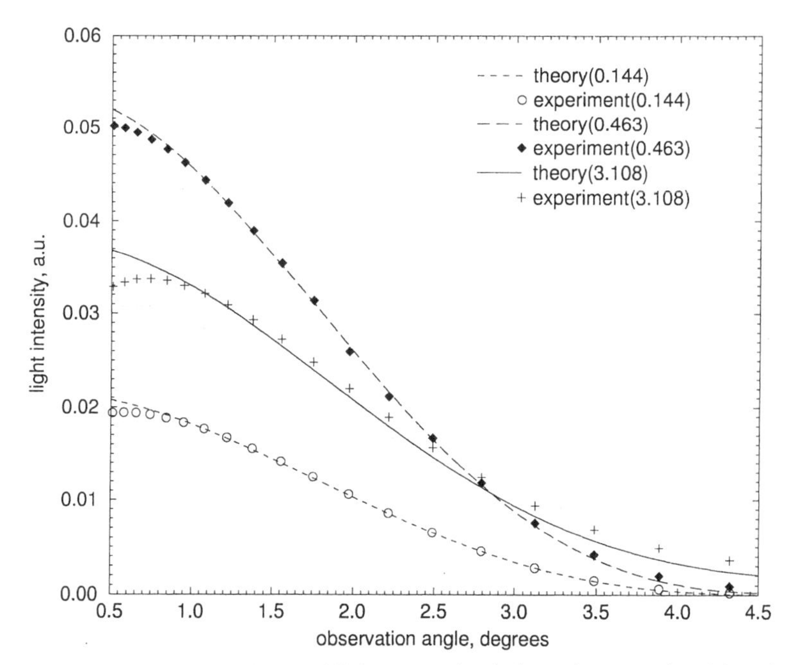

Figure 3.8. The angular dependence of light transmitted through a sample with polystyrene spherical particles according to experiment and theory at optical thickness equal to 0.144, 0.463, and 3.108. The diameter of spheres is equal to 9.685 μm.

of variation in particle size distribution was equal to 0.042 for large particles and 0.014 for smaller ones. Particles were composed of polystyrene (divinylbenzene, 4–8%) and had specific gravity 1.05 g ml^{-1}. The refractive index of particles at wavelength 0.59 μm was equal to 1.59.

The results of the experiment are presented in Figures 3.8, 3.9 for different optical thicknesses τ and sizes of particles. Diameters of particles were much

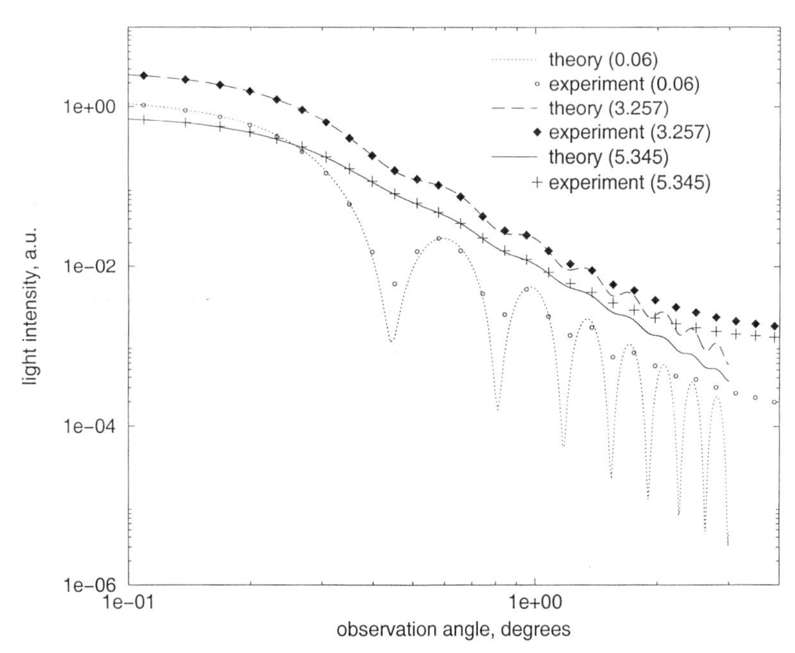

Figure 3.9. The same as Figure 3.8 but for $d_{ef} = 100\,\mu\text{m}$ and optical thickness equal to 0.06, 3.257, and 5.345.

larger than the wavelength of incident light. Note that as an accurate approximation for the extinction efficiency factor it follows that $Q_{ext} = 2$ in this case. Thus, optical thickness is approximately given by the following equation:

$$\tau = \frac{3Lc}{d_{ef}}$$

where $d_{ef} = 2a_{ef}$, a_{ef} is the effective radius of particles, L is the geometrical thickness of a sample and c is the volumetric concentration of particles in the suspension. It follows for the value of c that:

$$c = \frac{d_{ef}\tau}{3L}$$

We can see that the value of c is simply expressed via the optical thickness of medium τ. The value of τ was measured independently during the experiment. The geometrical thickness L was equal to 20 mm and $d_{ef} \approx d_0$ due to the high monodispersity of particles in the sample.

The experiment (symbols) shows that increase in concentration of particles leads to higher values of transmitted intensity in the region of small values of τ. However, the transmitted intensity reduces with further growth of the optical thickness. Such behaviour can easily be understood. Indeed, the intensity of the transmitted diffused light I is equal to 0 if there are no particles in a cuvette. It grows with the optical thickness initially. However, it is equal to 0 for semi-infinite media. Thus, it should be a maximum in the dependencies $I_d(\tau)$ at some value of τ. This is in correspondence with our experimental results. It should be pointed out that oscillations in Figure 3.9 confirm the high monodispersity of particles in the sample.

Another interesting feature is related to broadening of the diffraction pike due to the multiple light-scattering phenomenon. The larger optical thickness τ, the larger the half-width of the angular spectrum of the transmitted light. It should be pointed out that half-widths of small-angle peaks of scattered light also increase as particle sizes decrease. Thus, the presence of multiple light scattering in a sample could lead to underestimation of the size of particles in standard inversion procedures, which do not account for multiple light scattering.

Figures 3.8, 3.9 differ due to the absence of minima in Figure 3.8 for low concentrations of particles. This is due to limitation of observation angles within 4° in the experiment. The first minimum should appear at the scattering angle:

$$\theta_d = \frac{3.83 \times 180}{\pi x}$$

degrees according to the Fraunhofer theory (Born and Wolf, 1965). Here $x = (2\pi a)/\lambda$ is the size parameter. The value of θ_d is equal approximately to 0.4° at $a = 50\,\mu\text{m}$. But it is about 4° at $a = 5\,\mu\text{m}$, which is almost outside the measured angular range.

It should be pointed out that multiple light scattering reduces oscillations and makes diffraction patterns smoother. The same effect could be due to polydispersity of particles in a turbid medium.

Diffused transmitted intensity can be found by the solution of the RTE, as discussed in the previous section. It follows from equation (3.104) for diffused transmitted light that:

$$I_d(\tau, \vartheta) = \frac{I_0}{2\pi} \exp(-\tau) \int_0^\infty (\exp[\omega_0 P(\sigma)\tau] - 1) J_0(\sigma\vartheta)\sigma\, d\sigma \tag{3.117}$$

This equation can be simplified in the case of thin layers. It follows from equation (3.117) as $\tau \to 0$ that:

$$I_d(\tau, \vartheta) = \frac{I_0}{4\pi} \omega_0 \tau p(\vartheta) \tag{3.118}$$

where

$$p(\vartheta) = 2 \int_0^\infty P(\sigma) J_0(\sigma\vartheta)\sigma\, d\sigma \tag{3.119}$$

Thus, small-angle diffused intensity is determined by the angular distribution of singly scattered light, as it should be for small values of optical thickness.

Equation (3.117) allows us to study the dependence of diffraction patterns on the optical thickness and the concentration of particles. To do so we need to know the function $P(\sigma)$ in equation (3.117).

This function differs for particles of different shapes and internal structures. For instance, it follows in the framework of the Fraunhofer approximation for monodispersed spherical particles that:

$$\omega_0 = \tfrac{1}{2}, \qquad p(\theta) = \frac{4J_1^2(\theta x)}{\theta^2}$$

and (see Table 3.9):

$$P(\sigma) = \frac{2}{\pi}\left(\arccos\left(\frac{\sigma}{2x}\right) - \frac{\sigma}{2x}\sqrt{1 - \left(\frac{\sigma}{2x}\right)^2}\right)u\left(\frac{\sigma}{2x}\right) \tag{3.120}$$

where $u(\sigma/2x) = 1$ at $\sigma \leq 2x$ and $u(\sigma/2x) = 0$ at $\sigma > 2x$. Thus, the diffused intensity $I_d(\tau, \vartheta)$ can be presented in the following form:

$$I_d(\tau, \vartheta) = DF(\tau, \alpha) \tag{3.121}$$

where $D = (2x^2 I_0)/\pi$ is the constant which does not depend on the concentration of particles, and the function

$$F(\tau, z) = \exp(-\tau)\int_0^1 \left(\exp\left[\frac{\tau}{\pi}\left\{\arccos(y) - y\sqrt{1 - y^2}\right\}\right] - 1\right) J_0(yz) y\, dy \tag{3.122}$$

depends only on the optical thickness τ and the parameter $z = 2\vartheta x$. The function $F(z)$ at $\tau = 0.1$, 1, 3, and 5 is presented in Figure 3.10. We can see that the intensity of diffused transmitted light increases with the optical thickness, reaches maximal values and starts to decrease. The amplitude of oscillations decreases with optical thickness τ. On the other hand, the half-width of the small-angle peak increases with optical thickness. All these features are consistent with the experimental data presented in Figures 3.8, 3.9.

The dependence of light intensity (3.121) on optical thickness at $z = 1.5$ is presented in Figure 3.11. We can see that the intensity has a maximum at $\tau \approx 1$ in this case. It should be pointed out that the function $I_d(\tau, \vartheta)$ in Figure 3.11 was normalized to 1 at $\tau = 0.1$.

Let us compare the results of calculations with the experimental data obtained. To this end, results of calculations using equations (3.121), (3.122) were matched with experimental data at $\theta = 1^\circ$ for smaller particles (see Figure 3.8) and at $\theta = 0.1^\circ$ for larger particles (see Figure 3.9). The agreement between theory and experiment is excellent for all optical thicknesses except tails of angular distributions of transmitted light (see, in particular, Figure 3.9). Thus, simple equation (3.121) can indeed be used for the solution of both inverse and direct problems of light-scattering media optics.

The poor agreement between theory and experiment at $\vartheta > 2^\circ$ in Figure 3.9 is due to the fact that multiple scattering of light refracted and reflected by particles becomes important in this case. This is not accounted for in equation (3.121), where the geometrical optics component of scattered light is simply ignored. Accounting for geometrical optics scattering becomes important for larger particles at smaller

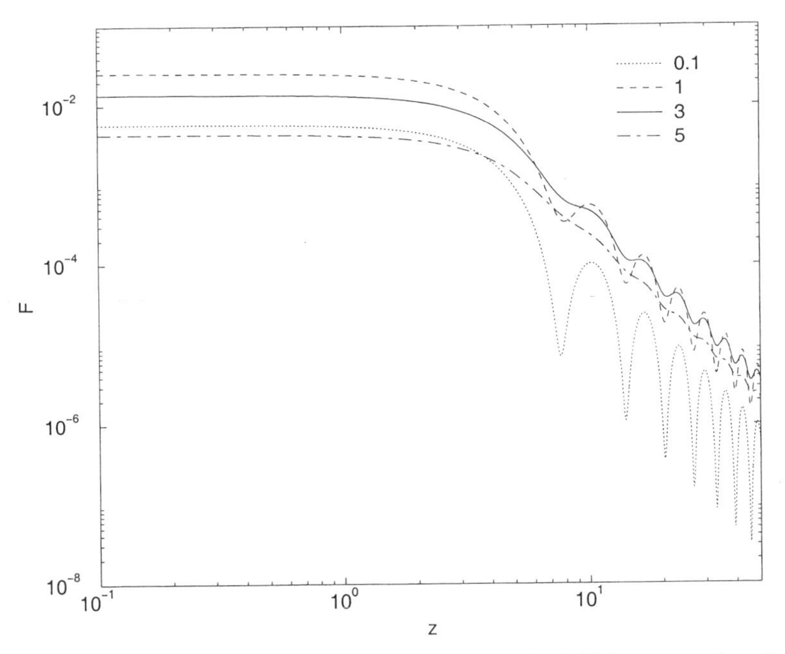

Figure 3.10. The dependence $F(z)$ at different values of optical thickness equal to 0.1, 1, 3, and 5.

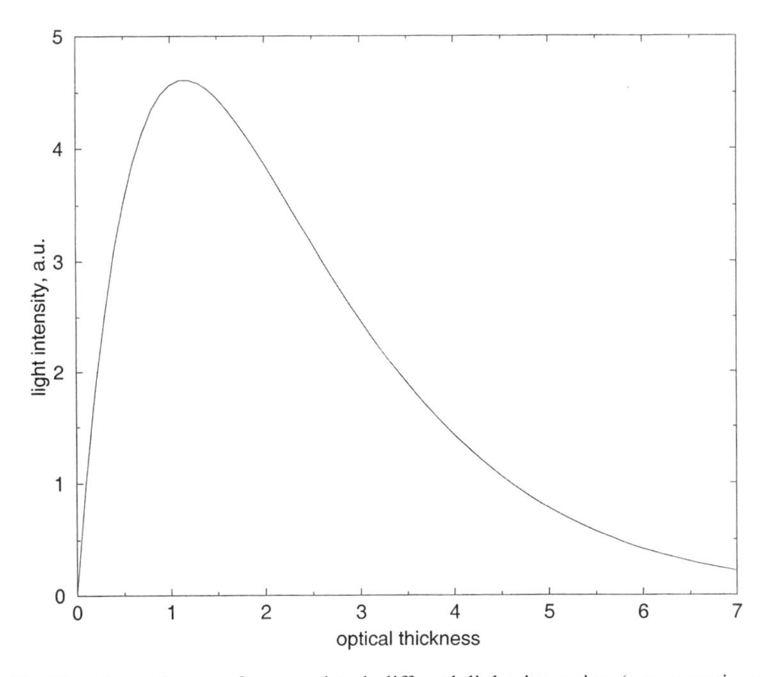

Figure 3.11. The dependence of transmitted diffused light intensity (see equations (3.121), (3.122)) on optical thickness at $z = 1.5$.

scattering angles than it is for smaller particles (compare Figures 3.8 and 3.9). This is due to smaller half-widths of Fraunhofer diffraction patterns for larger scatterers.

Accounting for geometrical optics scattering is simple. Indeed, ignoring interference between diffracted, reflected, and refracted beams we get:

$$p(\theta) = \frac{\sigma^d_{sca} p^d(\theta) + \sigma^g_{sca} p^g(\theta)}{\sigma^d_{sca} + \sigma^g_{sca}}$$

where σ^d_{sca} and σ^g_{sca} are parts of the total scattering coefficient $\sigma_{sca} = \sigma^d_{sca} + \sigma^g_{sca}$, related to diffraction and geometrical optics scattering processes, respectively.

It follows that $\sigma^d_{sca} = \sigma^g_{sca}$ for large transparent particles and

$$p(\theta) = \frac{p^d(\theta) + p^g(\theta)}{2}$$

where $p^d(\theta) = (4J_1^2(\theta x))/\theta^2$ and function $p^g(\theta)$ represents the geometrical optics contribution to the total phase function $p(\theta)$.

Calculations performed in the framework of the geometrical optics approximation show that function $p^g(\theta)$ can be approximated by the following simple formula:

$$p^g(\theta) = 4\beta e^{-\beta\theta^2} \tag{3.123}$$

where the value of β depends only on the refractive index of transparent particles, but not on their size. It follows for the Fourier–Bessel transform of phase function in this case

$$P(\sigma) = \frac{P^d(\sigma) + P^g(\sigma)}{2} \tag{3.124}$$

where the value of $P^d(\sigma)$ can be found from equation (3.120) and (see Table 3.9):

$$P^g(\sigma) = e^{-\sigma^2/4\beta} \tag{3.125}$$

Results of comparison of calculated diffused transmitted intensity (see equations (3.122), (3.124)) with experimental data for monodipersed spheres with diameter 100 μm are presented in Figure 3.12 at $\beta = 5$. We can see that taking the geometrical optics scattering into account (see equations (3.124), (3.125)) indeed provides higher accuracy at larger values of ϑ. Note that the value of $\beta = 5$ in equation (3.125) was obtained by fitting experimental and theoretical results.

It should be pointed out that the general equations presented here are valid also for randomly oriented nonspherical particles and for polydispersed media. However, these media are characterized by different spectra $P(\sigma)$ when compared with the simple case of monodispersed spheres. This results in different angular spectra $I_d(\tau, \vartheta)$ of diffused light. For instance, it follows for disperse media with polydisper-

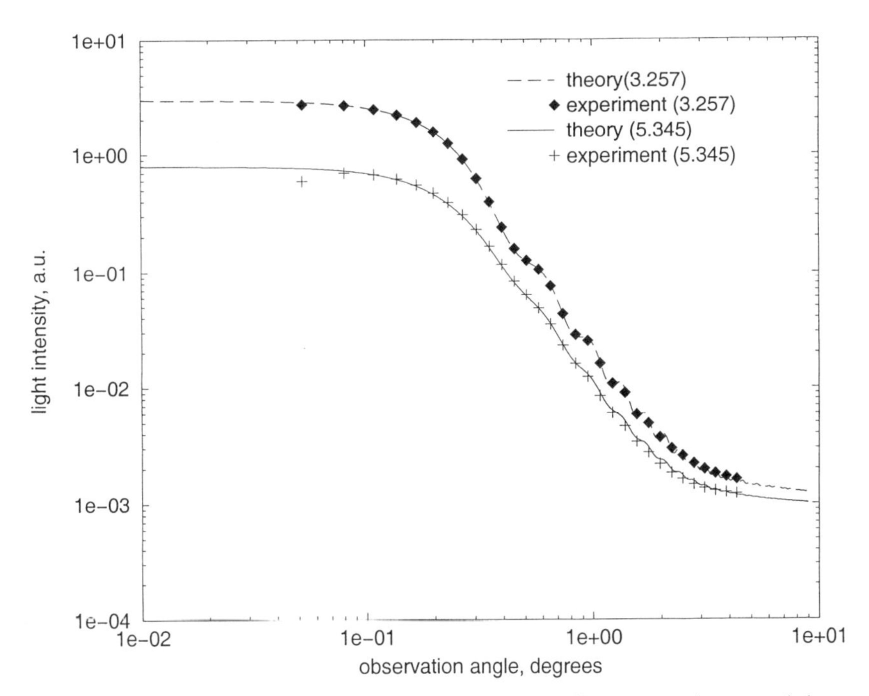

Figure 3.12. The angular dependence of light intensity according to experiment and theory for samples with polystyrene spheres of diameter 100 μm accounting for geometrical optical scattering at optical thickness equal to 3.257 and 5.345.

sions of spherical particles that:

$$\langle P(\sigma)\rangle = \frac{\int_0^\infty a^2 f(a) P(\sigma)\, da}{\int_0^\infty a^2 f(a)\, da} \tag{3.126}$$

where $f(a)$ is particle size distribution and $P(\sigma)$ is described by equation (3.124).

Experimental measurements of small-angle patterns of particulate media show that the influence of multiple light scattering causes broadening of the angular spectrum of transmitted light and reduces the intensity of the Fraunhofer rings for monodispersed particles. We can see that equations (3.122), (3.123) describe experimental data with high accuracy. Thus, they can be used as a theoretical base for the laser diffraction spectroscopy of turbid media.

The function $\langle P(\sigma)\rangle$ in equation (3.126) can be retrieved from equation (3.117) by the inverse Fourier–Bessel transform. On the other hand, the Fourier–Bessel transform of function $\langle P(\sigma)\rangle$ allows us to obtain the function $p(\theta)$ (see Table 3.9) and reduce the inverse problem for a multiply light-scattering medium to the well-studied case of single scattering.

3.6 STRONGLY ABSORBING MEDIA

Let us now consider the case of strong absorption of radiation by particles ($\omega_0 \to 0$). In this case the reflection and transmission function can be represented in the following form:

$$R(\mu, \mu_0, \phi) = \sum_{n-1}^{\infty} \omega_0^n R_n^s(\mu, \mu_0, \phi)$$

$$T(\mu, \mu_0, \phi) = \sum_{n=1}^{\infty} \omega_0^n T_n^s(\mu, \mu_0, \phi)$$

where the values of R_n^s, T_n^s are obtained from the RTE by the method of successive orders. In the framework of this approach intensities of reflected and transmitted light are first calculated with using equations (3.9), (3.10), where the value of $B(\tau', \eta, \xi, \varphi)$ is equal to $((\omega_0 I_0)/4)p(\theta)\, e^{-\tau'/\cos\vartheta_0}$ (singly scattered light). The derived values of $I_\uparrow$ and $I_\downarrow$ are substituted in equations (3.9), (3.10) as $B(\tau', \eta, \xi, \varphi)$ to obtain secondary scattered light, etc. The results for thin layers are (see equations (3.16), (3.17)):

$$R_1^s(\mu, \mu_0, \phi) = p(\theta) r_1^s, \qquad r_1^s = \frac{1}{4(\mu + \mu_0)} \left\{1 - e^{-\tau_0(1/\mu + 1/\mu_0)}\right\} \tag{3.127}$$

$$T_1^s(\mu, \mu_0, \phi) = p(\theta) t_1^s, \qquad t_1^s = \frac{1}{4(\mu - \mu_0)} \left\{e^{-\tau_0/\mu} - e^{-\tau_0/\mu_0}\right\} \tag{3.128}$$

in the first approximation. If we account for secondary scattering (Wauben, 1992) it follows that:

$$\begin{aligned} R_2^s(\mu, \mu_0, \phi) &= \frac{1}{8}\int_0^1 d\mu'\, p(-\mu, -\mu') p(-\mu', \mu_0) \xi_1(\mu', \mu, \mu_0) \\ &\quad + \frac{1}{8}\int_0^1 d\mu'\, p(-\mu, \mu') p(\mu', \mu_0) \xi_2(\mu', \mu, \mu_0) \end{aligned} \tag{3.129}$$

$$\begin{aligned} T_2^s(\mu, \mu_0, \phi) &= \frac{1}{8}\int_0^1 d\mu'\, p(\mu, -\mu') p(-\mu', \mu_0) \xi_3(\mu', \mu, \mu_0) \\ &\quad + \frac{1}{8}\int_0^1 d\mu'\, p(\mu, \mu') p(\mu', \mu_0) \xi_4(\mu', \mu, \mu_0) \end{aligned} \tag{3.130}$$

where

$$\xi_1(\mu', \mu, \mu_0) = \frac{4}{\mu_0 + \mu'} \left\{\mu_0 r_1^s(\mu, \mu_0) + \mu' \exp\left(-\frac{\tau_0}{\mu'} - \frac{\tau_0}{\mu_0}\right) r_1^s(\mu, -\mu')\right\}$$

$$\xi_2(\mu', \mu, \mu_0) = \frac{4}{\mu_0 - \mu'} \{\mu_0 r_1^s(\mu, \mu_0) - \mu' r_1^s(\mu, \mu')\}$$

$$\xi_3(\mu', \mu, \mu_0) = \frac{4}{\mu_0 + \mu'} \left\{\mu_0 t_1^s(\mu, \mu_0) + \mu' e^{-\tau_0/\mu - \tau_0/\mu_0} t_1^s(\mu, -\mu')\right\}$$

$$\xi_4(\mu', \mu, \mu_0) = \frac{4}{\mu_0 - \mu'} \{\mu_0 t_1^s(\mu, \mu_0) - \mu' t_1^s(\mu, \mu')\}$$

The value $p(\mu, \mu')$ is the phase function, averaged on the azimuth ϕ. For isotropic scattering and some other simple analytical forms of phase functions, integrals for values of $R_2(\mu, \mu_0, \phi)$, $T_2(\mu, \mu_0, \phi)$ can be calculated analytically. The expressions for successive orders of scattering are complicated. However, they can easily be calculated numerically.

3.7 WEAKLY ABSORBING MEDIA

Clearly, for weakly absorbing ($\beta = 1 - \omega_0 \to 0$) media, other expansions should be used. Namely, it follows that:

$$R(\mu, \mu_0, \phi) = \sum_{n=1}^{\infty} \beta^n R_n(\mu, \mu_0, \phi) \tag{3.131}$$

$$T(\mu, \mu_0, \phi) = \sum_{n=1}^{\infty} \beta^n T_n(\mu, \mu_0, \phi) \tag{3.132}$$

Unfortunately even at $\omega_0 = 1$ the RTE (3.2) is not simplified very much. Thus, this case is difficult to solve analytically. But expansions (3.131), (3.132) can be useful from the point of view of investigating the general properties of the RTE solutions as $\beta \to 0$. Rozenberg (1962) used very similar expansions to find that the ratio Ξ of reflection functions of semi-infinite nonabsorbing and weakly absorbing media. It is determined by the following equation:

$$\Xi = \exp\left(-\alpha D(\mu, \mu_0, \phi)\sqrt{\beta/(1-g)}\right) \tag{3.133}$$

where the value $\alpha = 4/\sqrt{3}$ and for the value of $D(\mu, \mu_0, \phi)$ we can approximately obtain (Zege *et al.*, 1991):

$$D(\mu, \mu_0, \phi) = \frac{K_0(\mu)K_0(\mu_0)}{R_0(\mu, \mu_0, \phi)} \tag{3.134}$$

Note that function $D(\mu, \mu_0, \phi)$ depends only on the phase function. This dependence is rather weak as shown in previous sections. To obtain equation (3.133) Rozenberg used expansions on the parameter $\sqrt{\beta}$, which is inversely proportional to the average number of scattering events N in weakly absorbing media, instead of expansions (3.131), (3.132). The value of N is infinite for semi-infinite nonabsorbing media.

3.8 DIFFUSION APPROXIMATION

The diffusion approximation is used for the investigation of light fields deep inside light-scattering media. In this case the angular distribution of the diffused intensity only weakly depends on the observation angle. Such conditions occur, for instance, in dense fogs and clouds.

The main feature of the diffusion domain is the independence of the diffused light intensity I on the azimuth. Thus, we obtain from equation (3.2) under this assumption:

$$\eta \frac{dI(\tau,\eta)}{d\tau} = -I(\tau,\eta) + \frac{\omega_0}{2} \int_{-1}^{1} p(\eta,\eta') I(\tau,\eta')\, d\eta' \tag{3.135}$$

where

$$p(\eta,\eta') = \frac{1}{2\pi} \int_0^{2\pi} p(\theta)\, d\varphi \tag{3.136}$$

is the phase function averaged over the azimuth. Let us multiply equation (3.135) by η and integrate it from -1 to 1 with respect to η. Then it follows that:

$$\frac{dK(\tau)}{d\tau} = -F(\tau) + \omega_0 G(\tau) \tag{3.137}$$

where

$$K(\tau) = \frac{1}{2} \int_{-1}^{1} \eta^2 I(\tau,\eta)\, d\eta \tag{3.138}$$

$$F(\tau) = \frac{1}{2} \int_{-1}^{1} \eta I(\tau,\eta)\, d\eta \tag{3.139}$$

$$G(\tau) = \frac{1}{4} \int_{-1}^{1} \eta\, d\eta \int_{-1}^{1} d\eta' p(\eta,\eta') I(\tau,\eta') \tag{3.140}$$

Function $G(\tau)$ is proportional to function $F(\tau)$. Let us show it. Function $p(\theta)$ in equation (3.136) can be presented as (see equation (2.70)):

$$p(\theta) = \sum_{l=0}^{\infty} x_l P_l(\cos\theta) \tag{3.141}$$

where

$$\cos\theta = \eta\eta' + \sqrt{(1-\eta^2)(1-\eta'^2)} \cos(\phi - \phi') \tag{3.142}$$

Note that it follows from the addition theorem for spherical functions that:

$$P_l(\cos\theta) = P_l(\eta) P_l(\eta') + 2 \sum_{m=1}^{l} \frac{(i-m)!}{(i+m)!} P_i^m(\eta) P_i^m(\eta') \cos m(\phi - \phi') \tag{3.143}$$

where functions $P_i^m(\eta)$ are associated Legendre functions. Substitution of equation (3.143) into equation (3.141) gives:

$$p(\theta) = \sum_{l=0}^{\infty} x_l P_l(\eta) P_l(\eta') + 2 \sum_{l=0}^{\infty} \sum_{m=1}^{l} \frac{(i-m)!}{(i+m)!} P_i^m(\eta) P_i^m(\eta') \cos m(\phi - \phi') \tag{3.144}$$

Thus, it follows from equations (3.136), (3.144) that:

$$p(\eta,\eta') = \sum_{l=0}^{\infty} x_l P_l(\eta) P_l(\eta') \tag{3.145}$$

This representation of the averaged phase function is of great importance in radiative transfer theory. We obtain from equations (3.140), (3.145):

$$G(\tau) = \frac{1}{4}\sum_{l=0}^{\infty} x_l \int_{-1}^{1} P_l(\eta)\eta\,d\eta \int_{-1}^{1} I(\tau,\eta')P_l(\eta')\,d\eta' \tag{3.146}$$

or

$$G(\tau) = \frac{1}{4}\sum_{l=0}^{\infty} x_l\delta_{l1}\left(l+\frac{1}{2}\right)^{-1} \int_{-1}^{1} I(\tau,\eta')P_l(\eta')\,d\eta' \tag{3.147}$$

where we accounted for the orthogonality relation:

$$\int_{-1}^{1} P_l(\eta)P_s(\eta)d\eta = \frac{\delta_{ls}}{l+\frac{1}{2}} \tag{3.148}$$

Indeed, it follows from equation (3.148) that:

$$\int_{-1}^{1} P_l(\eta)\eta\,d\eta = \frac{2\delta_{l1}}{2l+1}$$

Finally, we can obtain from equations (3.139), (3.147):

$$G(\tau) = gF(\tau) \tag{3.149}$$

where $g = x_1/3$ is the asymmetry parameter. Thus, equation (3.137) can be written in the following form:

$$\frac{dK(\tau)}{d\tau} = -(1-\omega_0 g)F(\tau) \tag{3.150}$$

Taking the second derivative, we get:

$$\frac{d^2K(\tau)}{d\tau^2} = -(1-\omega_0 g)S(\tau) \tag{3.151}$$

where $S(\tau) \equiv dF(\tau)/d\tau$. The function $S(\tau)$ can also be obtained from equation (3.135). Indeed, integrating equation (3.135) from -1 to 1 with respect to η we get:

$$S(\tau) = -E(\tau) + \omega_0\Pi(\tau) \tag{3.152}$$

where

$$E(\tau) = \frac{1}{2}\int_{-1}^{1} I(\tau,\eta)\,d\eta \tag{3.153}$$

and

$$\Pi(\tau) = \frac{1}{4}\int_{-1}^{1} d\eta \int_{-1}^{1} d\eta'\, p(\eta,\eta')I(\tau,\eta') \tag{3.154}$$

or

$$\Pi(\tau) = \frac{1}{4}\sum_{l=0}^{\infty} x_l \int_{-1}^{1} P_l\,d\eta \int_{-1}^{1} I(\tau,\eta')P_l(\eta')\,d\eta' \tag{3.155}$$

Note that equation (3.155) follows from equations (3.154), (3.145). It should be pointed out that (see equation (3.148)):

$$\Pi(\tau) = \frac{1}{4}\sum_{l=0}^{\infty} x_l \delta_{l0} \left(l + \frac{1}{2}\right)^{-1} \int_{-1}^{1} I(\tau, \eta') P_l(\eta') \, d\eta' \tag{3.156}$$

Thus, it follows (see equations (3.153), (3.156)):

$$\Pi(\tau) = E(\tau) \tag{3.157}$$

and

$$S(\tau) = -(1 - \omega_0) E(\tau) \tag{3.158}$$

where we accounted for equalities: $x_0 = 1$, $P_0 = 1$.

Finally, the differential equation (3.151) can be written in the following form:

$$\frac{d^2 K(\tau)}{d\tau^2} = (1 - \omega_0)(1 - \omega_0 g) E(\tau) \tag{3.159}$$

We can expect that the ratio $d = K(\tau)/E(\tau)$ will be constant in the diffusion domain far from boundaries and sources. For instance, assuming that

$$I(\tau, \eta) = a + b\eta \tag{3.160}$$

where a and b are arbitrary constants, we obtain (see equations (3.138), (3.153)): $E = a$, $K = a/3$ and $K/E = \frac{1}{3}$. The Eddington approximation is based on this assumption ($d = \frac{1}{3}$). It follows from equations (3.159), (3.160):

$$\frac{d^2 W(\tau)}{d\tau^2} - \xi W(\tau) = 0 \tag{3.161}$$

where

$$\xi = \frac{(1 - \omega_0)(1 - \omega_0 g)}{d} \tag{3.162}$$

We used the relation $W(\tau) = (c/4\pi)E(\tau)$ between the energy density $W(\tau)$ and the average intensity $E(\tau)$. Equation (3.161) is called the diffusion equation and can be solved by standard methods. It allows us to interpret light propagation in scattering media far from light sources and boundaries as a diffusion process. Equation (3.161) has the following solution at $\xi > 0$:

$$W(\tau) = C_1 \cosh \sqrt{\xi}\tau + C_2 \sinh \sqrt{\xi}\tau \tag{3.163}$$

where C_1 and C_2 are constants which depend on boundary conditions.

We can see that the energy density of diffused light decreases exponentially with optical thickness in absorbing media. The extinction of diffused light with optical thickness is more rapid for larger values of $1 - \omega_0$ and $1 - g$.

Let us consider a semi-infinite absorbing medium. It follows that $W(\tau) \to 0$ as $\tau \to \infty$ and

$$W(\tau) = C \exp(-\sqrt{\xi}\tau) \tag{3.164}$$

where $C = (C_1 - C_2)/2$ is constant. This constant can be found from boundary conditions. It should be underlined that the intensity of light field I and the value of W are proportional to $\exp(-k\tau)$ (k is the diffusion exponent) in deep layers of light-scattering absorbing media as shown earlier in this chapter. Thus, it follows that $k = \sqrt{\xi}$ and

$$d = \frac{(1-\omega_0)(1-\omega_0 g)}{k^2} \tag{3.165}$$

This formula allows us to express the value of the d via the single-scattering albedo ω_0, asymmetry parameter g, and the diffusion exponent k.

It follows from equation (3.161) at $\xi = 0$ that:

$$W(\tau) = A_1 + A_2\tau$$

where A_1 and A_2 are constants. Thus, light energy density is constant or changes linearly with optical thickness in conservatively ($k = 0$) light-scattering media. The actual behaviour depends on boundary conditions.

Note that the problem of boundary conditions is not trivial for the diffusion approximation. Indeed, solutions of the diffusion equation are only valid far from light-scattering media boundaries. Thus, boundary conditions may only be approximate ones. This problem was discussed by Ishimaru (1978) in detail.

Note that equation (3.161) is easily generalized for non-plane-parallel geometries:

$$\sigma_{ext}^{-2}\Delta W - \xi W = 0 \tag{3.166}$$

where $\xi = (1-\omega_0)(1-\omega_0 g)/d$, Δ is the Laplacian. Equation (3.166) can be used for studies of light transfer in multiply light-scattering bodies of a finite volume (e.g. isolated clouds of different shapes).

Note that it follows:

$$\Delta W = \frac{\partial^2 W}{\partial x^2} + \frac{\partial^2 W}{\partial y^2} + \frac{\partial^2 W}{\partial z^2} \tag{3.167}$$

in rectangular co-ordinates x, y, z

$$\Delta W = \frac{1}{r^2}\frac{\partial}{\partial r}\left(r^2\frac{\partial W}{\partial r}\right) + \frac{1}{r^2 \sin\vartheta}\frac{\partial}{\partial\vartheta}\left(\sin\vartheta\frac{\partial W}{\partial\vartheta}\right) + \frac{1}{r^2\sin^2\vartheta}\frac{\partial^2 W}{\partial\varphi^2} \tag{3.168}$$

in spherical co-ordinates r, ϑ, $\varphi (r \geq 0,\ 0 \leq \vartheta \leq \pi)$, and

$$\Delta W = \frac{1}{\rho}\frac{\partial}{\partial\rho}\left(\rho\frac{\partial W}{\partial\rho}\right) + \frac{1}{\rho^2}\frac{\partial^2 W}{\partial\rho^2} + \frac{\partial^2 W}{\partial z^2} \tag{3.169}$$

in cylindrical co-ordinates ρ, φ, $z (\rho \geq 0)$. Equations (3.167)–(3.169) can be used for studies of light diffusion in parallelepipeds, spheres, and cylinders, respectively.

Let us introduce the so-called diffusion coefficient:

$$D = \frac{1}{3(\sigma_{abs} + \sigma_{sca}(1-g))} \tag{3.170}$$

Then it follows from equation (3.166) that:

$$-D\Delta W + \sigma_{abs}W = 0 \tag{3.171}$$

where we assumed that $d = \frac{1}{3}$. Equation (3.171) can be generalized to account for time-dependent processes due to photon sources $\vec{S}(\vec{r}, t)$, where $\vec{r}$ is the radius- vector inside the medium and t is time (Patterson *et al.*, 1989):

$$\frac{1}{c}\frac{\partial W(\vec{r}, t)}{\partial t} - D\Delta W(\vec{r}, t) + \sigma_{abs} W(\vec{r}, t) = S(\vec{r}, t) \tag{3.172}$$

where c is light speed.

The solution of this equation for a short light pulse from an isotropic point source, $S(\vec{r}, t) = \delta(0, 0)$, where δ is the delta function, has the following form in the case of infinite media (Zege and Katser, 1978; Patterson *et al.*, 1989):

$$W(r, t) = \frac{A}{(4\pi Dct)^{3/2}} \exp\left(-\frac{r^2}{4Dct} - \sigma_{abs} ct\right) \tag{3.173}$$

where $A = \text{const}$. Note that function $tW(r, t)$ is proportional to the Gaussian distribution $\Phi(r)$ with dispersion $\Delta = \sqrt{2Dct}$:

$$\Phi(r) = \frac{1}{\sqrt{2\pi}\Delta} \exp\left(-\frac{r^2}{2\Delta^2}\right) \tag{3.174}$$

We can see that the energy density decreases with r more rapidly for media with smaller values of the diffusion coefficient D and time t.

As a matter of fact, equation (3.173) is a Green function of equation (3.172). It can be used for finding energy densities of diffusing waves $W(\vec{r}, t)$ for arbitrary functions $\vec{S}(\vec{r}, t)$. This equation has important applications in biomedical imaging (Schotland, 1997; Tuchin, 1998; Van Rossum and Nienwenhuizen, 1999) and cloud remote sensing (Davies *et al.*, 1999).

3.9 SMALL-ANGLE DIFFUSION APPROXIMATION

As was shown before, the RTE can be written in the following form:

$$\left(\frac{1}{c}\frac{\partial}{\partial t} + \vec{\Omega}\vec{\nabla} + \sigma_{ext}\right) I(\vec{r}, t, \vec{\Omega}) = Q(\vec{r}, \vec{\Omega}) + q(\vec{r}, t, \vec{\Omega}) \tag{3.175}$$

where

$$Q(\vec{r}, t, \vec{\Omega}) = \frac{\sigma_{sca}}{4\pi} \int_{4\pi} p(\vec{\Omega}, \vec{\Omega}') I(\vec{r}, t, \vec{\Omega}')\, d\vec{\Omega}' \tag{3.176}$$

and $q(\vec{r}, t, \vec{\Omega})$ is the source function due to internal light sources. It can be applied for studies of narrow (e.g. laser) beam propagation in multiply scattering media. Equation (3.175) cannot be solved analytically. Thus, it is often reduced to a simpler form, called the small-angle diffusion approximation. First of all, like in the framework of the small-angle approximation, the differential operator is presented in the following form:

$$\vec{\Omega}\vec{\nabla} = \vec{\Omega}_\perp \vec{\nabla}_\perp + \Omega_z \frac{\partial}{\partial z} \tag{3.177}$$

where $\vec{\Omega}_\perp$ is the projection of vector $\vec{\Omega}$ on the plane perpendicular to the direction of coherent (or direct) light propagation.

Secondly, like in the diffusion approximation, the integral term $Q(\vec{r}, t, \vec{\Omega})$ is transformed into the Laplacian:

$$Q(\vec{r}, t, \vec{Q}) \approx h\Delta_\perp \tag{3.178}$$

where the Laplacian $\Delta_\perp$ acts with respect to vector $\vec{\Omega}_\perp$ and

$$h = \frac{\sigma_{sca}}{4} \langle \theta^2 \rangle \tag{3.179}$$

The value of

$$\langle \theta^2 \rangle = \frac{1}{2} \int_0^\pi p(\theta)\theta^2 \, d\theta \tag{3.180}$$

is the average square of the scattering angle.

Approximate equality (3.178) is derived by expanding the intensity $I(\vec{r}, t, \vec{\Omega})$ in equation (3.176) in the Taylor series with respect to $\vec{\Omega} - \vec{\Omega}'$. Only three terms of the expansion are used (Sobolev, 1972). Thus, it follows from equations (3.175), (3.176), and (3.178):

$$\left[\frac{1}{c} \frac{\partial}{\partial t} + \Omega_z \frac{\partial}{\partial z} + \vec{\Omega}_\perp \vec{\nabla}_\perp \sigma_{ext} \right] I(\vec{r}, t, \vec{\Omega}) = h\Delta_\perp I(\vec{r}, t, \vec{\Omega}) + q(\vec{r}, t, \vec{\Omega}) \tag{3.181}$$

This equation is approximately valid only if the intensity weakly depends on the angle when rotating $\vec{\Omega}$ at an angle $\vartheta' \sim \sqrt{\langle \theta^2 \rangle}$. This means that the light field should be almost isotropic on the scale of the angular width of the phase function.

Equation (3.181) is simpler than equation (3.175). For instance it follows (Bremmer, 1964) from equation (3.181) assuming that $\Omega_z = 1$, which is a standard assumption in the framework of small angle approximations, and $q(\vec{r}, t, \vec{\Omega}) = \delta(\vec{r}_\perp)\, \delta(z)\, \delta(\vec{\Omega})\, \delta(t)$:

$$I(\vec{r}_\perp, \vec{\Omega}_\perp) = -\frac{P\delta\left(t - \dfrac{z}{c}\right)}{\pi^2 v} \exp\left[-\frac{\beta_0 r_\perp^2 - 2\beta_1 \vec{r}_\perp \vec{\Omega}_\perp + \beta_2 \Omega_\perp^2}{v} \right] \tag{3.182}$$

where z is the geometrical depth, $\tau = \sigma_{ext} z$ is the optical depth and (Dolin and Levin, 1991)

$$\beta_0 = 4zh, \qquad \beta_1 = 2z^2 h, \qquad \beta_2 = \tfrac{4}{3} z^3 h, \qquad v = \beta_0 \beta_2 - \beta_1^2, \qquad P = \exp(-\tau)$$

The intensity in equation (3.182) is the Green function of equation (3.181). It can be used to find the intensity under arbitrary light sources $q(\vec{r}, t, \vec{\Omega})$.

More accurate results for the function $I(\vec{r}, t, \vec{\Omega})$ can be obtained assuming that $\Omega_z = 1 - \Omega_\perp^2/2$ (Dolin and Levin, 1991).

4

Light scattering and radiative transfer in densely packed disperse media

4.1 SINGLE LIGHT SCATTERING BY DENSELY PACKED PARTICLES

It is evident that local optical characteristics of media with touching or nearly touching particles will be different from those obtained at low concentrations (Mishchenko *et al.*, 1995a). For instance, the differential scattering cross-section (DSCS) $C_{sca}(\theta)$ per particle in close-packed media can be approximately represented as:

$$C_{sca}(\theta) = C^0_{sca}(\theta) \cdot S(\theta)$$

where $S(\theta)$ is the static structure factor (Balescu, 1975; Rytov *et al.*, 1978; Al-Nimr and Arpaci, 1994; Lock and Chiu, 1994), θ is the scattering angle, $C^0_{sca}(\theta)$ is the DSCS without account for correlation effects. The factor $S(\theta)$ can be calculated by different methods (Ivanov *et al.*, 1988). One of the most popular and simple approximations to find the function $S(\theta)$ was proposed by Percus and Yevick (1958). The results of the calculation of $S(\theta)$ in the framework of this approximation is presented in Figure 4.1. We can see that the value of $S(\theta) \to 1$ and $C_{sca}(\theta) \to C^0_{sca}(\theta)$ as $\theta x \to \infty$ (x is the size parameter). Within good accuracy we can ignore correlation effects at $\theta \geq 7/x$. For small particles ($x \to 0$) correlation effects are important at any scattering angle. But for large particles we should account for them only in calculations of the diffraction component of scattered light ($\theta \leq 7/x$). It follows at $\theta = 0$:

$$C_{sca}(0) = C^0_{sca}(0) \cdot S(0) \tag{4.1}$$

where

$$S(0) = \frac{(1-c)^4}{(1+2c)^2} \tag{4.2}$$

and c is the volumetric concentration of particles. We can see that the value of $S(0) \to 0$ as $c \to 1$. The intensity of the scattered light decreases due to the correlation effects. The scattering cross-section C_{sca}, the asymmetry parameter g,

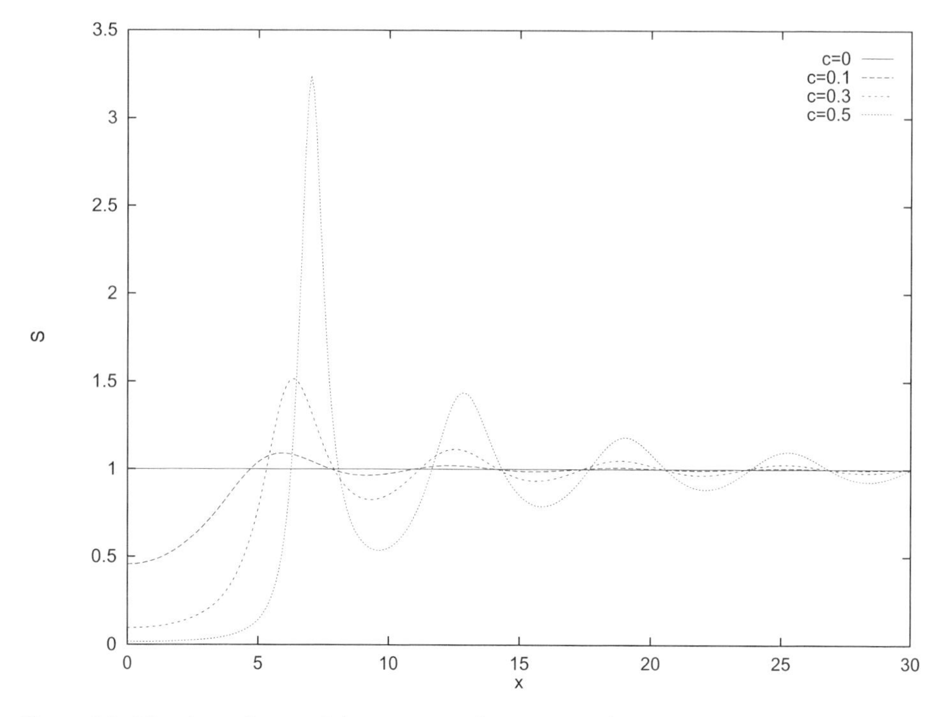

Figure 4.1. The dependence of the structure factor S on the value of $X = 2\theta x$ at different concentrations c.

and the phase function $p(\theta)$ of media with densely packed particles can be calculated with following equations:

$$C_{sca} = \int_0^\pi C^0_{sca}(\theta)S(\theta)\sin\theta\, d\theta \tag{4.3}$$

$$g = \frac{\displaystyle\int_0^\pi C^0_{sca}(\theta)S(\theta)\cos\theta\sin\theta\, d\theta}{\displaystyle\int_0^\pi C^0_{sca}(\theta)S(\theta)\sin\theta\, d\theta} \tag{4.4}$$

$$p(\theta) = \frac{C^0_{sca}(\theta)S(\theta)}{\displaystyle\frac{1}{2}\int_0^\pi C^0_{sca}(\theta)S(\theta)\sin\theta\, d\theta} \tag{4.5}$$

Calculations of functions $C_{sca}, g, p(\theta)$ for correlated distributions of particles are easily performed with equations (4.3)–(4.5) for particles of different sizes and internal structures. This was done in the book by Ivanov *et al.* (1988) for spherical uniform particles. The values of $C^0_{sca}(\theta)$ were calculated with the Mie theory. Equations (4.3)–(4.5) can be simplified if we apply the approximate solutions presented in Chapter 2.

To take one example, we will consider here the case of large particles ($\rho \gg 1$, $x \gg 1$, where $\rho = 2x|m-1|$, $m = n - i\chi$ is the refractive index, a is the radius of particles and $x = 2\pi a/\lambda, \lambda$ is the wavelength). As discussed earlier, the correlation effects are important only for the diffraction component in the case of large particles. Thus, we can write instead of equation (4.1):

$$C_{sca}(\theta) = C_{sca}^{D}(\theta)S(\theta) + C_{sca}^{GO}(\theta) \tag{4.6}$$

where the diffraction component

$$C_{sca}^{D}(\theta) = \frac{k^2a^4F(z)}{4}, \qquad F(z) = 4J_1^2(z)/z^2, \qquad z = \theta x \tag{4.7}$$

and the value of geometrical optics (GO) part of the DSCS $C_{sca}^{GO}(\theta)$ was studied in Chapter 2.

As follows from equation (4.6), we need to know the function $S(\theta)$ only at small angles θ in this case. This allows us to obtain the following simple approximation for the static structure factor (Ashcroft and Lekner, 1966; Ailawadi, 1973):

$$S(\theta) = \frac{1}{1 + H(\theta)} \tag{4.8}$$

where

$$H(\theta) = \frac{24c\psi(\theta)}{X^6}$$
$$\psi(\theta) = \alpha X^3(\sin X - X\cos X) + \beta X^2(2X\sin X + [2 - X^2]\cos X - 2)$$
$$+ \gamma([4X^3 - 24X]\sin X - [X^4 - 12X^2 + 24]\cos X + 24)$$

and

$$\alpha = \frac{(1+2c)^2}{(1-c)^4}, \qquad \beta = -6c\frac{(1+c/2)^2}{(1-c)^4}, \qquad \gamma = \alpha c/2, \qquad X = 2z, \qquad z = \theta x$$

Results of computation of the value of $A(z) = F(z)S(z)$ (see equations (4.7), (4.8)) are given in Figure 4.2 for different values of the concentration c. Note that:

$$A(0) = \frac{(1-c)^4}{(1+2c)^2} \qquad \text{and} \qquad A(0) \to 0 \quad \text{as } c \to 1$$

We can see that even small concentrations of particles (10%) produce the significant drop in the value of the phase function in the forward direction. It is worth pointing out that correlations change the phase function mostly in the region of the forward peak up to $z = 3.83$ and slightly move the maximum of scattered light intensity to larger angles.

Similar results were obtained by Lock and Chiu (1994) in the case of the dense distribution of condensation droplets on a window pane (see their Fig. 3). Note that droplets on a window pane form a monolayer and the function $H(\theta)$ in equation (4.8) is given by the following formula (Lock and Chiu, 1994):

$$H(\theta) = 4s\left[\frac{J_1(2z)}{z}\right] + 8s^2\left[\frac{J_0(z)J_1(z)}{z}\right] + 4(s^2 + 2s^3)\left[\frac{J_1(z)}{z}\right]^2$$

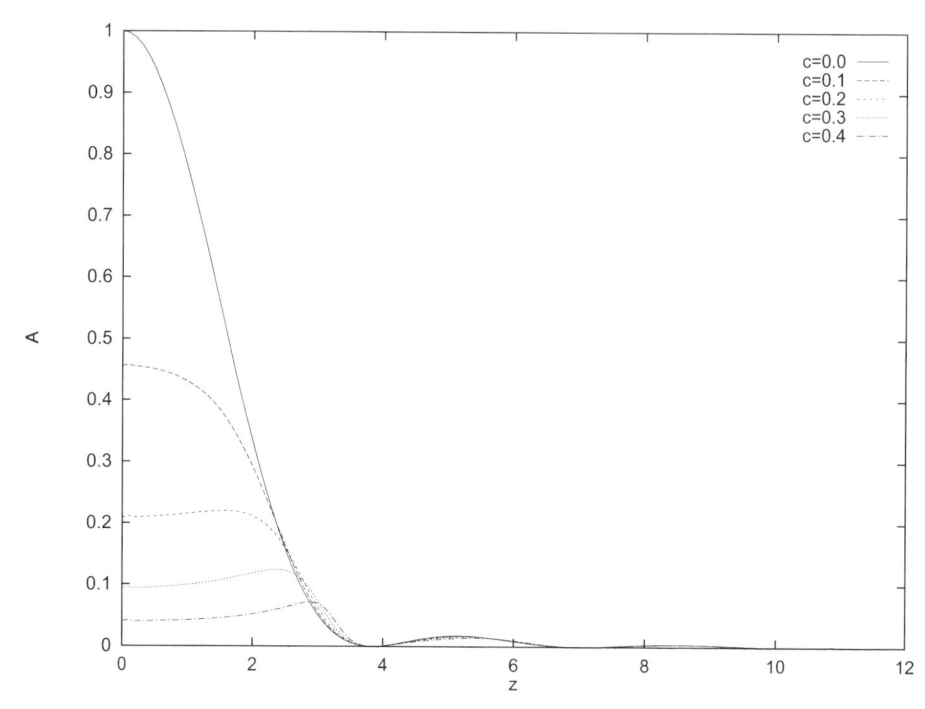

Figure 4.2. The dependence $A(z)$ at different concentrations c.

where $s = f/(1-f)$, f is the area fraction covered by particles and J_0 and J_1 are Bessel functions. This equation was used by Loiko and Konkolovich (2000) for studies of coherent transmittance quenching for surface ferroelectric liquid crystal droplets. Lock and Chiu (1994) considered very high values of $f \leq 0.7$ and observed a Bragg peak in the results of their calculations at $f = 0.7$ and $\theta = \lambda/d$, where d is the diameter of droplets. The origin of this peak is due to the short-range semiordering of the droplets. This semiordering is capable of near-forward direction scattering suppression and of producing the experimentally observed (see plate 44 in the paper by Lock and Chiu (1994)) bright ring of coloured light. The position of the Bragg maximum almost coincides with the minimum of the diffraction pattern for a single droplet, which is located at $\theta = 1.22\lambda/d$.

It follows from equations (4.6), (4.7) for the value of the diffraction part of the scattering cross-section $C_{sca}^{D} = C_{sca} - C_{sca}^{GO}$:

$$C_{sca}^{D} = 2\pi a^2 \int_0^{\infty} \frac{J_1^2(\theta x)}{\theta} S(\theta)\, d\theta \tag{4.9}$$

Equation (4.9) was obtained taking into account that $C_{sca}^{D}(\theta) \to 0$ at $\theta \gg 1$. Thus, it is allowed to change the upper limit of the integration from π to ∞. It follows as $c \to 0$ that $S(\theta) \to 1$ and $C_{sca}^{D} = \pi a^2$ as it should be.

Results of calculation of the efficiency factor $Q_{sca}^{D} = C_{sca}^{D}/\pi a^2$ are presented in

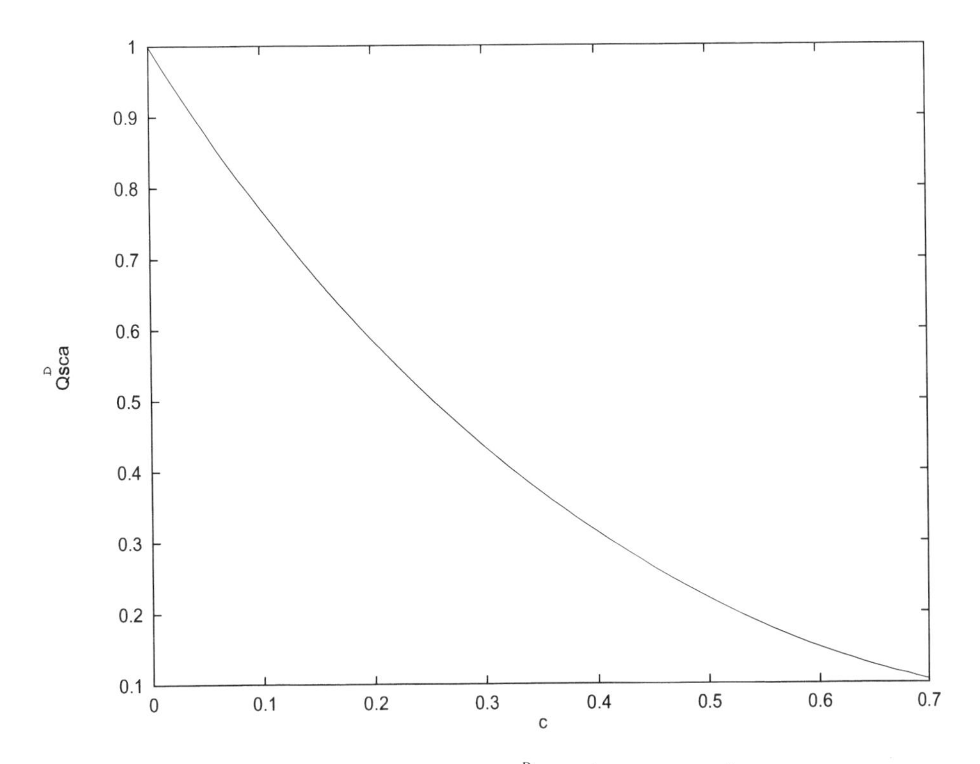

Figure 4.3. The dependence Q_{sca}^{D} on the concentration c.

Figure 4.3. It is seen that the efficiency factor changes considerably with the concentration of particles. It follows that the value of $Q_{sca}^{D} \to 0$ and $C_{sca} \to C_{sca}^{GO}$ as $c \to 1$. Note that the curve in Figure 4.3 can be represented by the following simple equation: $Q_{sca}^{D} = (1-c)^2$, within the framework of the first coarse approximation. Thus, the correlation effects in media with large particles decrease the value of the scattering coefficient (Hapke, 1981, 1993; Barabanenkov, 1982; Saulnier *et al.*, 1990; Mischenko, 1994; Gobel *et al.*, 1995).

Another important characteristics of the radiative transport in a scattering medium are the phase function $p(\theta)$ and the asymmetry parameter $g = \frac{1}{2}\int_0^{\pi} p(\theta)\sin\theta\cos\theta\,d\theta$. The phase function is normalized by the following equation:

$$\int_{4\pi} p(\theta)\,d\Omega = 4\pi \tag{4.10}$$

We can obtain from equations (4.3), (4.5):

$$p(\theta) = \frac{p^{D}(\theta)C_{sca}^{D} + p^{GO}(\theta)C_{sca}^{GO}}{C_{sca}^{D} + C_{sca}^{GO}} \tag{4.11}$$

where

$$p^D(\theta) = \frac{4\pi C^D_{sca}(\theta)}{C^D_{sca}} \tag{4.12}$$

$$p^G(\theta) = \frac{4\pi C^{GO}_{sca}(\theta)}{C^{GO}_{sca}} \tag{4.13}$$

It follows for the value of the asymmetry parameter:

$$g = \frac{g^D C^D_{sca} + g^{GO} C^{GO}_{sca}}{C^D_{sca} + C^{GO}_{sca}} \tag{4.14}$$

where $g^D \approx 1$ and (see equation (2.59)):

$$\left.\begin{aligned} g^{GO} &= \frac{1}{2\omega}\sum_{j=1}^{2}\int_0^{\pi/2} \frac{d_j(\tau)\sin 2\tau\, d\tau}{1 - 2r_j \exp(-b\xi)\cos(2\tau^*) + r_j^2 \exp(-2b\xi)} \\ d_j(\tau) &= \exp(-b\xi)(1 - r_j)^2 \cos(2(\tau - \tau^*)) + r_j(1 - \exp(-2b\xi))\cos 2\tau \\ &\quad + 2r_j^2(\exp(-b\xi) - \cos 2\tau^*)\exp(-b\xi)\cos 2\tau \end{aligned}\right\} \tag{4.15}$$

where $b = 4\chi x$, $\xi = \sqrt{1 - \sigma^2}$, $\tau^* = \arccos(\sigma)$, $\sigma = \cos\tau/n$, r_j are Fresnel coefficients. The value of ω is defined by equations (2.29), (2.30). The function $1 - g(c)$ at the refractive index $m = 1.333$ is presented in Figure 4.4. It can be represented by the following equation: $1 - g = 0.116 + 0.15c$ at $c < 0.6$. Thus, the asymmetry parameter g decreases due to correlation effects.

Note that the value of g for composite particles was studied by Mishchenko and Macke (1997). It was also found that the value of $1 - g$ increases with the concentration of particles.

Let us now consider the value of the transport mean free path length l_0 (for nonabsorning scatterers):

$$l_0 = \frac{1}{NC_{sca}(1 - g)} \tag{4.16}$$

where N is the number concentration of particles. The photon transport in the diffuse regime depends on the value of l_0 mostly. For instance, it follows for the spherical albedo r of a thick nonabsorbing layer (Zege *et al.*, 1991) that:

$$r = \frac{3L}{4l_0}\left[1 + \frac{3L}{4l_0}\right]^{-1}$$

where L is the geometrical optics thickness of a medium. We can obtain from equations (4.14), (4.16):

$$l_0 = \frac{1}{NC^{GO}_{sca}(1 - g^{GO})} \tag{4.17}$$

This equation holds for weakly absorbing media ($\sigma_{abs} \ll \sigma_{sca}$) as well.

It follows from equation (4.17) that the value of l_0 does not depend on the partially coherent diffracted part of the light field. This result is important.

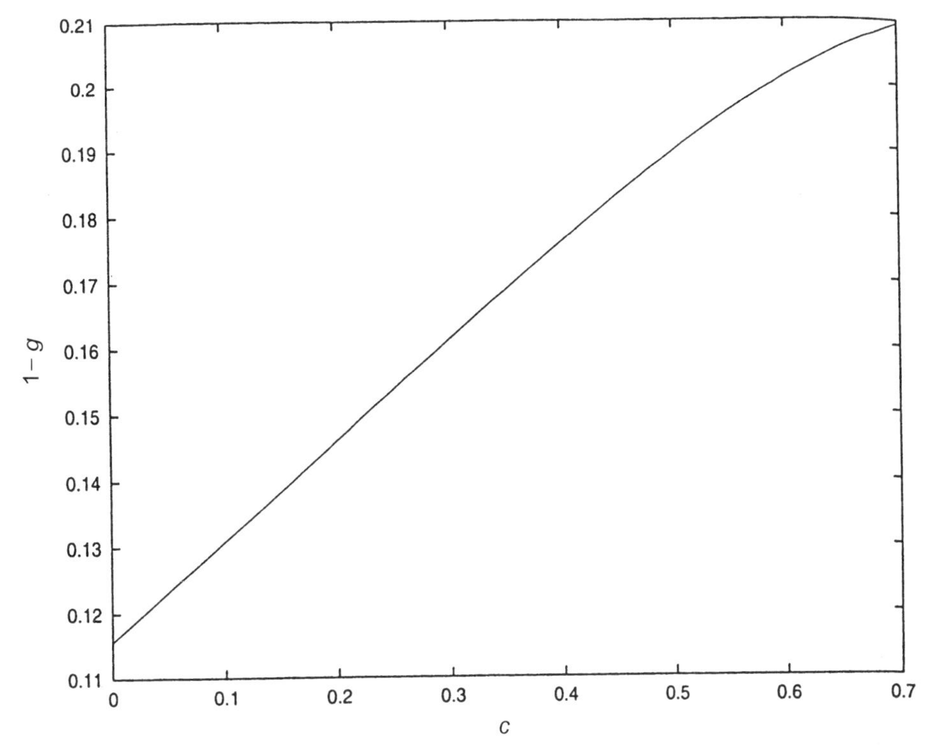

Figure 4.4. The dependence $1 - g$ on the concentration c at $m = 1.333$.

Indeed, it follows from equation (4.17) that we can ignore correlation effects in the diffuse regime for nonabsorbing (or weakly absorbing) media with large particles, because of their mutual cancellation. It is the genuine reason for the successful use of the standard RTE in blood (Dubova *et al.*, 1977), foam (Zege *et al.*, 1991), and snow (Grenfell *et al.*, 1994) optics. Of course, it is not true for small ($a \ll \lambda$) particles. But even in this case the correlation effects are more important for the scattering cross-section than for the transport mean free path length.

Another important local optical characteristic of light propagation in scattering media is the absorption coefficient $\sigma_{abs} = NC_{abs}$. Generally speaking, the value of C_{abs} increases with the concentration of particles (Dubova *et al.*, 1977, 1982; Ivanov *et al.*, 1987) due to correlation effects. However, this increase is rather weak and can be ignored at $c \leq 0.3$ in a first coarse approximation (Ishimaru, 1978; Gobel *et al.*, 1995; Mishchenko and Macke, 1997).

According to measurements of σ_{abs} in human blood (Dubova *et al.*, 1981), the value of σ_{abs} is only 20% larger due to correlation effects at $c \approx 0.5$. Loiko and Ruban (2000) found experimentally that function $\sigma_{abs}(c)$ has a nonlinear dependence for photolayers.

4.2 TRANSMITTANCE OF LIGHT THROUGH A MONOLAYER OF DENSELY PACKED LARGE PARTICLES

Let us now consider a monolayer of large ($x \gg 1$) densely packed particles, bounded by a diaphragm with area σ and illuminated by a normally incident plane wave. We will assume that multiple light scattering can be ignored. In this case, as shown in the previous section, we should account for dependent scattering effects only in the diffraction region (small angle scattering). In principle, it could be done using the static structure factor (Ivanov *et al.*, 1988). Here we will consider another approach, which is based on the amplitude-phase screen (APS) model. In this model the monolayer of particles is replaced by the APS with the transmission function V^*, which equals 1 in a region not shaded by particles, and a function V, depending on the shape, structure, optical constants, and sizes of the scatterers in a shaded region. We place this screen in the near field of a monolayer. Then the field u at any point behind the screen can be found by using Kirchhoff's formulation of the Huygens principle (Born and Wolf, 1965). Thus, one obtains in the framework of the scalar approximation:

$$u = C \int_{\sigma} e^{-ik(px+qy)}\, dx\, dy - C \sum_{n=1}^{N} e^{-ik(px_n+qy_n)} \int_{S_n} (1 - V_n)\, e^{-ik(px+qy)}\, dx\, dy \quad (4.18)$$

where $C = \text{const}$, $k = 2\pi/\lambda$; $p = l - l_0$; $q = f - f_0$; S_n is the projected area of the nth particle; x_n and y_n are the co-ordinates of its centre; l, f and l_0, f_0 are the direction cosines of diffracted and incident light; V_n is the transmission function of a particle (or the equivalent APS). The first term in equation (4.18) describes the diffraction on the diaphragm. The second one describes the diffraction of light on N screens with transmission functions V_n, while their sum describes the distribution of light in the Fraunhofer diffraction pattern for N particles.

Let us consider the central spot ($p = q = 0$) of the diffraction pattern. Then it follows from equation (4.18):

$$u = C\sigma \left\{ 1 - \frac{1}{\sigma} \sum_{n=1}^{N} \int_{S_n} (1 - V_n)\, dS_n \right\} \quad (4.19)$$

We can obtain from equation (4.19) at $V_n = 0$ (strong absorption):

$$u = u_0(1 - \eta) \quad (4.20)$$

where $u_0 = C\sigma$ is the value of the field in the absence of particles, and

$$\eta = \sum_{n=1}^{N} S_n/\sigma \quad (4.21)$$

is the overlap parameter. Thus, the intensity of the central spot of the diffraction pattern $I = |u|^2 = I_0(1 - \eta)^2$, where $I_0 = |u|^2$, decreases monotonically with parameter η. We obtain for the coherent transmission coefficient $T_k = I/I_0$ the following equation:

$$T_k = (1 - \eta)^2 \tag{4.22}$$

The discussion above is valid for polydispersed media as well. For simplicity, we assume now that the projections of particles onto the amplitude-phase screen are circles of equal areas S. Then it follows from equation (4.19) that:

$$T_k = \left| 1 - \eta + \frac{\eta}{S} \int_S V \, dS \right|^2 \tag{4.23}$$

Equation (4.23) is a generalization of equation (4.22) for the case of arbitrary phase shifts. For soft spherical particles, it is easy to show that (Van de Hulst, 1957):

$$V = e^{-i\Delta \cos \varphi} \tag{4.24}$$

where $\Delta = kd(m_2 - m_1)$ is the phase shift at diameter d of a particle, $k = 2\pi/\lambda$, λ is the wavelength, m_2 (m_1) is the refractive index of particles (a host medium), and φ is the angle of incidence of a ray on the surface of a particle. Substituting equation (4.24) into equation (4.23), we obtain after integration:

$$T_k = |1 - 2\eta K(i\Delta)|^2 \tag{4.25}$$

where $K(\omega) = \frac{1}{2} + e^{-\omega}/\omega + (e^{-\omega} - 1)/\omega^2$ is the Van de Hulst function. As follows from equation (4.25), $T_k \to (1 - \eta)^2$ not only as $\text{Im}(\Delta) \to \infty$, but also for arbitrary values of $\text{Im}(\Delta)$ if $\text{Re}(\Delta) \to \infty$. This means that equation (4.25) transforms into equation (4.22) not only in the region of strong absorption but also in the region of large phase shifts.

It is interesting that the coherent transmission T_k disappears both for $\eta = 1$ and $\eta < 1$ at some selected phase shifts Δ. Such behaviour appears when the field diffracted by the particles is equal in value and 180° out of phase with the field diffracted by the diaphragm. The general formulation of the conditions for obtaining $T_k = 0$ was given by Ivanov *et al.* (1988). We find the condition for the absence of coherent transmission ($T_k = 0$) from equation (4.25):

$$\text{Im}(K(i\Delta)) = 0, \qquad \eta = \frac{1}{2K(i\Delta)} \tag{4.26}$$

or (for nonabsorbing particles):

$$\tan \Delta = \Delta, \qquad \eta = \left(1 + \frac{2(1 - \cos \Delta)}{\Delta^2} - \frac{2 \sin \Delta}{\Delta}\right)^{-1} \tag{4.27}$$

Note that the maximal attainable overlap parameter is $\eta_{max} = 0.907$ for monodispersed spheres. Thus, the condition $T_k = 0$ can be obtained only for:

$$\text{Re}[K(i\Delta)] \geq \frac{1}{2\eta_{max}} \approx \frac{1}{2} \tag{4.28}$$

It is easy to find (see equation (4.27)) that the coherent transmission T_k disappears for a discrete set of phase shifts Δ_l, which are approximately described by the relation ($\Delta_l \gg 1$):

$$\Delta_l \approx \frac{\pi}{2}(2l+1) \tag{4.29}$$

where l is an integer [the first correction term to equation (4.29) equals $(\Delta_l)^{-1}$]. Thus, it follows from equation (4.27) accounting for equation (4.29):

$$\eta_l = \frac{1}{1+\dfrac{2}{\Delta_l}} \tag{4.30}$$

As follows from equation (4.30), the larger the parameter Δ, the larger the value of η at which condition $T_k = 0$ is satisfied.

Substituting equation (4.29) in equation (4.28), we obtain (with accuracy $\approx \Delta_l^{-2}$):

$$\frac{1}{2\eta_{max}} \leq \frac{1}{2} \pm \frac{1}{\Delta_l} \tag{4.31}$$

Since it is always true that $\eta_{max} < 1$, only the upper sign in equation (4.31) has physical meaning. Thus, only values of Δ_l with odd l should remain in equation (4.29). With this condition, it is easy to obtain from equation (4.31):

$$\Delta_l \leq \frac{2\eta_{max}}{1-\eta_{max}} \tag{4.32}$$

It follows that $\Delta \approx 19.5$ for $\eta_{max} = 0.907$. For larger values of phase shift Δ, the function $T_k(\eta)$ does not have a minimum.

Ivanov *et al.* (1988) derived that $T_k = 0$ if $\Delta_l/2 = 2.2, 5.7, 8.5, \ldots$ Their paper was based on the approximation of numerical calculations according to the Mie theory for spherical particles with $\Delta m = m_2 - m_1 \leq 0.552$, $(\pi d)/\lambda \leq 420$. With odd l, it follows from equation (4.29) that $\Delta_l/2 = 3\pi/4, 7\pi/4, 11\pi/4, \ldots$ or $2.4, 5.5, 8.6$. On the whole, the agreement is extremely close. This should be expected, since the results of calculations for the value of Δ_l using the Mie theory for $|\Delta m| \to 0$ should tend asymptotically to equation (4.29). Substituting equation (4.29) into equation (4.30), we obtain:

$$\eta_l = \frac{1}{1+\dfrac{4}{(2l+1)\pi}} \tag{4.33}$$

where $l = 1, 3, 5, \ldots$ Thus, the condition for the disappearance of the central spot of the diffraction pattern from N particles in a monolayer is satisfied only for a discrete set of phase shifts Δ_l and the values of the packing parameter corresponding to them, $\eta \approx 0.7, 0.85, 0.9$. This feature can be used in technological applications. For optically soft coarse particles, the value of T_k vanishes only for three values of the packing parameter η. The essential point is that the approach explained here, based on the model of an amplitude-phase screen, permits us to study nonspherical, anisotropic, gyrotropic, and structured particles, packed into a monolayer, without significant complications. The treatment of ellipsoidal and cylindrical particles (with the wave incident along the axis) thereby requires practically no calculations because of the well-known fact that absorption and extinction efficiencies of ellipsoidal and spherical particles have the same analytical form within the framework of the anomalous diffraction approximation (Lopatin and Sid'ko, 1988). In essence, equation (4.23) describes the transmission of a monolayer in the anomalous diffraction approximation and, in this way, can be extended to ellipsoidal particles without changes. Accordingly, in equation (4.27) we should use $\Delta^* = kh\Delta m$ (instead of Δ), where $h = 3V/2\Sigma$, V is the volume of an ellipsoid and Σ is the geometrical cross-section of an ellipsoid in the plane normal to the incident light beam. Then conditions (4.29) and (4.30) are rewritten in the form:

$$\Delta_l^* = \pi(2l+1)/2, \qquad \eta_1 = [1 + 2(\Delta_l^*)^{-1}]^{-1}$$

It follows from this that for ellipsoids, just as for spheres, the value of $T_k = 0$ at discrete values of the phase shift Δ^*.

The other case that allows for an elementary analysis is incidence of a plane wave along the axes of cylinders packed into a monolayer. For this case, independently of the shape of the transverse cross-section, we get:

$$V = e^{-ikl\Delta m} \tag{4.34}$$

where l is the length of a cylinder. Substituting equation (4.34) into equation (4.23) and bringing V out from inside the integral sign (because it is a constant), we have:

$$T_k = |1 - \eta(1 - e^{-ikl\Delta m})|^2 \tag{4.35}$$

Thus, it follows for nonabsorbing particles:

$$T_k = (1-\eta)^2 + 2\eta(1-\eta)\cos\Delta + \eta^2 \tag{4.36}$$

where $\Delta = kl\Delta m$. We can obtain at the minimum of this function:

$$\frac{dT_k}{d\Delta} = -2\eta(1-\eta)\sin\Delta = 0 \tag{4.37}$$

$$\frac{d^2T_k}{d\Delta^2} = -2\eta(1-\eta)\cos\Delta > 0 \tag{4.38}$$

We conclude from this that:

$$\Delta = \pi l$$

at the minimum value of T_k, where $l = 1, 3, 5, 7, \ldots$. This condition has a clear physical meaning: in this case rays passing through a cylinder and a host medium are found to be 180° out of phase, which leads to their destructive interference. As can be seen from equation (4.36), T_k has the form:

$$T_k = 4\eta^2 - 4\eta + 1 \tag{4.39}$$

at the minimum and vanishes at $\eta = \frac{1}{2}$ (i.e. when half the area of the equivalent screen is occupied by a field with the phase of the incident wave, while the other half is occupied by a field 180° out of phase with it, which guarantees the equality of the amplitudes of the interfering beams). We point out that it follows at the maximum that $T_k = 1$ ($l = 2, 4, 5, \ldots$) and a particulate medium with an ideal transparency can be obtained.

The properties of the small-angle scattering region can be investigated with equation (4.18). Further details on the optical properties of close-packed monolayers can be found in the book written by Ivanov *et al.* (1988).

4.3 MULTIPLE LIGHT SCATTERING IN DISPERSE MEDIA WITH DENSELY PACKED PARTICLES

Radiative transfer in random media with low concentrations of scatterers c (e.g. water and cirrus clouds, mists, aerosols, hydrosols, interstellar dust) has been studied by many authors (Chandrasekhar, 1960; Sobolev, 1956; Ambarzumian, 1961; Van de Hulst, 1980; Lenoble, 1985; Thomas and Stamnes, 1996; Natsyama *et al.*, 1998; Tsang *et al.*, 1985, 2000). As a result, at present there is no problem in solving the radiative transfer equation (RTE) for a plane-parallel light-scattering medium at any optical thickness at $c \ll 1$. A lot of analytical, approximate and numerical methods of RTE solutions have been developed. Some of them are presented in Chapter 3.

In case of intermediate or high concentrations of particles ($c > 0.1$) the situation is completely different. Generally speaking, the standard RTE cannot be used in this case and we should apply the T-matrix approach (Waterman, 1971) or different approximate techniques such as Foldy's (1945), Twersky's (1962), or quasicrystalline (Lax, 1952; West *et al.*, 1994) approximations. It is a straightforward matter to investigate light scattering and propagation in media with small particles ($x < 1$–10, $x = ka$, $k = 2\pi/\lambda$, λ is the wavelength, a is the radius of a particle) with these methods. In this case the scattering of radiation is weak. Moreover, different procedures of replacement of an actual inhomogeneous medium by a uniform one, but with the 'effective' refractive index (Tsang and Kong, 1983; Tsang *et al.*, 1985) can be used. Such approximations are very important in the microwave region of the electromagnetic spectrum.

For the optical band, the dimension of particles in many densely packed natural media (soils, snow fields, oceanic whitecaps) is very large in comparison with the wavelength of incident light and the scattering of radiation is extremely important. Theoretical methods (Tsang *et al.*, 1985), which appeared to be successful in the microwave region, encounter many obstacles. It is not possible to avoid difficulties with numerical algorithms of the dense medium theory at large values of size parameter ka, at least in the near future.

Since there is not an exact solution of the problem under discussion, very coarse approximations should be applied and experimentally verified. One of them is the use of the effective radiative transport equation (ERTE) to solve the problem (Hapke, 1993). This equation has the same structure as the ordinary RTE, but the relation of the phase function $p(\theta)$ (or the scattering matrix), the extinction coefficient, and the single scattering albedo ω_0 to the optical characteristics of individual particles differs from that for the case of disperse media with low concentrations of particles.

It is worth noting that it is possible to use the well-developed methods (Rozenberg, 1967; Ishimaru, 1978; Zege *et al.*, 1991) of the RTE solution within the framework of this approximation. Some recent experiments and theoretical investigations (Bohren and Barkstrom, 1974; Whitlock *et al.*, 1982; Bohren, 1987; Grenfell, 1994) show that there are grounds for such an approach (at least, for optically thick weakly absorbing light-scattering media).

Let us consider the case of weakly absorbing optically thick close-packed media in more detail. The spherical reflectance r and global transmittance r of such media can be found from the following simple equations (Rozenberg, 1962; Zege *et al.*, 1991):

$$r = \frac{\sinh yz}{\sinh[y(1+z)]}, \qquad t = \frac{\sinh y}{\sinh[y(1+z)]} \tag{4.40}$$

where

$$y = 4\sqrt{\frac{L_{tr}}{3L_{abs}}}, \qquad z = \frac{3L}{4L_{tr}} \tag{4.41}$$

$L_{abs} = 1/\sigma_{abs}$ and $L_{tr} = 1/(\sigma_{ext}(1-g))$, L is the geometrical thickness of a layer. At $L/L_{tr} \to \infty$ it follows from equations (4.41) that:

$$t = 0, \qquad r = \exp(-y) \tag{4.42}$$

We can see that the reflection and transmission of light by close-packed media depends on two parameters, namely transport L_{tr} and absorption L_{abs} lengths. Both of them can be calculated from the Mie theory in the case of large particles, because the dependence of values σ_{abs} and $\sigma_{ext}(1-g)$ on close-packed effects is weak in this case, as was found in Section 4.1. For small particles we must take account of the structure factor.

To summarize, we note that, generally speaking, there are two types of correlation effects in dense media:

A. the change of the effective electromagnetic field in a scattering medium;
B. interference effects.

For small particles ($a \ll \lambda$) the A-effects are the most important ones. Thus, the B-effects can be neglected because of very weak light scattering in such media. The main effect is the change in the effective field and the dielectric permittivity.

On the other hand, for large particles ($a \gg \lambda$) the B-effects are more important, but they take place only in a very narrow region of scattering angles near the forward direction. Interference effects for large particles reduce the scattering coefficient, the asymmetry parameter, and the extinction coefficient of media with large particles considerably, but they almost do not change the absorption coefficient and the transport mean free path length, which are responsible for the optical properties of light-scattering media in the diffuse regime (optically thick weakly absorbing light-scattering layers). Thus, it is possible to apply the standard RTE to study the optical properties of dense media with large particles.

For particles of intermediate sizes ($a \sim \lambda$) the A and B effects are both equally important and exact methods of the dense media transport theory (Tsang *et al.*, 1985; Zurk *et al.*, 1995) should be applied.

5

Applications

5.1 GEOPHYSICAL OPTICS

5.1.1 Cloudy media

5.1.1.1 Water clouds

Let us apply methods developed in the previous chapters to the investigation of optical properties of natural media. The first example will be optics of water clouds (Shifrin, 1955; Busygin *et al.*, 1973; Kondratyev and Binenko, 1984; Arking and Childs, 1985; Rossow, 1989; Rossow *et al.*, 1989; Liou, 1992; Cahalan *et al.*, 1994; Stuhlmann *et al.*, 1985; Spinhirne and Nakajima, 1994). The optical thickness τ of water clouds is large in most cases and the approximation of thick layers, considered in Chapter 3, can be applied. The single scattering albedo ω_0 of drops is close to 1 ($\omega_0 \geq 0.95$ for most cloud media at the visible and near infrared wavelengths $\lambda \leq 2.2\,\mu\text{m}$). Thus, the reflection and transmission functions of water clouds can be calculated with the following equations:

$$R(\mu, \mu_0, \phi) = R^0_\infty(\mu, \mu_0, \phi)\exp\left\{-y\frac{K_0(\mu)K_0(\mu_0)}{R^0_\infty(\mu, \mu_0, \phi)}\right\} - \frac{\sinh(y)\exp(-y-x)}{\sinh(x+y)}K_0(\mu)K_0(\mu_0) \tag{5.1}$$

$$T(\mu, \mu_0) = \frac{\sinh(y)}{\sinh(x+y)}K_0(\mu)K_0(\mu_0) \tag{5.2}$$

where

$$K_0(\mu) = \tfrac{3}{7}(1+2\mu), \qquad R^0_\infty(\mu, \mu_0, \phi) = \frac{1}{2} + \frac{\mu\mu_0}{\mu+\mu_0} + \beta\frac{p(\theta)}{4(\mu+\mu_0)} \tag{5.3}$$

$$\beta = 8 - 4.5\exp(-5(\pi-\theta)) - 5\exp(-5(\theta^*-\theta)) \tag{5.4}$$

$\mu = \cos\vartheta$, $\mu_0 = \cos\vartheta_0$, ϑ, and ϑ_0 are observation and incidence angles, ϕ is the azimuth, θ^* is the rainbow angle, $p(\theta)$ is the phase function (see Appendix 5), $x = \sqrt{3(1-g)\beta}\tau_0$, $y = 4\sqrt{\beta/(3(1-g))}$, $\tau = \sigma_{ext}L$, L is the geometrical thickness of a cloud, $\beta = 1 - \omega_0$, $\omega_0 = \sigma_{sca}/\sigma_{ext}$. The values of σ_{ext}, ω_0, g, and $p(\theta)$ can be obtained in the framework of the Mie theory, because drops in water clouds have a spherical shape. The average size of drops is larger than the wavelength. Thus, the geometrical optics approximation is a useful tool to calculate local optical parameters as well (see equations (2.93)–(2.95)):

$$\sigma_{ext} = \frac{1.5C_v}{a_{ef}}\left\{1 + \frac{1.1}{(ka_{ef})^{2/3}}\right\} \tag{5.5}$$

$$\sigma_{abs} = \frac{5\pi\chi(1-\alpha a_{ef})}{\lambda}C_v\left\{1 + 0.34\left(1 - \exp\left(-\frac{8\lambda}{a_{ef}}\right)\right)\right\} \tag{5.6}$$

$$1 - g = 0.12 + 0.5(ka_{ef})^{-2/3} - 0.15\alpha a_{ef} \tag{5.7}$$

where $\alpha = 2k\chi$, $k = 2\pi/\lambda$. These equations are valid in the visible and near infrared. Note that parameterizations of local optical characteristics, which are valid in the wider spectral range, including the far infrared were developed by Chylek *et al.* (1992, 1995) and Mitchell (2000).

It follows for parameters y and $z = x/y$ from equations (5.5)–(5.7) that:

$$y = 6\sqrt{\alpha a_{ef}}, \qquad z = \frac{0.12w}{a_{ef}\rho}\left(1 + \frac{6}{(ka_{ef})^{2/3}}\right) \tag{5.8}$$

where $w = C_v\rho L$ is the liquid water path, $\rho = 10^6\,\mathrm{g\,m^{-3}}$ is the density of water.

The total absorptance, plane albedo, spherical albedo, transmittance, and global transmittance of water clouds can be calculated from the formulae presented in Table 3.3. In most cases the reflection function of water clouds decreases with the solar angle ϑ_0 or the observation angle ϑ. The angular dependence of the transmittance is proportional to the function $1 + 2\cos\vartheta$ (see equation (5.3)). The reflection function of a semi-infinite cloud only weakly depends on the microstructure of a cloud in the visible. It can be parameterized as follows at the nadir observation geometry:

$$R^0_\infty(\vartheta_0) = 1.125 + 0.0036\vartheta_0 - 0.00013\vartheta_0^2 \tag{5.9}$$

where ϑ_0 is the solar angle in degrees. It follows that $R^0_\infty(\vartheta_0) = 1$ at $\vartheta_0 \in [30°, 60°]$ with an accuracy better than 10%.

It is important that the optical properties of water clouds depend mostly on two microphysical parameters (i.e. a_{ef} and C_v) (see equations (5.1), (5.2), (5.8)). The half-width and specific shape of the particle size distribution can be ignored in this problem. Thus, the sensitivity of reflection and transmission functions of clouds to the half-width of the particle size distribution is weak. The degree of polarization at the rainbow angle, however, does depend on the half-width parameter (mainly around the scattering angle 150 deg, see Figure 5.1). This is important for the formulation of the inverse problems of cloud optics.

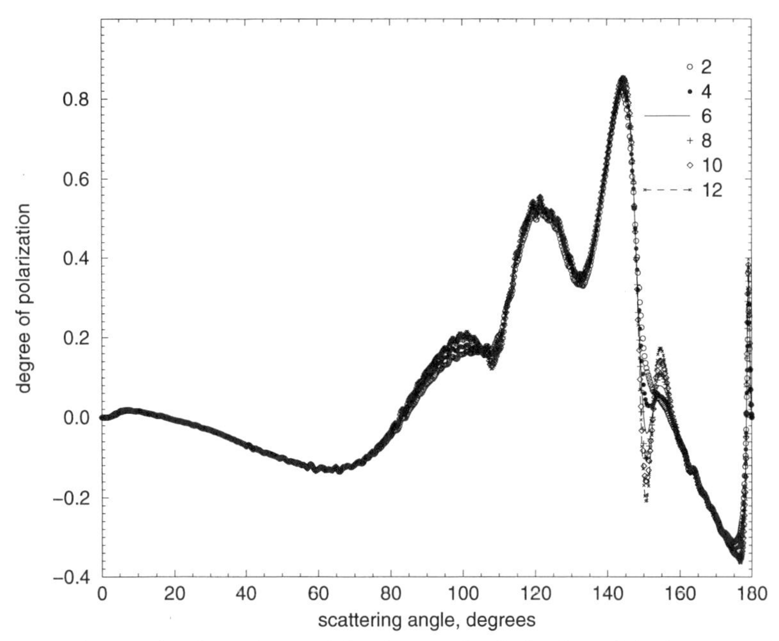

Figure 5.1. The angular dependence of the degree of polarization of singly scattered light for water clouds with the gamma particle size distribution at $a_{ef} = 6\,\mu\text{m}$ and the half-width parameter $\mu = 2$, 4, 6, 8, 10, 12. Calculations were performed with the Mie theory at $\lambda = 0.55\,\mu\text{m}$.

Note that optical characteristics of various atmospheric aerosols can be modelled very accurately from first moments of the particle size distribution as well (McGraw *et al.*, 1998). Thus, the particle size distribution contains more information than is generally required for the solution of geophysical problems.

Let us investigate the error of the approximate formula for the reflection function of a cloudy medium at nadir measurements ($\mu = 1$). This function can be expressed as follows (see equations (3.18), (5.1)):

$$R_*(1, \mu_0) = R(1, \mu_0) + \frac{r_s t^2 K_0(1) K_0(\mu_0)}{1 - r_s r} \tag{5.10}$$

where

$$R_*(1, \mu_0) = R^0_\infty(1, \mu_0) \exp\left(-y \frac{K_0(1) K_0(\mu_0)}{R^0_\infty(1, \mu_0, \phi)} \right) - t K_0(1) K_0(\mu_0) \exp(-y(1 + z))$$

$$r = \frac{\sinh(x)}{\sin((1 + z)y)}, \qquad t = \frac{\sinh(y)}{\sinh((1 + z)y)}, \qquad K_0(\mu_0) = \tfrac{3}{7}(1 + 2\mu_0),$$

$$R^0_\infty(1, \mu_0) = 0.49 \frac{1 + 4\mu_0}{1 + \mu_0} \qquad y = 6\sqrt{\alpha a_{ef}}, \qquad z = \frac{0.12w}{\rho a_{ef}} \left(1 + \frac{6}{(k a_{ef})^{2/3}} \right)$$

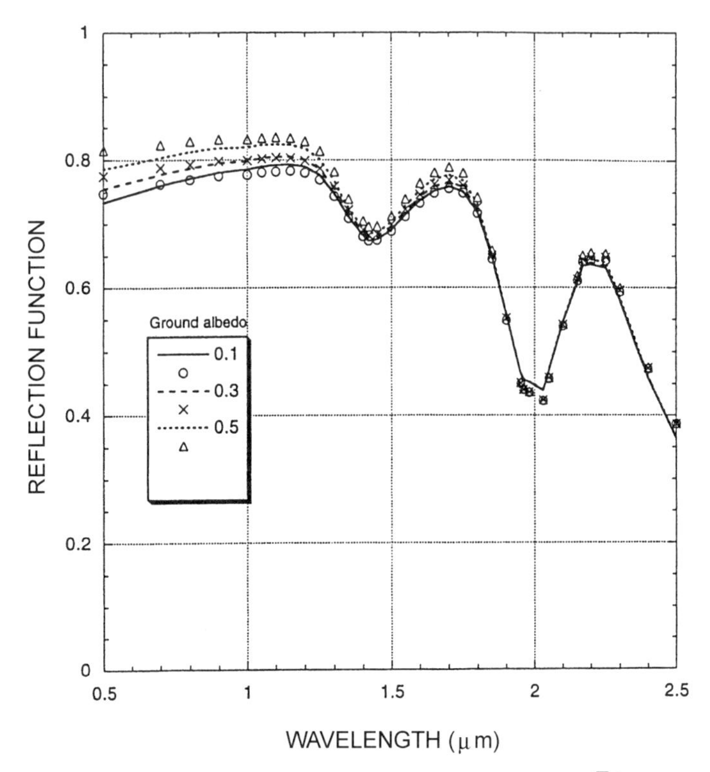

Figure 5.2. Reflection function of water clouds at $\tau = 30$, $\mu = 1$, $\mu_0 = \sqrt{2}/2$, $a_{ef} = 6\,\mu\text{m}$, and several values of the ground albedo obtained using the exact radiative transfer code (curves) and approximate formulae (symbols) for optically thick layers.

Here r_s is the albedo of an underlying surface. Note that we used the simplified equation for the value of $R^0_\infty(1, \mu_0)$ (see Table 3.3). The function $R_*(1, \mu_0)$ is important for satellite optical particle sizing (Kokhanovsky and Zege, 1996).

The accuracy of the calculation of the function $R_*(1, \mu_0)$ using these formulae was studied by Kokhanovsky *et al.* (1998) using the exact radiative transfer equation. The phase function and the single-scattering albedo in the exact code were calculated using the Mie theory for the gamma particle size distribution. The value of the effective radius of drops was changed from 4 to 16 μm, the wavelength was changed from 0.5 to 2.5 μm (see Appendix 1), the solar angle was changed from 30 to 60 deg, the ground albedo was changed from 0.1 to 0.5, and the liquid water path was changed from 50 to 300 $\text{g}\,\text{m}^{-2}$, which are representative values for water clouds. The coefficient of variance of the gamma particle size distribution was equal to $1/\sqrt{7}$ (Deirmendjian, 1969).

The results of computations are presented in Figures 5.2–5.5. They can be used to study the dependence of light reflectance from clouds on various parameters. The optical thickness in Figure 5.3 for different liquid water paths is in the range

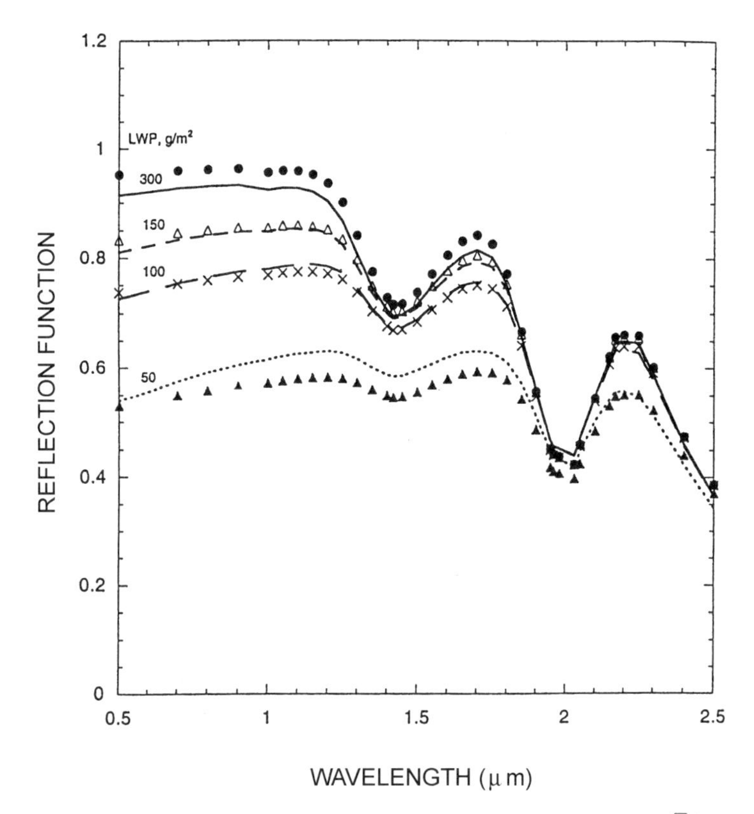

Figure 5.3. Reflection function of water clouds at $a_{ef} = 6\,\mu\text{m}$, $\mu = 1$, $\mu_0 = \sqrt{2}/2$, $r_s = 0$, and several values of the liquid water path obtained using the exact radiative transfer code (curves) and approximate formulae (symbols) for optically thick layers.

$\tau = 10$–70. The maximal probability of photon absorption was equal to 5% in our calculations. We can see that the accuracy of the formulae obtained is fairly good. Even at comparatively large absorption ($R_*(1, \mu_0) = 0.25$; see Figure 5.5 at $\lambda = 2\,\mu\text{m}$) the accuracy is high. The accuracy is not so good for thinner clouds (see Figure 5.3). This is due to the fact that the approximate theory presented here is valid only for optically thick media. It is interesting to see that the error has a minimum at intermediate (and the most frequently occurring) values of the liquid water path (see Figure 5.3). It results from the fact that the approximate equations presented here underestimate the reflection functions of thin water clouds and overestimate the values of reflection functions of semi-infinite light-scattering layers.

The accuracy of the approximations is high at the solar angle of 45 deg (see Figure 5.4), but it decreases for smaller and larger solar angles. The accuracy decreases with ground albedo r_s as well (see Figure 5.2). However, the influence of ground albedo on the cloud reflection function at $\tau > 30$ is important only at $\lambda \leq 1.3\,\mu\text{m}$ (see Figure 5.2).

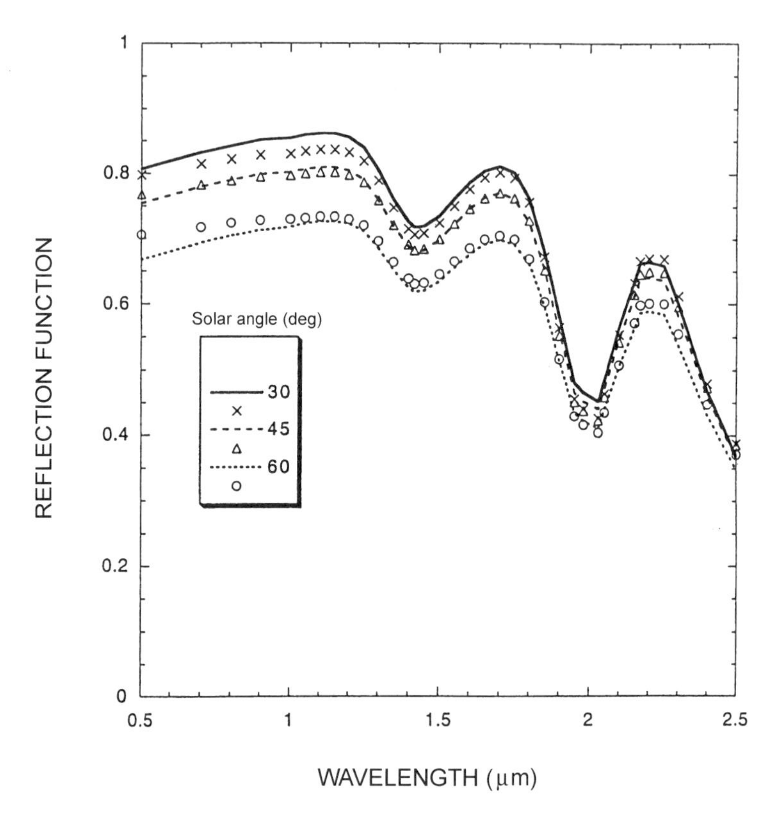

Figure 5.4. Reflection function of water clouds at $a_{ef} = 6\,\mu\text{m}$, $\tau = 30$, $\mu = 1$, $r_s = 0$, and several values of the solar zenith angle obtained using the exact radiative transfer code (curves) and approximate formulae (symbols) for optically thick layers.

Thus, we come to the conclusion that we can use the approximate equations presented here to estimate the reflection function $R(1, \mu_0)$ of water clouds with an error of less than 15% in the visible and near-infrared regions of the electromagnetic spectrum at $\tau \geq 10$ and $\omega_0 > 0.95$.

The accuracy of equation (5.1) is increased if special correction terms are used as shown by Kokhanovsky and Rozanov (2003). Then it follows:

$$R(\mu, \mu_0, \phi) = R_\infty^0(\mu, \mu_0, \phi)\exp(-(1 - 0.05y)u)$$
$$- \left[\frac{\sin h(y)\exp(-x)}{\sinh(x + \alpha y)} - \frac{a - b\mu\mu_0 + c\mu\mu_0}{\tau^3}\right]\exp(-y)K_0(\mu_0)K_0(\mu)$$

where $a = 4.86$, $b = 13.08$, $c = 12.76$, and

$$u = \frac{K_0(\mu_0)K_0(\mu)}{R_\infty^0(\mu_0, \mu, \phi)}$$

The function $R_\infty^0(\mu, \mu_0, \phi)$ can be calculated using exact radiative transfer code,

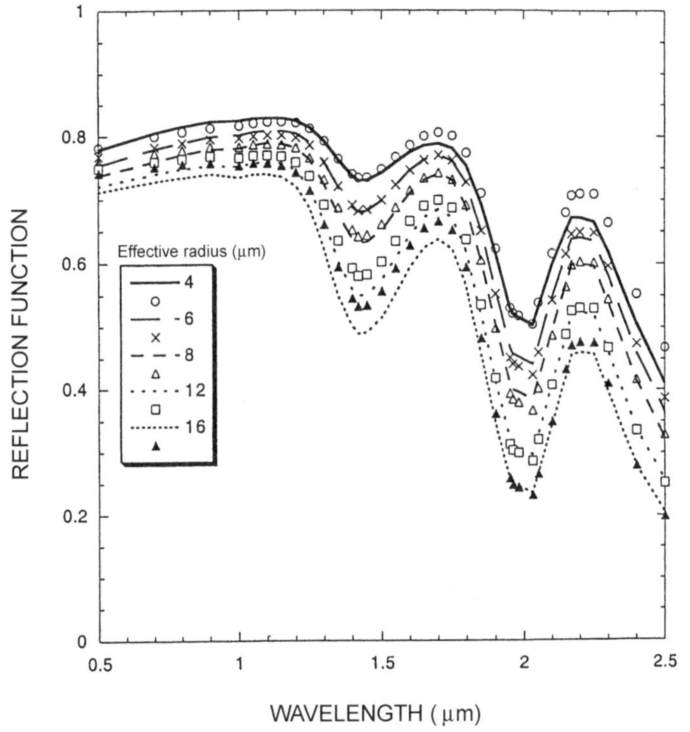

Figure 5.5. Reflection function of water clouds at $\tau = 30$, $\mu = 1$, $\mu_0 = \sqrt{2}/2$, $r_s = 0$, and several values of the effective radius of droplets obtained using the exact radiative transfer code (curves) and approximate formulae (symbols) for optically thick layers.

look-up tables or approximations (see, e.g. equation (5.3), and also results presented by Kokhanovsky and Rozanov (2003) and Kokhanovsky (2004)). Kokhanovsky and Rozanov (2004) generalized equation (5.1) for the case of the gaseous absorption band.

5.1.1.2 Ice clouds

Calculation of the optical properties of ice clouds is a much more complex task than it is for water clouds. This is mainly related to the fact that the shape of particles in ice clouds is not spherical. Moreover, there is no single shape of particles in ice clouds. We can find hexagonal plates and columns, stellar and fernlike crystals, needles, bullets, etc. (Auer and Veal, 1970) in crystalline clouds. Particles are characterized by different size distributions and can in principle have a preferential orientation.

Clearly, we need a statistical approach to calculate the local optical characteristics of ice clouds. In particular, the phase function of an elementary volume of a crystalline cloud in the visible can be modelled by the phase function of a single crystal with a very complex shape, reproducing the complexity of shape distributions in ice clouds (Mishchenko *et al.*, 1996b).

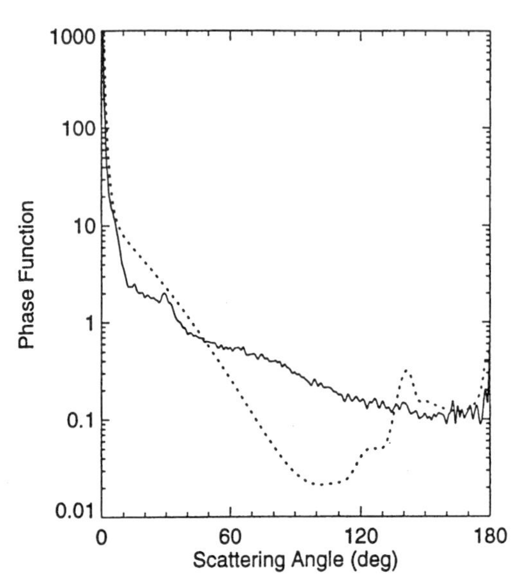

Figure 5.6. Phase functions of fractal ice particles (solid line) and water droplets (dashed line) with gamma particle size distribution at $a_{ef} = 10\,\mu\text{m}$ and $\lambda = 0.63\,\mu\text{m}$ (Mishchenko *et al.*, 1995b).

Such models were developed by Macke and Tzchihholz (1992), Macke (1994), Mishchenko *et al.* (1995b), and Macke *et al.* (1996). Namely, they proposed to use the phase function of fractal particles for modelling purposes (see Figure 5.6). This function is characterized by stronger scattering at side angles when compared with spheres, which is in correspondence with observations in natural ice clouds. The asymmetry parameter of the phase function in Figure 5.6 is equal to 0.75, which is close to experimentally observed asymmetry parameters in ice clouds.

The extinction coefficient of light in ice clouds can be found from the following equation:

$$\sigma_{ext} = N \int_0^\infty f(\vec{a}) C_{ext}(\vec{a})\, d\vec{a}$$

where $C_{ext}(\vec{a})$ is the extinction cross-section of a single crystal with geometrical parameters, defined by the vector-parameter $\vec{a}$, N is the number of particles in a unit volume and $f(\vec{a})$ is the function which describes the size and shape distributions of particles. We will assume that particles are randomly oriented. The size of crystals is much larger than the wavelength of incident light. Thus, it follows in a good approximation (Van de Hulst, 1957) that $C_{ext} = 2S$, where S is the cross-section of a particle in the plane perpendicular to the incident wave. Therefore, we can obtain:

$$\sigma_{ext} = \frac{2C_v \int_0^\infty f(\vec{a}) S(\vec{a})\, d\vec{a}}{\int_0^\infty f(\vec{a}) V(\vec{a})\, d\vec{a}}$$

where we accounted for the equality:

$$N = \frac{C_v}{\int_0^\infty f(\vec{a})V(\vec{a})\,d\vec{a}}$$

Here C_v, is the volumetric concentration of particles and $V(\vec{a})$ is the volume of a single crystal.

It is known that the average value of cross-section $\langle S\rangle$ for convex particles in random orientation is equal to $\langle\Sigma\rangle/4$, where $\langle\Sigma\rangle$ is the average surface area of particles.

The majority of particles in ice clouds have a convex form. Thus, it follows approximately that:

$$\sigma_{ext} = \frac{3C_v}{2a_{ef}}$$

where a_{ef} is the effective size of particles, defined as:

$$a_{ef} = \frac{3\langle V\rangle}{\langle\Sigma\rangle}$$

It follows for monodispersed spheres of the radius a: $\langle\Sigma\rangle = 4\pi a^2$, $\langle V\rangle = (4\pi a^3)/3$, and $a_{ef} = a$. We can see that the extinction coefficient does not depend on the wavelength in the visible. This is, of course, not a case in the infrared (Arnott *et al.*, 1995). Also it decreases with a_{ef} at $C_v = \text{const}$. Note that the value of $\langle S\rangle$ could be related not only to the value of $\langle\Sigma\rangle$, but also to the maximal dimension of a crystal (Mitchell *et al.*, 1996). The absorption of light by ice is negligible in the visible. However, it cannot be ignored in the infrared. The value of the absorption coefficient of particles in random orientation is determined by the following equation:

$$\sigma_{abs} = N\int_0^\infty f(\vec{a})C_{abs}(\vec{a})\,d\vec{a}$$

In the case of the near-infrared spectral region the absorption of light by crystals is weak and

$$C_{abs}(\vec{a}) \sim V(\vec{a})\alpha$$

where $\alpha = 4\pi\chi/\lambda$. Thus, it follows that:

$$\sigma_{abs} = B\alpha C_v$$

where B depends on the real part of the refractive index of ice and the shape of particles. It is difficult to find the value of B theoretically due to the complex shapes of crystals in clouds. Thus, there is a task to derive the value of B from experimental data. It follows for ice spheres in the visible that $B \approx \frac{5}{4}$. Thus, we can obtain for the probability of photon absorption $1 - \omega_0 = \sigma_{abs}/\sigma_{ext}$:

$$1 - \omega_0 = \tfrac{5}{6}\alpha a_{ef}$$

where we used $B = \frac{5}{4}$. The multiplier B for nonspherical particles is larger than in the case of spheres (see Figure 2.11). The case of strongly absorbing ice crystals (e.g., in the infrared) can be treated using equations (2.83) and (2.84). It should be pointed out that the phase function presented in Figure 5.6 does not account for haloes, which often appear for crystal clouds. Thus, the better model of ice crystal cloud phase function could be the combined phase function for random particles, presented in Figure 5.6, and the phase function of regular shaped particles, say, hexagonal columns (see Table A5.1 in Appendix 5). Note that the account for inhomogenity of ice crystals due to air bubbles, soot and dust aerosols also produce featureless, smooth phase functions (Macke, 2000). Parameterization schemes in ice cloud optics were developed by many authors (see, e.g., Mitchell and Arnott, 1994; Mitchell, 2000; Yang *et al.*, 2000).

5.1.2 Snow

5.1.2.1 Snow radiative characteristics

Snow covers a large portion of the terrestrial surface during the winter season. Therefore, snow fields have potentially significant effects on the planetary albedo and climate. The main parameters of the influence are snow cover and snow albedo r. Snow is highly reflective in the visible spectrum, where the solar spectrum maximum is located. Then the albedo r is close to 1. However, the albedo decreases in the UV and IR spectral regions. The albedo decrease is the most pronounced in the near-IR spectral region with values of r close to 0 at 1.5, 2.0, and 2.5 μm, where ice absorption maxima are located. Snow is essentially black then.

The snow albedo controls the surface energy budget and the rate of melting. Therefore, knowledge of r helps us to understand the change in snow cover over the seasons. This explains a huge amount of both experimental and theoretical research devoted to studies of snow optical properties. Recent comprehensive reviews (Nolin and Liang, 2000; Massom *et al.*, 2001) contain around 300 references related to the subject. One of the most important results found is the fact that snow spectral albedo can be fitted using radiative transfer calculations under the assumption of the spherical shape of snow grains (Bohren and Barkstrom, 1974; Wiscombe and Warren, 1980; Warren, 1982; Aoki *et al.*, 1998, 2000) (see, e.g. figure 10 in Wiscombe and Warren (1980) and figure 9 in Aoki *et al.* (2000)). This allowed the development of algorithms for the snow grain size retrieval from airborne and spaceborne optical instruments (Nolin and Liang, 2000; Li *et al.*, 2001).

It is known, however, that snow *in situ* has an extremely complex structure and consists of irregularly shaped ice grains in contact with one another (see, e.g. figure 5 in Massom *et al.* (2001)). Optical properties of particles are controlled not only by their size but also by their shape (Mishchenko *et al.*, 2002; Kokhanovsky, 2003b). Therefore, it is clear that the model of spherical particles has very limited applications for snow optics. Another question, which is also of importance in snow optics, is the examination of the possible influence of close-packed effects on snow reflective properties.

The size d of snow grains is usually close to 1 mm (Massom *et al.*, 2001). It means that the methods of geometrical optics can be applied to studies of snow local optical characteristics in the visible and near-IR spectra. It is easy to show that the photon free path length in snow l, in the spectral range considered, is close to the value of d:

$$\lambda \ll l \sim d$$

The radiative transfer equation can not be applied if $\lambda \gg l$ (Tsang *et al.*, 1985). Then, the simple geometrical optics scheme laying at the very heart of radiative transfer theory is not applicable anymore. As we see, this is fortunately not the case for snow in the visible and near-IR. Therefore, we will assume that the radiative transfer equation can be used to calculate snow radiative characteristics.

In particular, we will use the following approximate analytical solution of this equation, valid in the limit of small absorption (Rozenberg, 1962; Zege *et al.*, 1991):

$$R(\vec{q}) = R_0(\vec{q}) \exp(-\alpha f(\vec{q})) \tag{5.11}$$

Here R is the bi-directional reflectance (or the reflection function) of a semi-infinite snow layer. The vector-parameter $\vec{q}$ has co-ordinates ϑ_0, ϑ, φ, which are the incidence zenith angle, the observation zenith angle, and the relative azimuth, respectively.

The function $f(\vec{q})$ in equation (5.11) is given by the following ratio:

$$f(\vec{q}) = \frac{K_0(\vartheta_0)K_0(\vartheta)}{R_0(\vartheta_0, \vartheta, \varphi)} \tag{5.12}$$

where R_0 equals to the value of R at zero absorption and

$$K_0(\vartheta_0) = \frac{3}{4}\cos\vartheta_0 + \frac{3}{4\pi}\int_0^{2\pi} d\varphi \int_0^{\pi/2} R_0(\vartheta_0, \vartheta, \varphi)\cos^2\vartheta \sin\vartheta \, d\vartheta \tag{5.13}$$

The function $K_0(\vartheta_0)$ is called the escape function in radiative transfer theory. It determines the angular distribution of light escaping the semi-infinite nonabsorbing medium in the framework of the Milne problem (with sources located at infinity inside a medium (van de Hulst, 1980)).

We also have for α in equation (5.11):

$$\alpha = 4\sqrt{\frac{l_{tr}}{3l_{abs}}} \tag{5.14}$$

where l_{abs} is the absorption path length and l_{tr} is the transport path length. Studies of the accuracy of equation (5.11) as compared to the solution of the radiative transfer equation have been performed by Kokhanovsky (2002a, 2002b). Equation (5.11) can be applied with a high accuracy for values of α smaller than unity.

Note that the phase coherence of scattered light is lost over length scales that are longer than l_{tr}. The path of a scattered photon can then be described as a random walk with the step size l_{tr}. On length scales that are large compared with l_{tr} it is useful to regard this as a photon diffusion with the diffusion coefficient $D = vl_{tr}/3$ where v is the speed of light in the medium. Therefore, l_{tr} is an important parameter in our theory. The value of l_{abs} is the length scale, where absorption effects become important. This parameter is large in the visible range of the electromagnetic spectrum. However, it rapidly decreases with the wavelength due to stronger light absorption by ice in the near-IR region. Actually, only these two parameters of a scattering medium enter the light diffusion equation.

We have from equation (5.11) as $\alpha \to 0$:

$$R(\vec{q}) = R_0(\vec{q}) - \alpha K_0(\vartheta_0)K_0(\vartheta) \tag{5.15}$$

which is a familiar result of the radiative transfer theory (van de Hulst, 1980). Equation (5.15) allows us to establish the physical meaning of α. Namely, performing the integration with respect to ϑ_0, ϑ, φ, we obtain from equation (5.15) for the spherical albedo (Kokhanovsky *et al.*, 2003):

$$r = 1 - \alpha \tag{5.16}$$

where we used equation (5.13) and the energy conservation law. Let us remember that the spherical albedo r is defined as (van de Hulst, 1980):

$$r = \frac{2}{\pi}\int_0^{2\pi} d\phi \int_0^1 \mu_0 d\mu_0 \int_0^1 \mu d\mu R(\mu_0, \mu, \phi) \tag{5.17}$$

where we introduced the cosines $\mu_0 = \cos\vartheta_0$ and $\mu = |\cos\vartheta|$.

The value of r gives the fraction of light energy reflected from a semi-infinite medium under diffused illumination. The value of α is, therefore, the fraction of the absorbed energy in the same conditions (and r close to unity). It follows (see, equation (5.14)) that media with more absorbing particles (smaller l_{abs}) do not necessarily give the largest absorption by a medium as a whole. Instead, it is the ratio $\mathbb{Z} = l_{tr}/l_{abs}$, which is of importance. In particular, different semi-infinite weakly absorbing disperse media, having the same values of $\mathbb{Z}$, have close values of the reflectance $R(\vec{q})$, providing that functions $R_0(\vec{q})$ are close as well. The physical reason behind this is quite clear. Indeed, this is not only the absorption by a local volume which matters, but also the type of scattering diagram. In particular, scattering diagrams which are highly extended in the forward direction with large values of l_{tr} also lead to larger numbers of scattering/absorption events prior to the escape of photons from the medium (hence, a larger probability of absorption by a medium as a whole).

Let us consider the escape function $K_0(\vartheta)$ (see equation (5.13)) in more detail now. It follows from the general radiative transfer theory that the function $K_0(\vartheta)$ describes the angular distribution of light emerging from a semi-infinite nonabsorbing layer with light sources placed at infinite depth inside the medium (the Milne problem) (Minin, 1988: Yanovitskij, 1997). It means, in particular, that this function should not depend strongly on the type of scatterers. This is really the case as derived from comparisons with numerical calculations (Yanovitskij, 1997; Kokhanovsky *et al.*, 2003). In particular, the following approximation can be used for all types of scatterers (Sobolev, 1972):

$$K_0(\mu_0) = \tfrac{3}{7}(1 + 2\mu_0) \tag{5.18}$$

if $\mu_0 \geq 0.2$ [12, 15, 22]. Equation (5.18) is not influenced by a particular single scattering law and can be used for media of various types, including snow.

We underline that equation (5.18) also describes the angular distribution of light transmitted by thick layers of snow. This suggests a simple way for the experimental verification of equation (5.18) (e.g. performing angular measurements of transmitted light in a cold chamber). Such data are not available to us. So we have chosen an indirect way to check equation (5.18) as applied to snow optics.

Namely, one can obtain from equation (5.11) for a plane albedo $r_p(\mu_0)$:

$$r_p(\mu_0) = \exp[-\alpha K_0(\mu_0)] \tag{5.19}$$

where $r_p(\mu_0)$ is defined as:

$$r_p(\mu_0) = \frac{1}{\pi}\int_0^{2\pi} d\phi \int_0^1 R(\mu_0, \mu, \phi)\mu d\mu. \tag{5.20}$$

A similar equation can be derived for the spherical albedo given by equation (5.17). Namely, it follows: $r = \exp(-\alpha)$.

The quantity $r_p(\xi)$ has been measured by many authors (Hubley, 1955; Rusin, 1961). It follows from equation (5.19):

$$K_0(\mu_0) = \frac{1}{\alpha}\ln\frac{1}{r_p(\mu_0)} \tag{5.21}$$

The precise value of α is usually unknown for a given experiment. However, α does not depend on ξ.

Let us define the normalized escape function $K_{0n}(\mu_0) = K_0(\mu_0)/K_0(\mu_0^*)$. Then it follows from equation (5.18):

$$K_{0n}(\mu_0) = \frac{1 + 2\mu_0}{1 + 2\mu_0^*} \tag{5.22}$$

where μ_0^* is a cosine of the selected incidence angle. We assume that $\mu_0^* = 0.7$. Clearly, any other value of $\mu_0^* > 0.2$ (see equation (5.18)) can be used for normalization. It follows from equation (5.21):

$$K_{0n}(\xi) = \frac{\ln r_p(\mu)}{\ln r_p(\mu_0^*)} \tag{5.23}$$

Therefore, measurements of $r_p(\mu)$ can be used to establish the relative angular dependence of the function $K_0(\mu)$. The remaining proportionality constant can be easily found from the condition:

$$2\int_0^1 K_0(\mu_0)\mu_0 d\mu_0 = 1 \tag{5.24}$$

which follows from the energy conservation law.

The comparison of the function $K_{0m}(\mu_0)$ derived using equation (5.22) at $\mu_0^* = 0.7$ and measurements of $r_p(\mu_0)$ (see equation (5.23)) is given in Figure 5.7. We see that experimental data closely follow the approximation in the range of $\mu_0 \geq 0.4$. The difference between curves can be attributed in particular to the inaccuracy related to the procedure of taking experimental data from plots $r_p(\mu_0)$

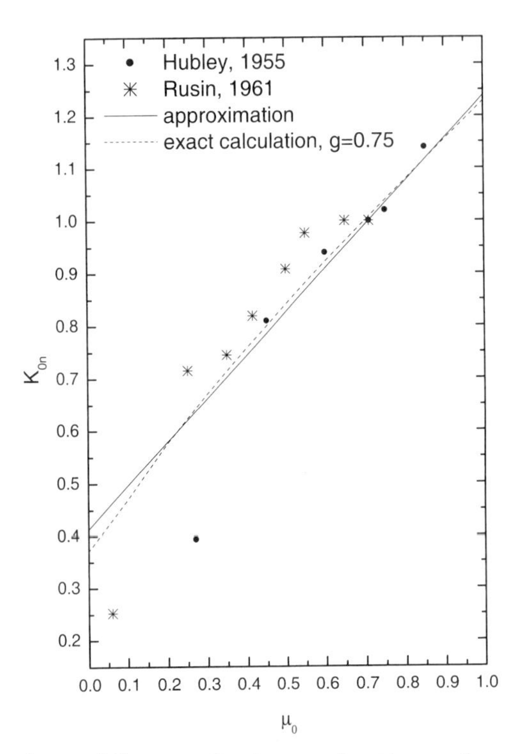

Figure 5.7. The dependence of the normalized escape function on the cosine of the incidence angle according to measurements (Hubley, 1955; Rusin, 1961), equation (5.22), and exact radiative transfer calculations for the Heney–Greenstein (HG) phase function with $g = 0.75$.

(especially at $\xi < 0.4$, where r_p is close to 1). We also show the results of the exact calculation (Yanovitskij, 1997) of the function $K_{0n}(\mu_0)$ for media characterized by the HG phase function $p(\theta) = \sum_{l=0}^{\infty} g^l P_l(\cos\theta)$ in Figure 5.7. Here $P_l(\cos\theta)$ is the Legendre polynomial. It was assumed that $g = 0.75$. Note that $p(\theta)$ gives the probability of photon scattering in the direction specified by the scattering angle θ. $g = \frac{1}{2}\int_0^{\pi} p(\theta) \sin\theta \cos\theta \, d\theta$ is the asymmetry parameter.

Aoki *et al.* (2000) found that the HG phase functions are more suitable for snow optical property studies as compared to Mie phase functions. It should be stressed, however, that the dependence $K_0(\mu_0)$ is very robust to the phase function variation (at least for $\mu_0 \geq 0.2$). The exact result and the approximation in Figure 5.7 closely correspond to each other at $\mu_0 \geq 0.2$ (the error is less than 2%). Therefore, we conclude that equation (5.18) can be used in conjunction with equations (5.11) and (5.12) to study the influence of absorption on the snow bi-directional reflectance. Note that $K_0(\mu_0)$ increases with μ_0. This leads to the decrease of r_p with μ_0 (see equation (5.19)), which has been confirmed by field measurements (Hubley, 1955; Rusin, 1961; Wiscombe and Warren, 1980).

We see, therefore, that the snow plane albedo is larger for the Sun at the horizon (small values of μ_0) as compared to the Sun overhead. The physics behind this effect

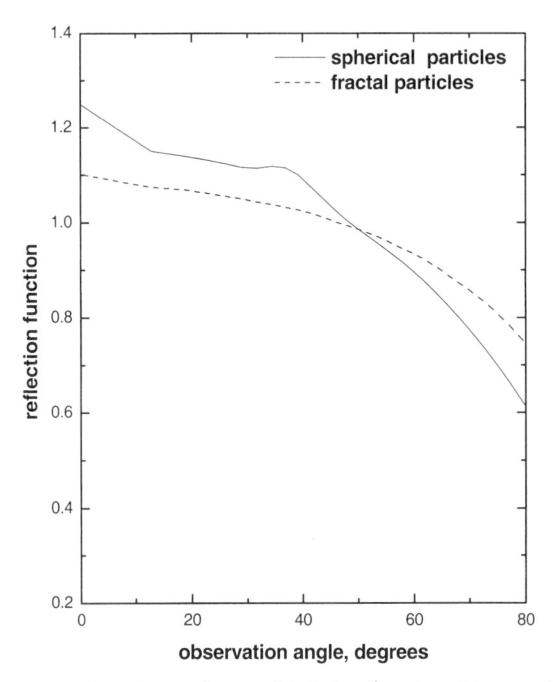

Figure 5.8. The reflection functions of a semi-infinite nonabsorbing medium with spherical and fractal grains at the nadir illumination as the function of the observation angle. Calculations were performed using the radiative transfer code described by Mishchenko *et al.* (1999) for the refractive index equal to 1.31 and the wavelength equal to 0.55 μm. The fractal particle model used is described by Macke *et al.* (1996). The size distribution of spherical particles is given by the model Cloud C.1 with the effective radius $a_{ef} = 6\,\mu m$ (Deirmendjian, 1969) and the coefficient of variance equal to $7^{-1/2}$.

is quite transparent. Indeed, photons injected in the medium along slant paths have more chances of survival compared with the case of the perpendicular incidence.

To model the bi-directional reflectance function of a plane-parallel semi-infinite snow layer $R(\vec{q})$ we should also know the function $R_0(\vec{q})$ for nonabsorbing snow (see equations (5.11) and (5.12)). It depends on the phase function $p(\theta)$ only. Various assumptions of the function $p(\theta)$ (see, e.g. Macke, 1994; Macke *et al.*, 1996; Muinonen *et al.*, 1996; Yang and Liou, 1998; Nousiainen and Muinonen, 1999; Barkey *et al.*, 2002; Liou, 2002) lead to diverse results as far as the function $R_0(\vec{q})$ is concerned. This was clearly demonstrated by Mishchenko *et al.* (1999) (see figure 5 therein) for three shapes of particles including spheres, hexagonal cylinders, and fractals (see also Figure 5.8). Therefore, it is of importance to make the right selection of the phase function in the modelling of the snow bi-directional reflectance.

Several theoretical phase function models are given in Figure 5.9 together with the experimental data of Barkey *et al.* (2002) for a crystalline medium. The spherical model should be rejected. In particular, if this model is valid, then rainbows and

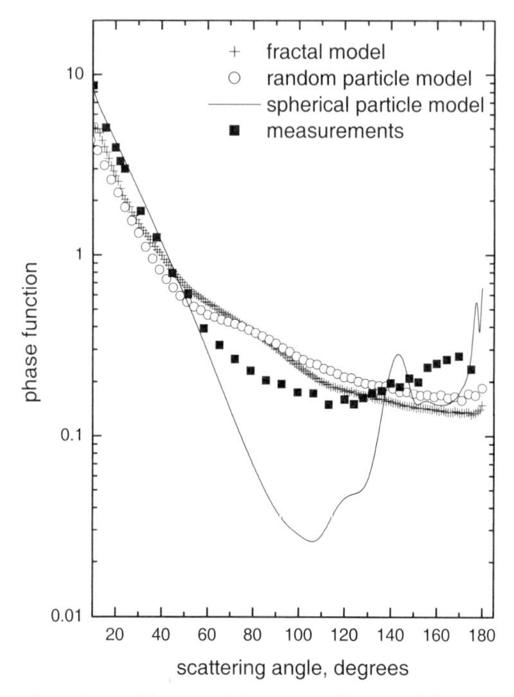

Figure 5.9. The phase function of ice particles according to different theoretical models and measurements (Barkey *et al.*, 2002). Calculations were performed using the geometrical optical approximation at the 0.55 μm wavelength, where light absorption by ice is extremely weak. The fractal particle model is described by Macke *et al.* (1996). The random particle model is described by Muinonen *et al.* (1996). It was assumed that the standard deviation $\sigma = 0.2$ and the correlation angle $\Gamma = 5^\circ$ in calculations according to the random particle model (Muinonen *et al.*, 1996). The size of particles for fractal and random particle models does not influence results given in this figure. The phase function for spherical particles was calculated for the same spherical polydispersion as in Figure 5.8 but for the effective radius $a_{ef} = 15\,\mu\text{m}$.

glories should be observed (see the solid line in Figure 5.9 at the scattering angle θ close to 145° and 180° and the solid line in Figure 5.8 at 35° and 0°). Clearly, this is not the case for snow. The hexagonal cylinders model given by Mishchenko *et al.* (1999) should also be rejected because such shapes do not appear in snow in large numbers.

The random particle model (Muinonen *et al.*, 1996) and the fractal particle model (Macke *et al.*, 1996) are good candidates, however. The first model is based on the distortion of a spherical shape using the Gaussian distribution of its surface with varied parameters. The fractal particle is constructed in the following way. The initial shape is a regular tetrahedron. A part of its surface is replaced by a reduced version of the same tetrahedron. The repetition of this procedure leads to a fractal surface with the fractal dimension: $\ln 6/\ln 2 \approx 2.58$ (Macke *et al.*, 1996).

We would like to underline, that functions $p(\theta)$ for fractals and random particles are close to each other (see Figure 5.9). Therefore, functions $R_0(\vec{q})$ for both of them do not differ very much. This underlines the fact that what is of importance is not the particular stochastic particle model but rather the level of the randomness in a given model (Kokhanovsky, 2003a).

Here we choose the fractal model because it has no free parameters for nonabsorbing ice particles and closely corresponds to the random particle model at extreme values of its randomness parameters. The final decision, however, should be made when measurements of the snow phase function *in situ* are performed.

It should be stressed that the poor correspondence of measurements and the theory for fractals around 90° and 170° (see Figure 5.9) could be due to the presence of small crystals or even ice spheres in the experiment performed by Barkley and Liou (2002), who have been able to describe their experimental curve with a high precision using the unified theory of light scattering by ice crystals assuming the following shapes of crystals: irregular bullet rosettes and plates with rough surfaces. Clearly, the combination of different shapes can be used to model observed phase functions of crystalline media with a high accuracy (Liou, 2002). However, here the problem of uniqueness arises. Therefore, we prefer to use a model of a single randomly oriented fractal particle, which has no free parameters attached.

The comparison of measurements performed by Middleton and Mungall (1952) of $R(\vartheta)$ at $\vartheta_0 = 0°$ and calculations using equation (5.10) with account of equation (5.18) and the function R_0 (given in Figure 5.8 for fractal particles) is shown in Figure 5.10. The value of α was not specified in the experiment discussed. So we used α as a fitting parameter at the angle 30° (see Figure 5.10).

We see that our model is capable of describing the experimental snow bidirectional reflectance accurately. It has no free parameters except the value of R at $\vartheta = 30°$. This is in contrast with Hapke's model (Hapke, 1993) as applied to snow. In particular, it is assumed in the framework of Hapke's model (Domingue *et al.*, 1997):

$$R = \frac{\omega_0}{4(\mu_0 + \mu)} \{(1 + s(g))p(\theta) + H(\mu_0)H(\mu) - 1\} \tag{5.25}$$

where ω_0 is the single scattering albedo, $p(\theta)$ is the phase function, and

$$s(g) = \frac{B_0}{1 + (1/h)\tan(g/2)} \tag{5.26}$$

Here h and B_0 are free parameters of the theory and g is the asymmetry parameter. The function $H(\mu)$ can be calculated approximately via

$$H(\mu) = \frac{1 + \mu}{1 + \mu\sqrt{1 - \omega_0}} \tag{5.27}$$

In Figure 5.10 we used the following values of parameters in equations (5.25) and (5.26): $B_0 = 1.0$, $\omega_0 = h = 0.995$, and the Heney–Greenstein phase function $p(\theta)$ with $g = 0.449$. Domingue *et al.* (1997) found that this set of parameters gives best fits of measurements presented in figure 5.10 by Hapke's model.

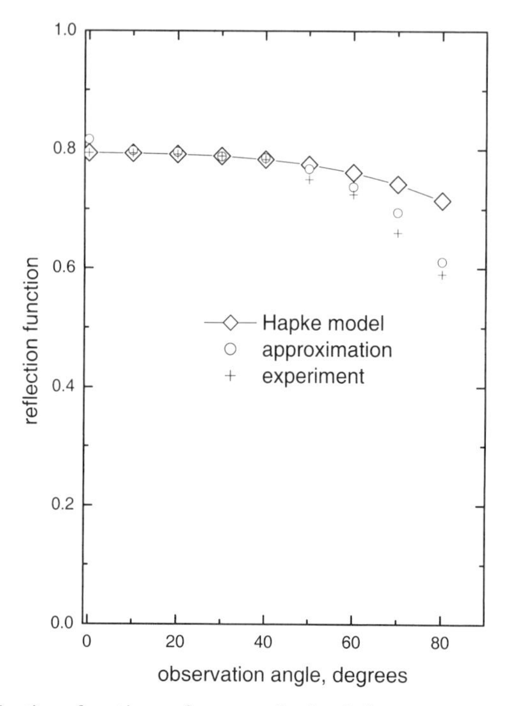

Figure 5.10. The reflection function of snow obtained from measurements performed by Middleton and Mungall (1952), according to our approximation (equation 5.11) and Hapke's model.

It follows that both our snow model and that according to Hapke's theory explain the almost constant value of the reflection function in the range of viewing angles smaller than 50° (near-nadir observations). This feature is also confirmed by experiments performed by Middleton and Mungall (1952). However, our theory gives better agreement with measurements at the range of angles 45°–80°. Note that the use of a more complex model of the phase function in equation (5.25) fails to improve the accuracy of Hapke's model considerably (Domingue *et al.*, 1997).

The parameter α obtained from fitting measurement data (Middleton and Mungall, 1952) using equation (5.11) and that calculated from values of g and ω_0 of Hapke's model ($\alpha = 4\sqrt{(1-\omega_0)/3(1-g)}$) are close to each other. In particular, it is equal to 0.2 in the framework of our model and it is only 10% larger (0.22) as derived from the fitting techniques based on Hapke's equation.

5.1.2.2 *Local optical characteristics of snow*

Until now we have mostly considered the global optical characteristics of snow, like its bi-directional reflectance and albedo. Let us turn to the description of snow local optical characteristics related to equation (5.11). In particular, we would like to establish the relation of parameters l_{abs} and l_{tr} to the microphysical characteristics of snow. This will also lead to the relationship between α (see equation (5.14)) and the grain size d.

Let us start from l_{tr}. This parameter is defined by the following equation:

$$l_{tr} = \frac{1}{\sigma_{ext}(1-g)} \tag{5.28}$$

where σ_{ext} is the extinction coefficient and g is the asymmetry parameter. For large particles considered here, σ_{ext} and g are given as:

$$\sigma_{ext} = (\langle C^D_{ext}\rangle + \langle C^G_{ext}\rangle)N \tag{5.29}$$

$$g = \frac{\langle C^D_{sca} g^D\rangle + \langle C^G_{sca} g^G\rangle}{\langle C^D_{sca}\rangle + \langle C^G_{sca}\rangle} \tag{5.30}$$

where N is the number of grains in a unit volume. Symbols G and D signify the geometric optics (G) and the diffraction (D) contributions to the extinction cross section C_{ext}, the scattering cross section C_{sca}, and the asymmetry parameter g. Brackets around symbols represent the averaging of the snow grains geometrical characteristics (size, shape, and orientation).

It follows from equations (5.28)–(5.30) and the assumption that $C_{sca} \approx C_{ext}$, which is valid for weakly absorbing media considered here:

$$l_{tr} = \frac{1}{N(1 - \mathbb{Q}\langle g^G\rangle)\langle C^G_{sca}\rangle} \tag{5.31}$$

where $\mathbb{Q} = \langle C^G_{sca} g^G\rangle / \langle C^G_{sca}\rangle\langle g^G\rangle$. Note that we also used: $g^D \approx 1$.

Equation (5.31) shows that the transport path length in snow is defined by the geometrical part of the scattered field. This field is almost unaffected by the close-packed effects (in contrast to the diffraction contribution). Therefore, we can neglect the consideration of the close packed effects in the calculation of l_{tr} due to their mutual cancellation. This is an important point.

It is known that:

$$\langle C^G_{sca}\rangle = \frac{\langle\Sigma\rangle}{4} \tag{5.32}$$

for nonspherical convex particles in a random orientation at any values of the complex refractive index $m = n - i\chi$ of grains. Here $\langle\Sigma\rangle$ is the average surface area of particles.

It follows from equations (5.31) and (5.32):

$$l_{tr} = \frac{4}{N\langle\Sigma\rangle(1 - \langle g^G\rangle)} \tag{5.33}$$

where we assume that $\mathbb{Q} = 1$. This is exactly the case for nonabsorbing large spherical particles. Then g^G does not depend on the size of particles and $\langle C^G_{sca} g^G\rangle = \langle C^G_{sca}\rangle\langle g^G\rangle$. We will assume that this equality approximately holds for snow as a whole.

Equation (5.33) can be written in a slightly different form introducing the volumetric concentration of snow grains C_v:

$$C_v = N\langle V\rangle \tag{5.34}$$

where $\langle V\rangle$ is the average volume of grains. Then we have:

$$l_{tr} = \frac{2d}{3(1 - \langle g^G\rangle)C_v} \tag{5.35}$$

where

$$d = 6\frac{\langle V\rangle}{\langle \Sigma\rangle} \tag{5.36}$$

Equation (5.36) could be taken as a definition of the effective grain size d for such a complex medium as snow. Note that for monodispersed spheres, d is equal to their diameter. It follows for polydispersed media composed of spherical particles: $d = 2a_{ef}$, where a_{ef} is the effective radius defined as the ratio of the third to the second moment of the particle size distribution.

The value of C_v is close to 1/3 for snow (Massom *et al.*, 2001). This means that $l_{tr} \approx 2d/(1 - \langle g^G\rangle) \gg \lambda$ because $d \gg \lambda$ and $0 < g^G < 1$. This also points to the possibility of the application of the radiative transfer equation to the problems of snow optics.

It follows from equation (5.35) that the ratio M of transport lengths for media with spheres to that of media with nonspherical particles, with the same values of C_v and d, is given by the following formula:

$$M = \frac{1 - \langle g_n^G\rangle}{1 - \langle g_s^G\rangle} \tag{5.37}$$

where s and n here and below denote spherical and nonspherical scattering, respectively. Note that $g_n^G \approx 1/2$ for fractals (Kokhanovsky, 2003a). The value of the asymmetry parameter for large monodispersed spherical ice particles as functions of the size parameter $x = \pi d/\lambda$ is presented in Figure 5.11. We see that $g \approx 0.89$ for large ice spheres. It means that $g_s^G = 2g - 1 \approx 0.78$. This gives, with an account of equation (5.37), $M = 2.27$. Therefore, the transport path length in the media, composed of fractal particles, is approximately two times smaller than in the case of media with spherical ice particles having the same values of C_v and d. In particular, we have: $l_{tr}^s = 2.27 l_{tr}^n$. This has important consequences as far as the snow grain size retrieval techniques are concerned.

Let us study the absorption path length l_{abs} now. It is defined as the inverse value of the absorption coefficient σ_{abs}: $l_{abs} = 1/\sigma_{abs}$, where $\sigma_{abs} = N\langle C_{abs}\rangle$ and $\langle C_{abs}\rangle$ is the average absorption cross section per particle. In principle, $\langle C_{abs}\rangle$ can also depend on the number of particles N for close-packed media such as snow fields. However, *in situ* experiments (Bohren and Beschta, 1979) show that snow albedo does not depend on N. The value of $\alpha \sim \sqrt{l_{abs}/l_{tr}}$ is independent of N if $\langle C_{abs}\rangle$ is not a function of N.

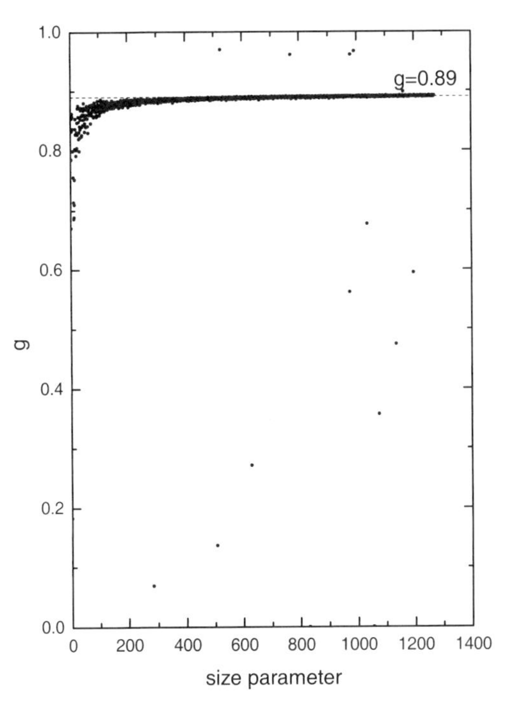

Figure 5.11. The dependence of the asymmetry parameter g on the size parameter x, obtained using the Mie theory for monodispersed spherical particles, having the refractive index of $1.31\text{–}10^{-7}i$.

Therefore, we neglect close-packed media effects in the calculation of l_{abs} and write:

$$l_{abs} = \frac{\langle V \rangle}{C_v \langle C_{abs} \rangle} \tag{5.38}$$

where we accounted for equation (5.34).

Let us derive an approximate expression for the value of C_{abs} for a particle of an arbitrary shape as it is needed in snow optics. For this we use the definition of C_{abs}:

$$C_{abs} = \frac{k}{|\vec{E}_0|^2} \int_V \varepsilon''(\vec{r}) \vec{E}(\vec{r}) \vec{E}^*(\vec{r}) d^3\vec{r} \tag{5.39}$$

where $k = 2\pi/\lambda$, λ is the wavelength, V is the volume of a particle, $|\vec{E}_0|$ is the absolute value of the incident electric field vector $\vec{E}_0$, $\varepsilon'' = 2n\chi$ is the imaginary part of the relative dielectric permittivity of a particle, and $\vec{E}(\vec{r})$ is the value of the electric field inside a particle at the point with the radius vector $\vec{r}$. We will assume:

$$\vec{E}(\vec{r}) = \vec{E}_0 \sum_{s=0}^{\infty} a_s(\vec{r}) \exp\{i\varphi_s(\vec{r})\} \tag{5.40}$$

which corresponds to the representation of $\vec{E}(\vec{r})$ by a linear combination of simple waves with amplitudes $a_s\vec{E}_0$ and phases φ_s. It follows from equation (5.40):

$$|\vec{E}|^2 = |\vec{E}_0|^2 \sum_{s=0}^{\infty} a_s^2 \exp(-2\varphi_s'') \tag{5.41}$$

where $\varphi_s'' = \mathrm{Im}(\varphi_s)$ and where we have neglected the phenomenon of interference, which is of no importance for snow grains with $d \gg \lambda$. Then we obtain – using the fact that φ_s'' is a small parameter for weakly absorbing grains:

$$|\vec{E}(\vec{r})|^2 = \beta(\vec{r})|E_0|^2 \tag{5.42}$$

where $\beta(\vec{r}) = \sum_{s=0}^{\infty} a_s^2(\vec{r})$:

It follows from equations (5.39) and (5.42):

$$C_{abs} = k\beta(\vec{r}_0)\int_V \varepsilon''(\vec{r})d^3\vec{r} \tag{5.43}$$

where we use the equality:

$$\int_V \psi(\vec{r})\varphi(\vec{r})d\vec{r} = \psi(\vec{r}_0)\int_V \varphi(\vec{r})d\vec{r} \tag{5.44}$$

The value of $\vec{r}_0$ is generally not known.

We neglect the internal inhomogeneity of grains. Then it follows from equation (5.43):

$$C_{abs} = n\beta\gamma V \tag{5.45}$$

where $\gamma = 4\pi\chi(\lambda)/\lambda$ is the ice absorption coefficient. Also, we have:

$$\langle C_{abs}\rangle = B\gamma\langle V\rangle \tag{5.46}$$

where $B = n\langle\beta\rangle$ and we assume that $\langle\beta V\rangle \approx \langle\beta\rangle\langle V\rangle$. One can obtain from equations (5.38) and (5.46):

$$l_{abs} = \frac{1}{B\gamma C_v} \tag{5.47}$$

Let us check the applicability of equation (5.45) for a particular case of spherical particles. For this, we have calculated the ratio $B = C_{abs}/\gamma V$ with Mie theory for monodispersed spheres with a refractive index $m = 1.31 - 10^{-7}i$ at $\lambda = 0.55\,\mu m$ as a function of the size parameter $x = \pi d/\lambda$. The results are given in Figure 5.12. We see that indeed B only weakly depends on x, if we neglect the effects of interference, which is of no importance for snow optics due to the nonspherical shape of large grains and their polydispersity. The numerical data for the value of B scatter between 1.27 and 1.31 for cases relevant to snow optics problems ($x = 500$–1200). This gives variability below 3%. We take the value $B \approx 1.27$ for estimations in this section, which is similar to that used by Bohren (1983) ($B \approx 12.6$). It means (see equation (5.42)) that $\beta = B/n \approx 1.0$ (with an error smaller than 3%) for ice spheres in the visible spectrum, where $n \approx 1.31$. Also we have on physical grounds: $\beta \to 1$ as $n \to 1$ (or $C_{abs} \to \gamma V$). This means that β is very robust against variations of the refractive index for the range of n relevant to snow optics problems.

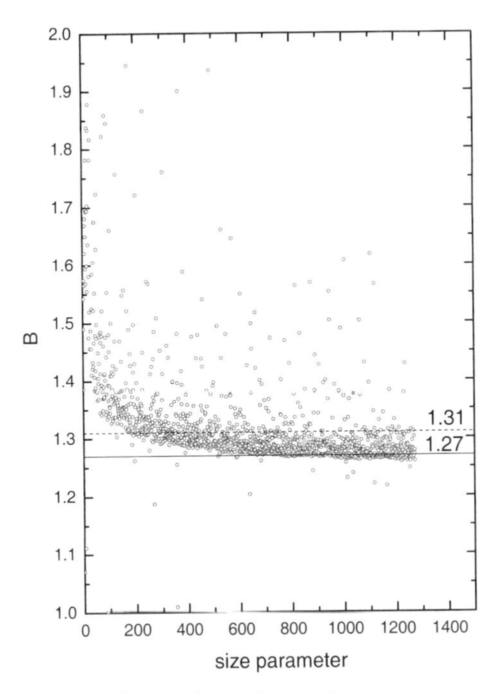

Figure 5.12. The dependence of the absorption enhancement parameter B on the size parameter x, obtained using Mie theory for monodispersed spherical particles, having the refractive index of $1.31–10^{-7}i$.

The value of B for spheroids and hexagonal cylinders with various aspect ratios are tabulated by Kokhanovsky and Macke (1997). In particular, it was found that $B = 1.2–2.1$ for spheroids and $B = 2.2–2.5$ for hexagonal cylinders, if it is assumed that the ratio of their axes is in the range 0.5–2.0. The average of B for the interval 1.2–2.5 is equal to 1.85, which is close to our estimation of $B \approx 1.84$ for fractal particles obtained using geometrical optics Monte Carlo calculations. The details of the code used to find B for fractals are given elsewhere (Macke, 1994). The close correspondence of B for fractals and a mixture of particles having various shapes is not surprising taking into account the fact that chaotic scattering/absorption by media with particles of diverse shapes should approach that of a single extremely irregular particle such as a fractal.

Therefore, we propose to use the value of $B = 1.84$ in snow optics instead of $B = 1.27$ as for spheres. This proposal, however, should be checked against carefully planned experiments for natural snow. We note that the value of B may also depend on the age of the snow and the ambient temperature.

Our estimations allow us to conclude that large weakly absorbing nonspherical ice particles have generally larger absorption cross sections compared with spheres of the same volume ($B_n > B_s$). This also means that $l^s_{abs} > l^n_{abs}$.

In particular, for ice particles we have: $l^s_{abs} \approx 1.45 l^n_{abs}$, if the volume concentration of snow grains is fixed. We remember that $l^s_{tr} = 2.27 l^n_{tr}$. This means that the ratio

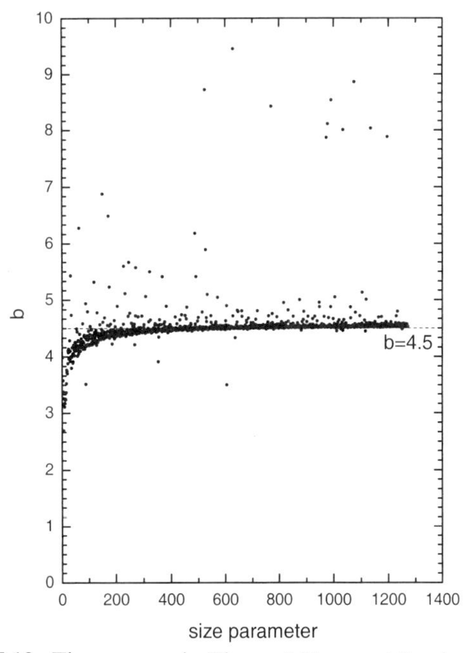

Figure 5.13. The same as in Figure 5.12 except for the value of b.

l_{tr}/l_{abs}, which is of importance for reflectance problems (see equations (5.11) and (5.14)) is 1.57 times larger for spheres compared with fractal particles with the same values of the product γd.

As seen from equation (5.11) and (5.14), this ratio can be easily determined from reflectance measurements. It is proportional to the size of grains (see equations (5.35) and (5.47)):

$$\frac{l_{tr}}{l_{abs}} = \frac{2B\gamma d}{3(1 - \langle g^G \rangle)} \tag{5.48}$$

In particular, we have $l_{tr}/l_{abs} \approx 2.45\gamma d$ for fractal particles and $l_{tr}/l_{abs} \approx 3.85\gamma d$ for spheres. It follows from equation (5.48): $d = \Im l_{tr}/\gamma l_{abs}$, where $\Im = 3(1 - \langle g^G \rangle)/2B$. Let us suppose that the ratio $l_{tr}/\gamma l_{abs}$ is exactly known (the ratio l_{tr}/l_{abs} is determined from the snow reflectance and the ice absorption coefficient γ is known to a high accuracy). Then the error of the grain size estimation due to the spherical assumption is given by $\varepsilon_d = 1 - \Im_s/\Im_n$. Taking into account that $\Im_s \approx 0.26$ and $\Im_n \approx 0.408$, we obtain $\varepsilon_d \approx 0.4$. This allows us to conclude that the uncertainty in the determination of the grain size due to shape effects could reach 40%. Therefore, snow grain size retrievals based on Mie theory may greatly underestimate the value of d.

Note that *in situ* measurements also confirm that the optically equivalent (in terms of the spectral albedo fitting) diameter of the snow grain size is smaller than that measured *in situ*. This problem has been addressed (e.g. Aoki *et al.*, 2000), suggesting that the spherical approximation gives not the value of d defined in equation (5.36) but rather a dimension of a narrower portion of broken crystals or the brunch width of dendrites.

The value of $\alpha \sim \sqrt{l_{tr}/l_{abs}}$ is approximately 1.25 times larger for spheres as compared with fractals. We conclude, therefore, that although values of l_{abs} and l_{tr} differ considerably for spheres compared with fractals (see above), the difference is more moderate for the value of α, which determines the snow albedo. Nevertheless, we underline that the use of the Mie model for the calculation of α with equation (5.14) leads to the overestimation of α by 25% compared with the fractal model. This gives overestimation of the snow heating (larger α) if the spherical particle model is used. This also gives the same error for the value of $1 - r$ as $r \to 1$ (see equation (5.16)).

The expression for α can be written in the following simple form using equations (5.14) and (5.48): $\alpha = B\sqrt{\gamma d}$, where $B = \frac{4}{3}\sqrt{2B/(1 - \langle gG \rangle)}$. The dependence $B(x)$ for spheres at $m = 1.31 - 10_i^{-7}$ is shown in Figure 5.13. The value of B approaches 4.5 for large spheres (as compared to 3.62 for fractals).

5.1.3 Aerosols

Unlike snow and clouds, atmospheric aerosol layers (Ivlev, 1986; d'Almeida *et al.*, 1991) are very thin ($\tau = 0.05 - 0.3$). Thus, we can obtain for the values of transmission and reflection functions of thin aerosol layers (see Chapter 3):

$$R = \frac{\omega_0 p(\theta)}{4(\mu + \mu_0)}\left(1 - e^{-\tau(1/\mu + 1/\mu_0)}\right) \tag{5.49}$$

$$T = \frac{\omega_0 p(\theta)}{4(\mu - \mu_0)}\left(e^{-\tau/\mu} - e^{-\tau/\mu_0}\right) \tag{5.50}$$

where

$$\theta = \arccos\left(\beta\mu\mu_0 + \sqrt{(1-\mu)^2(1-\mu_0^2)}\cos\phi\right)$$

is the scattering angle, $\beta = 1$ for the value of T and $\beta = -1$ for the value of R, $\mu = \cos\vartheta$, $\mu_0 = \cos\vartheta_0$, and ϑ_0 and ϑ are the angles of incidence and observation, and ϕ is the azimuth. The reflection of light from the underlying surface can be accounted for with equations (3.18)–(3.19).

The value of ω_0 is about unity in these equations for the background aerosol. The atmospheric phase function $p_a(\theta)$ is determined mostly by molecular scattering and two aerosol components, namely dust particles (or oceanic particles over oceans) and water soluble (WS) aerosols (see Appendix 5):

$$p_a(\theta) = \frac{p(\theta) + \xi p_m(\theta)}{1 + \xi} \tag{5.51}$$

where $p(\theta)$ is the aerosol phase function, $p_m(\theta) = \frac{3}{4}(1 + \cos^2\theta)$ is the molecular scattering phase function, and ξ is the ratio of scattering coefficient of molecular and aerosol scattering. Dust aerosols have approximately ten times larger average radii than WS aerosols and determine the phase function at small scattering angles.

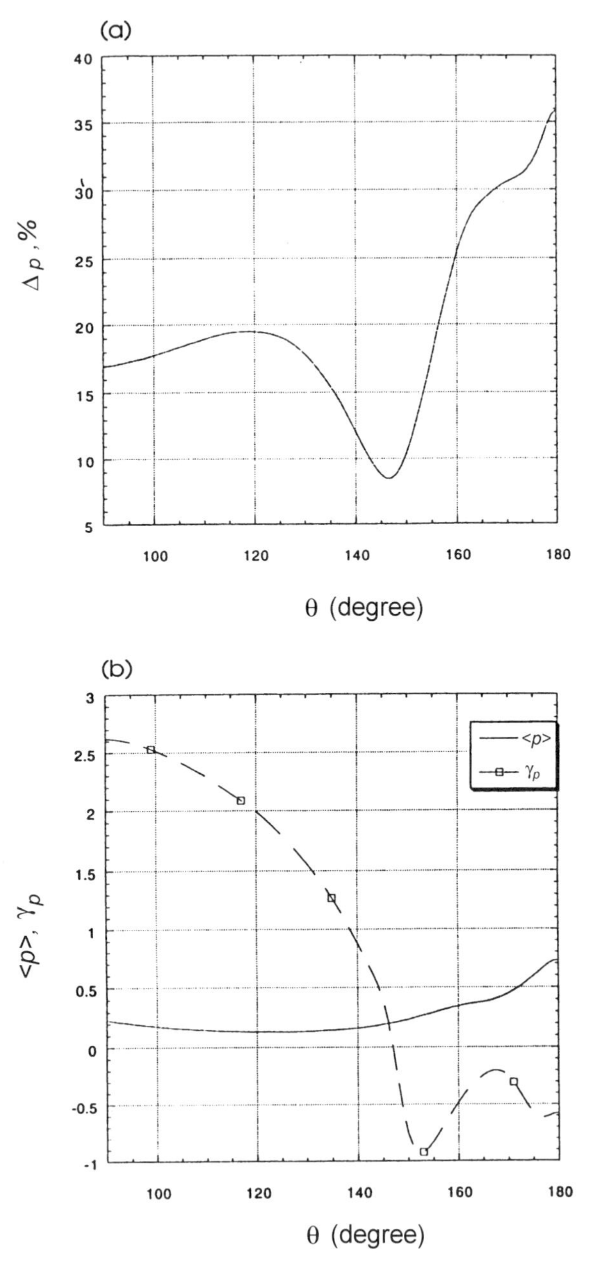

Figure 5.14. The dependence of the coefficient of variance of the phase function Δ_p (a), the average aerosol phase function $\langle p \rangle$, and skewness γ_p (b) on the scattering angle θ at $\lambda = 0.7\,\mu\text{m}$, $m = 1.53 - 0.006i$, and the lognormal particle size distribution with the coefficient of variance $\Delta = 0.7$. Statistical characteristics of the ensemble of phase functions were calculated with the Mie theory for 300 values of the effective radii, ranging from $0.1\,\mu\text{m}$ to $1.5\,\mu\text{m}$ (Kokhanovsky, 1998).

WS particles have greater concentration than dust particles. Thus, they determine the behaviour of the phase function $p_a(\theta)$ at large scattering angles. If we know the value of $p(\theta)$ in equations (5.13), (5.14), it can be used to find the value of τ from equations (5.13), (5.14), which is an important parameter for global climatology of the Earth. Generally speaking, the value of $p(\theta)$ is not known a priori. However, statistical analysis shows that the variability in the phase function $p(\theta)$ of the background aerosol $\Delta_p = s_p(\theta)/\langle p(\theta)\rangle$ is low at $\theta \approx 150^\circ$. Here $\langle p(\theta)\rangle = \sum_{i=1}^{N} p_i(\theta)/N$ is the average phase function, $s_p^2(\theta) = \sum_{i=1}^{N}(p_i(\theta) - \langle p(\theta)\rangle)/(N-1)$ is the variance, $p_i(\theta)$ is the ensemble of phase functions of atmospheric aerosols with different microstructure parameters. Kokhanovsky (1998) found that the aerosol phase function at $\theta = 150^\circ$ is $p(150^\circ) \approx 0.22 \pm 0.03$ for the ensemble of phase functions of atmospheric aerosols with different microstructure parameters. Functions Δ_p, $\langle p(\theta)\rangle$, and skewness

$$\gamma_p(\theta) = \frac{1}{(N-1)s_p^3(\theta)} \sum_{i=1}^{N} (p_i(\theta) - \langle p(\theta)\rangle)^3 \tag{5.52}$$

are presented in Figure 5.14 (Kokhanovsky, 1998). Thus, we can see that aerosol phase functions hardly change at the scattering angle 150°.

The demand for detailed information on the spatial distributions, microstructure parameters and chemical composition of atmospheric aerosols is increasing. The information on aerosol size distributions contained in solar reflected spectral radiances were studied by Tanre *et al.* (1996). Sasano (1996) investigated aerosol extinction profiles in the troposphere. The special case of oceanic aerosols was considered by Wang and Gordon (1993, 1994). The algorithm for the determination of the complex index of refraction of aerosol particles from scattering measurements of polarized light was proposed by Zhao *et al.* (1997).

Aerosol particles affect the climate of our planet both by directly interacting with solar and terrestrial radiation and indirectly by their effect on cloud microphysics, precipitation, and snow albedo. Past and future launches of different statellites (e.g. *ADEOS* and *ADEOS-2* in Japan, *EOS-AM1* and *EOS-PM1* in the USA, *ENVISAT* in Europe) with numerous modern multispectral, multiangle, and polarization instruments, including MODIS (King *et al.*, 1992), POLDER (Deschamps *et al.*, 1994), and SCIAMACHY (Bovensmann *et al.*, 1999), will provide humankind with important knowledge about aerosol characteristics on a global scale.

5.1.4 Ocean optics

5.1.4.1 Reflection of light from the ocean

The open ocean is a classical example of a semi-infinite light-scattering medium. The absoption of light in the ocean (Shifrin, 1988; Bricaud and Stramski, 1990) is rather high ($\omega_0 \approx 0.7$ in the visible) and the reflection function of ocean can be calculated in the framework of the quasi-single scattering approximation (Gordon, 1973):

$$R_\infty(\mu, \mu_0, \phi) = \frac{\omega_0 p(\theta)}{4(1-\omega_0\Phi)(\mu+\mu_0)} \tag{5.53}$$

Table 5.1. Typical phase function of oceanic water at $\lambda = 0.514\,\mu m$ (see table 3.10 in Mobley, 1994).

θ (deg)	$p(\theta)$	θ (deg)	$p(\theta)$	θ (deg)	$p(\theta)$	θ (deg)	$p(\theta)$
0	3.764E+04	1.995	2.866E+02	45.0	3.854E−01	115.0	3.515E−02
0.100	2.220E+04	2.512	1.905E+02	50.0	2.859E−01	120.0	3.375E−02
1.126	1.629E+04	3.162	1.259E+02	55.0	2.135E−01	125.0	3.231E−02
0.158	1.194E+04	3.981	8.269E+01	60.0	1.650E−01	130.0	3.111E−02
0.200	8.785E+03	5.012	5.397E+01	65.0	1.314E−01	135.0	2.987E−02
0.251	6.459E+03	6.310	3.527E+01	70.0	1.067E−01	140.0	2.927E−02
0.316	4.730E+03	7.943	2.286E+01	75.0	8.766E−02	145.0	2.907E−02
0.398	3.472E+03	10.0	1.449E+01	80.0	7.341E−02	150.0	2.972E−02
0.501	2.750E+03	15.0	6.149E+00	85.0	6.224E−02	155.0	3.149E−02
0.631	1.815E+03	20.0	3.071E+00	90.0	5.393E−02	160.0	3.345E−02
0.794	1.284E+03	25.0	1.850E+00	95.0	4.753E−02	165.0	3.563E−02
1.000	8.999E+02	30.0	1.082E+00	100.0	4.278E−02	170.0	3.809E−02
1.259	6.230E+02	35.0	7.453E−01	105.0	3.916E−02	175.0	3.886E−02
1.585	4.266E+02	40.0	5.290E−01	110.0	3.659E−02	180.0	3.963E−02

where

$$\Phi = \frac{1}{2}\int_0^{\pi/2} p(\theta)\sin\theta\,d\theta \tag{5.54}$$

Note that the value of Φ is close to 1 for most oceanic phase functions. Natural waters range from absorption dominated ($\omega_0 \approx 0.25$ at $\lambda = 0.514\,\mu m$) to scattering dominated ($\omega_0 \approx 0.83$ at $\lambda = 0.514\,\mu m$) waters (Mobley, 1994). The typical phase function of oceanic water is presented in Table 5.1. This table was obtained by Mobley (1994) by averaging scattering functions, measured by Petzold (1972). We changed the data in Table 3.10 of Mobley (1994) to satisfy the normalization condition:

$$\frac{1}{2}\int_0^{\pi} p(\theta)\sin\theta\,d\theta = 1 \tag{5.55}$$

The value of $p(0)$ was obtained by interpolation. It strongly depends on the size of particles in the water probe. This is not the case for phase functions in the backward hemisphere.

The error of simple equation (5.53) is less than 5% at μ, $\mu_0 > 5$ and it increases with the solar and observation angles (Zege *et al.*, 1991). This formula can be used to estimate the plane and spherical albedo of the ocean. We need to know the phase function $p(\theta)$ to apply equation (5.53). This function, as in the case of atmospheric aerosols, depends both on molecular scattering in water and on scattering of light by particles. The phase function at small angles is determined by particles. The fine component of oceanic suspensions is responsible for side and backward light scattering. The refractive indices of particles in the ocean are close to the refractive index of water. Thus, the theory of soft particles can be used to estimate the phase function of particulate matter in the ocean. In particular, the shape of the forward peak can be estimated in the framework of Van de Hulst or Perelman approximations (Aas, 1984). The phase functions of fine particles can be described in the framework of the Rayleigh–Gans approximation (Kerker, 1969).

Note that the phase function for ocean water is almost constant at $\theta > \pi/2$ (Shifrin, 1988). It is often assumed that $p(\theta) \in [0.029, 0.1]$ in the backward hemisphere (Golubitsky *et al.*, 1974). This assumption is supported by data, presented in Table 5.1, where $p(\theta) \in [0.029, 0.06]$ at $\theta \geq \pi/2$. Note that the diffuse reflectance of ocean waters was studied by Morel and Gentili (1991, 1993).

It follows from equation (5.18) that:

$$R_\infty(\lambda) = \xi \frac{\sigma_{sca}(\theta, \lambda)}{\sigma_{abs}(\lambda)} \tag{5.56}$$

where the value of ξ does not in practice depend on the wavelength. The dependence of the volume- scattering coefficient $\sigma_{sca}(\theta, \lambda) = (\sigma_{sca} p(\theta))/4\pi$ on the wavelength is also weak in the backward hemisphere. Thus, the spectral dependence of the reflection function (and the colour of the ocean) is determined mostly by the absorption coefficient of water. Oceanic deserts are characterized by a deep blue colour (Shifrin, 1988). The colour shifts from blue to green as chlorophyll concentration increases. Equation (5.56) is used to estimate the value of the absorption coefficient $\sigma_{abs}(\lambda)$ of oceanic waters from satellite measurements. In turn, the absorption coefficient σ_{abs} depends on the concentration of chlorophyll c in oceanic water (Shifrin, 1988; Sathyendranath *et al.*, 1989; Sturm, 1981; Mobley, 1994). In many cases there is linear correlation between values of $\lg c$ and $\lg(R(443\,\text{nm})/R(500\,\text{nm}))$ (Mobley, 1994). This correlation can be used to find the chlorophyll concentration from satellite measurements of the reflection function of water at these two wavelengths.

For a sake of completeness, we present a model for the oceanic water absorption coefficient σ_{abs}, given by Morel and Maritorena (2001). Also, we present the experimental data of Preiur and Sathyendranath (1981) for the spectral pigment absorption coefficient $p(\lambda)$ normalized at the wavelength of 440 nm and the data of Pope and Fry (1997) for the pure water absorption coefficient $\alpha_w(\lambda)$ in a tabular form (see Table 5.2). Morel and Maritorena (2001) have introduced the following parameterization for σ_{abs}:

$$\sigma_{abs} = \alpha_w + C_p \alpha_p + C_y \alpha_y \tag{5.57}$$

where we used a slightly different notation. $C_p \alpha_p$ gives the product of pigments concentration and their absorption coefficient and $C_y \alpha_y$ is the product of the concentration of yellow substance and its absorption coefficient (Morel and Maritorena (2001)):

$$C_y \alpha_y(\lambda) = 0.06 p(\lambda) \tag{5.58}$$

$$C_p \alpha_p(\lambda) = 0.2(\alpha_w(\lambda) + 0.06\mathbb{C}^{0.65}) e^{-0.014(\lambda - \lambda_0)} \tag{5.59}$$

where λ is given in nm, $\lambda_0 = 440\,\text{nm}$, and $\mathbb{C}$ is the chlorophyll concentration (in $\text{mg}\,\text{m}^{-3}$). Note that $\mathbb{C}$ is usually in the range 0.03–3 $\text{mg}\,\text{m}^{-3}$. We show the values of

Table 5.2. Functions $\alpha_w(\lambda)$ (Pope and Fry, 1997) and $p(\lambda)$ (Prieur and Sathyendranath, 1981).

λ (nm)	α_w (m^{-1})	p	λ (nm)	α_w (m^{-1})	p
0.4000E+03	0.6630E−02	0.6870E+00	0.5550E+03	0.5960E−01	0.3210E+00
0.4050E+03	0.5300E−02	0.7810E+00	0.5600E+03	0.6190E−01	0.2940E+00
0.4100E+03	0.4730E−02	0.8280E+00	0.5650E+03	0.6420E−01	0.2730E+00
0.4150E+03	0.4440E−02	0.8830E+00	0.5700E+03	0.6950E−01	0.2760E+00
0.4200E+03	0.4540E−02	0.9130E+00	0.5750E+03	0.7720E−01	0.2680E+00
0.4250E+03	0.4780E−02	0.9390E+00	0.5800E+03	0.8960E−01	0.2910E+00
0.4300E+03	0.4950E−02	0.9730E+00	0.5850E+03	0.1100E+00	0.2740E+00
0.4350E+03	0.5300E−02	0.1001E+01	0.5900E+03	0.1351E+00	0.2820E+00
0.4400E+03	0.6350E−02	0.1000E+01	0.5950E+03	0.1672E+00	0.2490E+00
0.4450E+03	0.7510E−02	0.9710E+00	0.6000E+03	0.2224E+00	0.2360E+00
0.4500E+03	0.9220E−02	0.9440E+00	0.6050E+03	0.2577E+00	0.2790E+00
0.4550E+03	0.9620E−02	0.9280E+00	0.6100E+03	0.2644E+00	0.2520E+00
0.4600E+03	0.9790E−02	0.9170E+00	0.6150E+03	0.2678E+00	0.2680E+00
0.4650E+03	0.1011E−01	0.9020E+00	0.6200E+03	0.2755E+00	0.2760E+00
0.4700E+03	0.1060E−01	0.8700E+00	0.6250E+03	0.2834E+00	0.2990E+00
0.4750E+03	0.1140E−01	0.8390E+00	0.6300E+03	0.2916E+00	0.3170E+00
0.4800E+03	0.1270E−01	0.7980E+00	0.6350E+03	0.3012E+00	0.3330E+00
0.4850E+03	0.1360E−01	0.7730E+00	0.6400E+03	0.3108E+00	0.3340E+00
0.4900E+03	0.1500E−01	0.7500E+00	0.6450E+03	0.3250E+00	0.3260E+00
0.4950E+03	0.1730E−01	0.7170E+00	0.6500E+03	0.3400E+00	0.3560E+00
0.5000E+03	0.2040E−01	0.6680E+00	0.6550E+03	0.3710E+00	0.3890E+00
0.5050E+03	0.2560E−01	0.6450E+00	0.6600E+03	0.4100E+00	0.4410E+00
0.5100E+03	0.3250E−01	0.6180E+00	0.6650E+03	0.4290E+00	0.5340E+00
0.5150E+03	0.3960E−01	0.5820E+00	0.6700E+03	0.4390E+00	0.5950E+00
0.5200E+03	0.4090E−01	0.5280E+00	0.6750E+03	0.4480E+00	0.5440E+00
0.5250E+03	0.4170E−01	0.5040E+00	0.6800E+03	0.4650E+00	0.5020E+00
0.5300E+03	0.4340E−01	0.4740E+00	0.6850E+03	0.4860E+00	0.4200E+00
0.5350E+03	0.4520E−01	0.4440E+00	0.6900E+03	0.5160E+00	0.3290E+00
0.5400E+03	0.4740E−01	0.4160E+00	0.6950E+03	0.5590E+00	0.2620E+00
0.5450E+03	0.5110E−01	0.3840E+00	0.7000E+03	0.6240E+00	0.2150E+00
0.5500E+03	0.5650E−01	0.3570E+00			

σ_{abs}, found with this model in Figure 5.15 for various values of $\mathbb{C}$. Figure 5.16 shows the relative contribution of various components to the value of σ_{abs} with $\mathbb{C}$ equal to 0.03, 0.3, and 3.0 mg m^{-3}. Note that the contribution of yellow substance and pigments generally decreases with the wavelength and can be neglected for wavelengths larger than approximately 700 nm. The contribution of pigments peaks around the 430 nm wavelength. Note also the nonnegligible contribution of pigments in the spectral range 600–675 nm at $\mathbb{C} = 3.0$ mg m^{-3}.

The reflection of light from the ocean is comparatively low. Thus, atmospheric correction (Kaufman and Sendra, 1988) is the key issue of satellite oceanography. It should be pointed out that information on the distribution of chlorophyll in oceanic water is of a paramount value for fisheries.

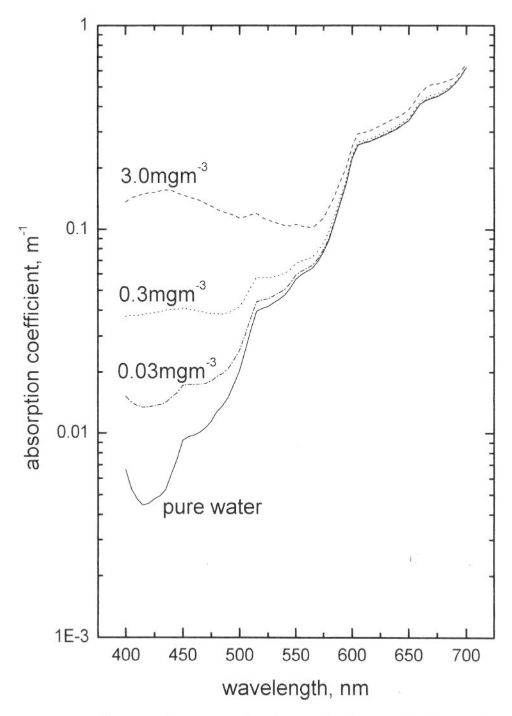

Figure 5.15. Oceanic water absorption coefficient (adapted from figure 13 of Morel and Maritorena, 2001) for various chlorophyll concentrations. The absorption spectrum of pure water (Pope and Fry, 1997) is also displayed.

Note that equation (5.53) can be applied only at low wind speed $\vec{v}$ over ocean. At large values of $\vec{v}$ we need to account for the reflection of light from the rough surface and whitecaps (Monahan and O'Muircheartaigh, 1986; Gordon and Wang, 1994). In this case the average reflectance $\langle r \rangle$ of the oceanic surface can be estimated using the following approximate equation (Mobley, 1994):

$$\langle r \rangle = \Xi r_f + (1 - \Xi) r_0 \tag{5.60}$$

where r_f and r_0 are average reflectances of foam and smooth oceanic surface. The fractional area $\Xi \in [0, 1]$ of the wind-blown ocean surface that is covered by foam depends on wind speed and air–sea temperature difference (Monahan and O'Muircheartaigh, 1986).

New methods in optical oceanography lie in the use not just of the intensity, but also the polarization (Mishchenko and Travis, 1999) of light fields in the ocean. Unpolarized light from the Sun becomes partially elliptically polarized in the underwater environment (Waterman, 1954). The characteristics of the polarization ellipse can be used for retrieval of the microphysical characteristics of oceanic water. This is a complex question, which is further complicated by the necessity to account for light scattering by chiral biological particles in water. These particles can rotate the

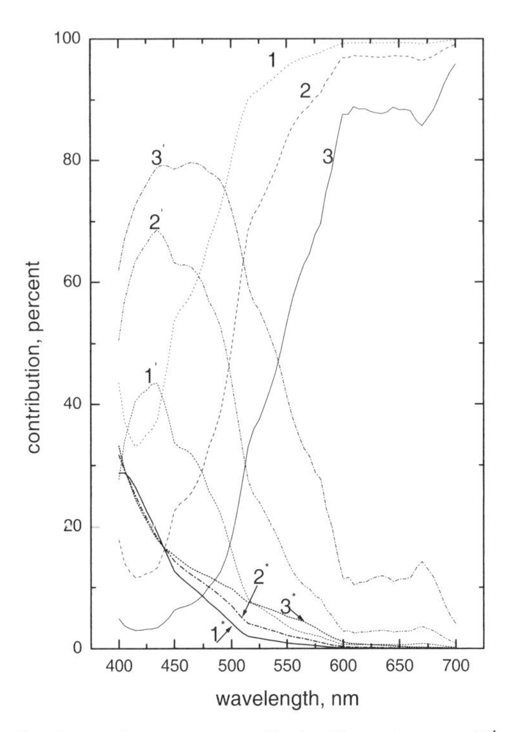

Figure 5.16. The contribution of pure water (1, 2, 3), priments (1′, 2′, 3′), and yellow substance (1*, 2*, 3*) to the total absorption coefficient of oceanic water according to the model of Morel and Maritorena (2001). Numbers 1, 2, and 3 correspond to different chlorophyll concentrations ($0.03\,\mathrm{mg\,m^{-3}}$, $0.3\,\mathrm{mg\,m^{-3}}$, and $3.0\,\mathrm{mg\,m^{-3}}$, respectively).

polarization plane of incident linearly polarized light. However, neither dispersion of the optical rotation nor circular dichroism of light in oceanic water has been studied so far. Some theoretical results in this area were obtained by Zege *et al.* (1987). Measurements of the Mueller matrices of oceanic waters were performed by Kadyshevich *et al.* (1976), Voss and Fry (1984), and others (Mobley, 1994).

Marine organisms use underwater polarization for orientation and navigation (Waterman, 1988). Underwater visibility can be improved using special polarization techniques (Gilbert and Pernicka, 1966).

5.1.4.2 Oceanic whitecaps

Spectral and radiative properties of whitecaps have attracted considerable attention in recent years. In particular, the spectral reflectance of whitecaps was measured in a broad spectral range from visible to IR (Frouin *et al.*, 1996; Moore *et al.*, 1998, 2000; Nicolas *et al.*, 2001). It was found that whitecaps may exert a cooling influence on the planet by increasing surface albedo (Frouin *et al.*, 2001). They influence aerosol and colour remote sensing from space (Koepke and Quenzel, 1981; Whitlock *et al.*, 1982; Koepke, 1984; Gordon, 1997; Husar *et al.*, 1997; Deuze *et al.*, 1999; Goloub *et al.*, 1999).

Theory did not follow the progress in experimental studies of optical properties of whitecaps, however. This was mostly due to the fact that a foam is a very unusual medium from the point of view of radiative transfer theory (Chandrasekhar, 1950), which is a standard tool for studies of light propagation in natural media with low concentrations of scatterers. Indeed, the volumetric concentration of bubbles in whitecaps is larger than 70% in most of cases. The radiative transfer equation (Chandrasekhar, 1950), however, is usually applied to the case of media with low concentrations of scatterers (smaller than 1% by volume). The main aim of this section is to propose a simple physically based parameterization for the spectral reflectance of whitecaps, which can be used, in particular, in ocean satellite retrieval techniques. We demonstrate that the parameterization developed is consistent with the available measurements of spectral reflectance of whitecaps.

The spherical albedo (van de Hulst, 1980) is defined as the integral of the azimuthally averaged reflection function with respect to the incidence and observation angles. By definition, it is equal to 1 for semi-infinite nonabsorbing media.

Let us find a simple analytical equation for the spectral spherical albedo of whitecaps, which can be used in oceanic sections of global circulation models and also in satellite aerosol and ocean remote sensing. We assume that whitecaps can be presented as a semi-infinite medium. However, this assumption is not crucial. Indeed, the theory of light reflection by a semi-infinite medium can be used (after simple modifications to the main equations) for the case of finite but optically thick layers (van de Hulst, 1980; Zege *et al.*, 1991a,b; Kokhanovsky, 2001). This issue, however, is out of the scope of this section. We start from the representation of the spherical albedo r as the Maclauren series with respect to the single scattering albedo ω_0:

$$r = \sum_{i=1}^{\infty} a_n \omega_0^n \tag{5.61}$$

We have for nonabsorbing media: $r = \omega_0 = 1$ and, therefore:

$$\sum_{i=1}^{\infty} a_n = 1 \tag{5.62}$$

Equation (5.61) is an exact formula. Clearly, coefficients a_n give us the contribution of n – scattering into the value of albedo r. Whitecaps are weakly absorbing media (ω_0 is close to 1). This means that one should account for a great number of terms in equation (5.61). To avoid this difficulty we introduce the probability of photon absorption $\beta = 1 - \omega_0$, which is a small number for whitecaps. Then it follows from equation (5.61):

$$r = \sum_{i=1}^{\infty} a_n (1 - \beta)^n \tag{5.63}$$

or

$$r = \sum_{i=1}^{\infty} a_n (1 - n\beta + 0.5n(n-1)\beta^2 - \ldots) \tag{5.64}$$

This can be rewritten as $r = 1 - \langle n \rangle \beta + 0.5\langle n(n-1)\rangle\beta^2 - \ldots$. Here $\langle n \rangle = \sum_{i=1}^{\infty} n a_n$, $\langle n(n-1)\rangle = \sum_{i=1}^{\infty} n(n-1)a_n$, etc. Whitecaps belong to a broad class of strongly multiply scattering media. It means that $\langle n \rangle$ is a large number and $\langle n(n-1)\rangle \approx \langle n^2 \rangle$, $\langle n(n-1)(n-2)\rangle \approx \langle n^3 \rangle$, etc. So we have approximately: $r = 1 - \langle n \rangle \beta + 0.5\langle n^2 \rangle \beta^2 - \ldots r = \langle \exp(-\beta n)\rangle \approx \exp(-\beta n_0)$, where we used the mean value theorem. The constant n_0 is generally not known. We find n_0 using the fact that $r \approx 1 - 4\sqrt{\beta/3(1-g)}$ for small β (van de Hulst, 1980). This means: $n_0 = 4/\sqrt{3\beta(1-g)}$, where g is the asymmetry parameter, which describes the asymmetry of angular light distribution in the single scattering event (van de Hulst, 1980).

Therefore, it follows:

$$r = \exp\left(-4\sqrt{\frac{\beta}{3(1-g)}}\right) \tag{5.65}$$

or

$$r = \exp\left(-4\sqrt{\frac{l_{tr}}{3l_{abs}}}\right) \tag{5.66}$$

where we introduced the transport photon path length $l_{tr} = l/1 - g$ and the absorption path length $l_{abs} = 1/\beta$. Here l is the average distance between scattering events, which is given by the inverse value of the total scattering coefficient in a medium. We assume that extinction and scattering path lengths are almost equal in whitecaps due to small light absorption by water in the visible and near-IR regions of the electromagnetic spectrum.

The next step is to relate the parameters in equation (5.66) to the microphysical characteristics of foams (e.g. the average diameters of bubbles d and the liquid fraction c). For this we will use an empirically established relation $l_{tr} = d/\sqrt{c}$ (Vera *et al.*, 2001), and also the fact that the absorption coefficient of weakly absorbing media (e.g. whitecaps) $\sigma_{abs} = 1/l_{abs}$ is proportional to the concentration of liquid and its bulk absorption coefficient α. Therefore, it follows: $l_{abs} = 1/(A\alpha c)$, where A is a generally unknown (but spectrally neutral for all practical purposes) constant. This constant depends only on the real part $\mathrm{Re}(m)$ of the refractive index of liquid in foam. Clearly, it follows at values of $\mathrm{Re}(m)$ close to that of air: $A = 1$. Although the value of A for water in the visible spectrum differs from that of air by more than 30%, we can neglect the difference of A from 1 as a first rough approximation. The exact value of this constant should be established using laboratory experiments.

It follows from equation (5.66) and taking into account considerations given above:

$$r = \exp(-Bc^{0.25}\sqrt{\alpha d}) \tag{5.67}$$

where $B = 4\sqrt{A/3} \approx 2.3$ at $A = 1$.

Equation (5.67) can be written in a slightly different form for dry foams. Then the value of d is inversely proportional to the concentration of liquid c. Namely, we have: $d = H(\delta/c)$, where δ is the average film thickness and the proportionality constant H is in the range 1.35–1.64 (Kruglyakov and Ekserova, 1990), depending on the form of the foam cell. This allows us to represent equation (5.67) in a more transparent form:

$$r = \exp(-G\sqrt{cp}) \tag{5.68}$$

where $G = B\sqrt{H}$ ($\approx$3.5 at $A = 1$, $H = 1.5$) and $p = \alpha\delta$. Note that $\exp(-p)$ gives the light attenuation on its passage though a single film.

We would like to underline that equation (5.68) can be used to find the plane albedo $r_p(\theta_0)$ and the reflection function $R(\theta_0, \theta, \psi)$ of whitecaps, using the following results valid for weakly absorbing strongly scattering media (Zege *et al.*, 1991): $r_p = r^p$ and $R = R_0 r^q$, where we omitted arguments for simplicity. Note that we have (Zege *et al.*, 1991): $p(\theta_0) = 3(1 + 2\cos(\theta_0))/7$ and $q = (p(\theta_0)p(\theta))/R_0$. The value of $R_0(\theta_0, \theta, \psi)$ is the reflection function of a nonabsorbing foam. Angles θ_0, θ, and ψ are the illumination angle, the observation angle, and the relative azimuth, respectively (van de Hulst, 1980). The theory is also easily generalized to account for a finite thickness L of whitecaps providing that $L/l \gg 1$.

Due to a number of assumptions involved in the calculations, a thorough study of the validity of our technique is clearly needed. For this, however, a controlled laboratory experiment should be performed. Meanwhile, we can check the consistency of equation (5.68) with the available spectral measurements of foam spectral reflectance (Frouin *et al.*, 1996, 2001; Moore *et al.*, 1998, 2000; Nicolas *et al.*, 2001). In particular, Frouin *et al.* (2001) found that there is a considerable dependence of the foam reflectance on the wavelength λ. They introduced the spectral factor $f(\lambda)$ such that the product $r_f f(\lambda)$ gives the spectral albedo of whitecaps. Here r_f is the so-called effective albedo of whitecaps. In particular, Koepke (1984) gives: $r_f = 0.22$ with a standard deviation of 0.11. Frouin *et al.* (2001) used the following parameterization for the spectral factor at wavelengths larger than 0.6 µm: $f = \exp(-v(\lambda - 0.6)^w)$, where $v = 1.75$ with a standard deviation of 0.48 and $w = 0.99$ with a standard deviation 0.05. It was assumed that $f = 1$ for smaller wavelengths. We believe that equation (5.67) allows for a more correct description of the spectral factor. Namely, we have: $f = \exp(-\sqrt{\alpha s})$, which accounts for the water spectral absorption. Here the constant $s = B^2 d\sqrt{c}$ depends on the type of foam.

Experimental measurements of the foam spectral reflectance were performed by Whitlock *et al.* (1982). In particular, they found that the measured spectral reflecton function can be approximated by the following polynomial at wavelengths larger than 0.8 µm (see Figure 5.17):

$$R = \sum_{s=0}^{4} (-1)^s \xi_s (\ln \alpha_w)^s \tag{5.69}$$

where $\xi_0 = 60.063$, $\xi_1 = 5.127$, $\xi_2 = 2.799$, $\xi_3 = 0.713$, and $\xi_4 = 0.044$. Note that the bulk water absorption coefficient $\alpha_w = (4\pi\kappa)/\lambda$ is expressed in m^{-1}. Here κ is the imaginary part of the refractive index m of water.

Equation (5.69) has the correlation coefficient 0.979 and standard error equal to

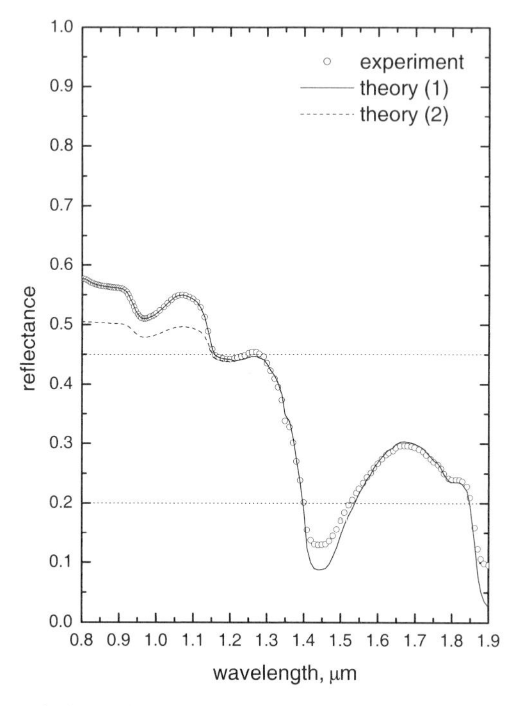

Figure 5.17. The spectral dependence of the foam reflectance coefficient calculated using equation (5.23) (from Whitlock *et al.*, 1982) and equation (5.25) with $b = 1.72$ mm (theory). Theoretical results differ due to a different choice of the absorption coefficient of additives, which was not specified in the experiment.

4.36 at wavelengths larger than 0.8 μm compared with actual measurements. It was found that foam is almost spectrally neutral for smaller wavelengths.

Whitlock *et al.* (1982) stated that the polynomial regression equations usually give few clues to actual physical relationships of the parameters being correlated.

Let us compare now the physically based parameterization (equation 5.68) with experimental data. For this we rewrite equation (5.68) in the following form:

$$r = \exp(-\sqrt{\alpha s}) \tag{5.70}$$

where s is defined above. Whitlock *et al.* (1982) actually measured not the spherical albedo but the reflection function R. Taking into consideration the development given above, we establish the following simple formula for the reflection function of whitecaps:

$$R = R_0 \exp(-\sqrt{\alpha b}) \tag{5.71}$$

where $b = sq^2$. We account for the fact that the absorption coefficient in the experiment was close to that of water for large wavelengths λ. This is due to an appreciable absorption by water at wavelengths larger than approximately 1.2 μm (see Figure

5.17). For shorter wavelengths, light absorption by dissolved substances can play a role. Therefore, we used: $\alpha = \alpha_w + c^*\alpha_a$, where α_w is the absorption coefficient of pure water and α_a is the absorption coefficient of additives (e.g. detergents), which presented in water during the experiment. The value of c^* gives the relative volumetric concentration of the additives.

We found that equation (5.71) fits experimental measurements pretty well (see Figure 5.17), which confirms the general applicability of our derivations and assumptions. The dashed line corresponds to calculations using equation (5.71) at $R_0 = 1$ and spectrally neural value of $c^*\alpha_a$, which was set equal to $0.000\,27\,\mu\text{m}^{-1}$. The value of α_w was taken from Segelstein (1981). We see that the assumption of an additional spectrally neutral absorber does not work for wavelengths smaller than $1.2\,\mu\text{m}$. Whitlock *et al.* (1982) did not specify the value of $c^*\alpha_a$ for their experiment. So we obtained the function $c^*\alpha_a(\lambda)$ at λ smaller than $1.2\,\mu\text{m}$ from experimental data itself, using equation (5.71). This is shown in Figure 5.17 by a solid line. Note that solid and dashed lines coincide at $\lambda \geq 1.2\,\mu\text{m}$.

Overall, equation (5.71) can be used for the interpretation of the experimental reflectance given in Figure 5.17 in the range $R = 0.2$–0.45 (without account of spectrally varying additional absorbers). For larger values of R, a spectral variation of additives should be taken into account. Equation (5.71) has larger errors for smaller values of R (see Figure 5.17) due to the violation of a weak absorption assumption in this case. Then β is not close to zero and, correspondingly, equation (5.71) is not valid.

The value of b in equation (5.71) was taken to be equal to 1.72 mm. This number was found using a fitting procedure. It is of importance to estimate values of R_0 and b not by applying fitting techniques but from the physics of the problem. Unfortunately, Whitlock *et al.* (1982) did not give all the details of their experiment, so we can make only a rough estimate of these parameters, which may be biased because of the unknown density of air bubbles and microstructure of the foam used in the experiment.

The value of R_0 coincides with the reflection function of a semi-infinite non-absorbing medium. We can conclude from data given by Whitlock *et al.* (1982) that the measurements are performed at near-nadir observation and illumination conditions. Then R_0 should be close to 1 (Kokhanovsky, 2001) as used in Figure 5.17. The value of b is given by the product $B^2q^2\sqrt{c}d$. We obtain the following estimates for multipliers: $q \approx 1.65$ (for nadir illumination and observation conditions) and $B \approx 2.3$ as specified above. Taking into account that $b \approx 1.72\,\text{mm}$ for Figure 5.17, we obtain: $d\sqrt{c} = 0.12\,\text{mm}$, which is a reasonable estimate (e.g. $d = 1.2\,\text{mm}$ and $c = 0.01$).

Therefore, the foam spectral reflectance is governed by the function: $\exp(-\sqrt{b\alpha(\lambda)})$ where $\alpha(\lambda)$ is the spectral reflectance of liquid in foam and b is the spectrally neutral constant.

Let us consider now the results of measurements of sea foam spectral reflectance performed at the Scripps Institute of Oceanography Pier, La Jolla, California. Frouin *et al.* (1996) measured the spectral reflectance of sea foam by viewing the sea surface radiometrically in a region of breaking waves near the pier. They found that the foam reflectance decreases substantially with the wavelength in the near-IR, contrary to the findings of previous studies (both theoretical and experimental). In particular, it follows from Figure 5.17 that the foam reflectance does not change considerably at wavelengths smaller than 1.2 μm (Whitlock *et al.*, 1981). This was not confirmed by Frouin *et al.* (1996) for measurements *in situ*.

This discrepancy can be understood using our simple model described above. It follows from equation (5.71) that the spectral reflectance of whitecaps depends strongly on the spectral absorption coefficient α of water. We underline that values of the natural water absorption coefficient can vary considerably due to the presence of dissolved and particulate matter (Bricaud *et al.*, 1998). This causes a variation of the foam spectral reflectance across the ocean. Clearly, filtered-deionized tap water used by Whitlock *et al.* (1982) differs considerably from surf water in the experiment of Frouin *et al.* (1996) (e.g. in terms of their spectral absorption coefficient). This is the main reason behind differences obtained in these two experiments.

Our theory can be applied to the experiment of Frouin *et al.* (1996) as well. For this we need to know the value of the water absorption coefficient. This unfortunately was not measured during the experiment, so we used the parameterization of oceanic water absorption coefficient, proposed by Morel and Maritorena (2001) (see above) to describe the experiment of Frouin *et al.* (1996).

The comparison of experimental data (Frouin *et al.*, 1996) and the theory for the normalized foam reflectance $R_n(\lambda) = R(\lambda)/R(0.44\,\mu\text{m})$ at the visible range of the electromagnetic spectrum is shown in Figure 5.18. We see that equation (5.71) allows us to explain the foam spectral selectivity in the visible spectrum. Note that the value of $b = 17.1$ mm used in Figure 5.18 was found using the fitting procedure. This is approximately 10 times larger than in the experiment of Whitlock *et al.* (1982) and can be attributed to larger sizes of bubbles near the pier compared with the laboratory experiment of Whitlock *et al.* (1982).

We also compared the measurements of Frouin *et al.* (1996) with results of our theoretical model in the IR region of the electromagnetic spectrum (see Figure 5.19), using the same value of b as specified above. The pure water absorption coefficient was taken from data of Segelstein (1981). The influence of additional absorbers like yellow substance and pigments was neglected at wavelengths larger than 0.7 μm. The experiment of Frouin *et al.* (1996) showed that the foam spectral reflectance typically reduced by 40% at the 0.85 μm wavelength, 50% at 1.02 μm, and 85% at 1.65 μm. We also found a 50% reduction at 1.02 μm (see Figure 5.18). However, the theory gives only a 25% reduction at 0.85 μm. It is 95% at 1.65 μm. The last number could be in error because our approximation has a poor accuracy in the region of large absorption. The discrepancy at 0.85 μm could be due to various reasons. One of them is the possible presence of an additional absorber during the experiment, which is always the case near a pier. This is supported by data obtained by Moore *et al.* (2000) who found that the value of the normalized reflectance is only 0.8 at the wavelength 0.86 μm in the open ocean for wind speeds of 9–12 $\text{m}\,\text{s}^{-1}$. This closely corresponds to the theoretical results given in Figure 5.18 at this wavelength. Moore *et al.* (2000) suggest, however, another reason for the discrepancy, namely the less

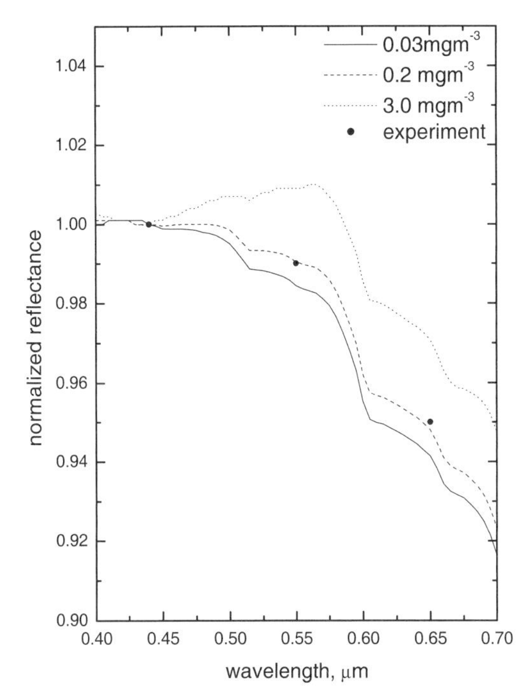

Figure 5.18. Normalized spectral reflectance of foam in the surf zone calculated using equation (5.25) at $b = 17.1$ mm with σ_{abs} given by equation (5.57) for different chlorophyll contents. Experimental data are taken from Frouin *et al.* (1996).

violent wave breaking in the open ocean compared with the surf zone. This leads to the decrease of d, and therefore, b, which in turn enhances R (see equation (5.71)).

Nicolas *et al.* (2001) found that the normalized reflectance at 0.855 μm is close to 0.65 with a standard deviation 0.15 for 15 cases measured in the open ocean. This contradicts, at first glance, data of Moore *et al.* (2000). However, note that Nicolas *et al.* (2001) have used the push-boom radiometer with the spectral bandwidth 0.19 μm (see Figure 5.19) at the central wavelength 0.855 μm. Moore *et al.* (2000) used a 6-channel radiometer with the nominal 0.01 μm spectral width. Due to a high variation of the normalized reflectance (0.4–0.8) in the range 0.76–0.95 μm (see Figure 5.19), the result should be highly sensitive to the radiometer spectral response function and the bandwidth in this particular case. This explains the differences obtained.

For the sake of completeness, we would like to mention the experiment of Moore *et al.* (1998), which was performed for ship-induced foam. The normalized reflectance obtained from this experiment and equation (5.71) using the values of α reported in the paper of Moore *et al.* (1998) (0.11, 0.1, 0.065, 0.08, 0.475, and 4.3 m^{-1} at wavelengths 0.41, 0.44, 0.51, 0.55, 0.67, and 0.865 μm, respectively) are shown in Figure 5.20. The value of b was taken to be equal to 85.5 mm to fit the experimental

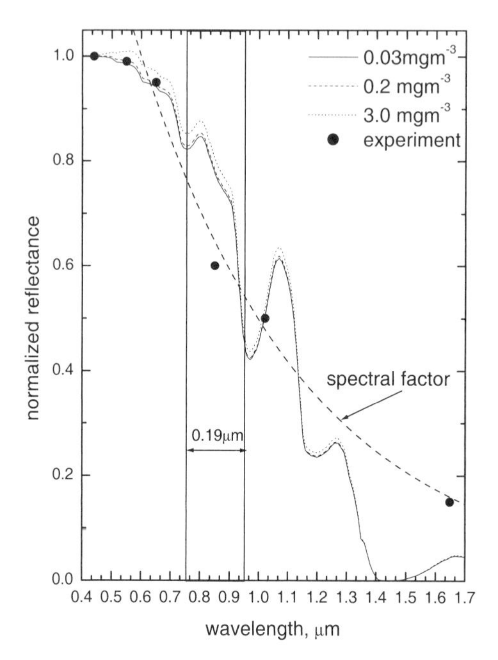

Figure 5.19. The same as in Figure 5.18 but for the broader spectral range. Also the spectral factor f (Frouin *et al.*, 2001) is shown. Vertical lines show the bandwidth (0.19 μm) of channel 3 of the radiometer used by Nicolas *et al.* (2001) for measurements performed in the open ocean.

data. Again we see that equation (5.71) can be used as an accurate base for the fitting procedure, which explains the foam spectral dependence.

5.2 IMAGE TRANSFER

From the point of view of vision theory a light-scattering medium is a high frequency filter. The transmission of high frequency signals (in time and space domains) is reduced considerably in comparison with low frequency signals as a result of light scattering. In this section we will consider the main concepts and ideas of the image transfer theory.

Any diffused source of light can be considered as a superposition of point light sources. Thus, in linear optical systems the image of such an object with intensity $I_0(\vec{r}')$ is a linear superposition of images of point sources:

$$I(\vec{r}) = \int_{-\infty}^{\infty}\int_{-\infty}^{\infty} d\vec{r}'\, I_0(\vec{r}')S(\vec{r}',\vec{r}) \tag{5.72}$$

where the point spread function (PSF) $S(\vec{r}',\vec{r})$ describes the process of transformation of the object intensity $I_0(\vec{r}')$ in the initial plane to the image intensity $I(\vec{r})$ in the

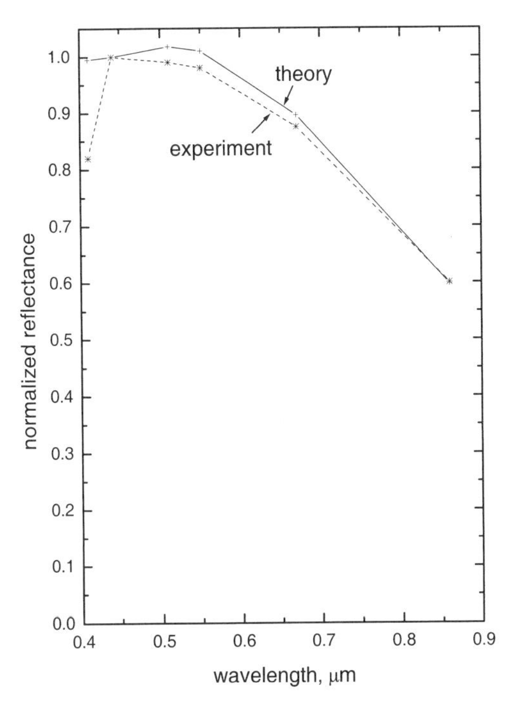

Figure 5.20. Foam normalized reflectance calculated from equation (5.71), using values of σ_{abs} given by Moore *et al.* (1998) and $b = 85.5$ mm (theory). Experimental data are taken from figure 10 of Moore *et al.* (1998).

image plane. The PSF is the main notion of the image transfer theory. Equation (5.72) has a simpler form in a frequency domain:

$$I(\vec{v}) = S(\vec{v}) I_0(\vec{v}) \tag{5.73}$$

where

$$I(\vec{v}) = \int_{-\infty}^{\infty} \int_{-\infty}^{\infty} I(\vec{r}')\, e^{-i\vec{v}\vec{r}'}\, d\vec{r}' \tag{5.74}$$

$$I_0(\vec{v}) = \int_{-\infty}^{\infty} \int_{-\infty}^{\infty} I_0(\vec{r}')\, e^{-i\vec{v}\vec{r}'}\, d\vec{r}' \tag{5.75}$$

$$S(\vec{v}) = \int_{-\infty}^{\infty} \int_{-\infty}^{\infty} S(\vec{r}')\, e^{-i\vec{v}\vec{r}'}\, d\vec{r}' \tag{5.76}$$

and $\vec{v}$ is the space frequency. One of the main problems of the image transfer theory is to determine the value of the Fourier transform of the PSF, namely the optical transfer function (OTF) $S(\vec{v})$. After that the value of $I(\vec{v})$ can be determined from equation (5.73) and it follows from equation (5.74), for light intensity in the image plane, that:

$$I(\vec{r}') = \frac{1}{4\pi^2} \int_{-\infty}^{\infty} \int_{-\infty}^{\infty} I(\vec{v})\, e^{i\vec{v}\vec{r}}\, d\vec{v} \tag{5.77}$$

The OTF depends on the properties of media between an object and an image (Wells, 1969; Volnistova and Drofa, 1986; Zege *et al.*, 1991). Different methods of calculation of this function and the modulation transfer function (MTF) $T(\vec{v}) = |S(\vec{v})|/|S(0)|$ are described by Ishimaru (1978) and Zege *et al.* (1991).

Here we will only consider the case of image transfer through a light-scattering layer with large particles at normal incidence of a light beam. In this case the OTF is a real function. It does not depend on the azimuth for randomly oriented particles.

It follows from equation (3.112) that:

$$S(v) = \exp(-\tau_0 + \omega_0 \tau_0 \eta(v^*)) \tag{5.78}$$

where $v^* = vL$ is the dimensionless frequency, L is geometrical thickness of a layer, $v = |\vec{v}|$ and

$$\eta(v^*) = \int_0^1 p(v^* y)\, dy, \qquad p(v^* y) = \frac{1}{2}\int_0^\infty p(\theta) J_0(\theta v^* y)\theta\, d\theta \tag{5.79}$$

We can see that the OTF can be obtained from equations (5.78), (5.79) if we have information about the optical thickness of a medium (≤ 7 in approximations (5.78), (5.79)), single scattering albedo ω_0, and phase function $p(\theta)$.

For media with complex microstructure (e.g. oceanic waters), the crux of the difficulty is that the phase function is not known a priori. In this case different models to describe this function can be used. For instance, if we suppose that the phase function is described by the Gaussian:

$$p(\theta) = \frac{2}{b^2} e^{-\theta^2/2b^2} \tag{5.80}$$

where b is constant, θ is the scattering angle, it follows that:

$$p(v^* y) = e^{-(b^2 (v^* y)^2)/2} \tag{5.81}$$

and

$$S(v) = \exp\left(-\tau + \frac{\omega_0 \tau}{b v^*}\sqrt{\frac{\pi}{2}}\,\mathrm{erf}\left(\frac{b v^*}{\sqrt{2}}\right)\right) \tag{5.82}$$

where $\mathrm{erf}(bv^*/\sqrt{2})$ is the probability integral:

$$\mathrm{erf}(x) = \frac{1}{\sqrt{2\pi}} \int_{-x}^{x} e^{-t^2/2}\, dt \tag{5.83}$$

We can obtain at large values of frequency v^*:

$$\mathrm{erf}\left(\frac{b v^*}{\sqrt{2}}\right) = 1, \qquad S(v^*) = \exp(-\tau_0) \tag{5.84}$$

Thus, the OTF is determined by the nonscattered light alone at $v^* \gg 1$. This conclusion follows directly from equations (5.78), (5.79) as well.

The number of nonscattered photons is low at large optical thickness. Thus, the value of the OTF is small and the contrast of the image is low. It follows at small values of frequency υ^* that:

$$\operatorname{erf}\left(\frac{b\upsilon^*}{\sqrt{2}}\right) = b\upsilon^*\sqrt{\frac{2}{\pi}}\left(1 - \frac{(b\upsilon^*)^2}{6}\right) \tag{5.85}$$

and

$$S(\upsilon^*) = \exp\left\{-\tau_0\left[1 - \omega_0\left(1 - \frac{(b\upsilon^*)^2}{6}\right)\right]\right\} \tag{5.86}$$

The modulation transfer function in this case is:

$$T(\upsilon) = e^{-((b\upsilon^*)^2/6)\omega_0\tau_0} \tag{5.87}$$

Note that the value of b is inversely proportional to the average size of particles (see equation (5.80)). Thus, the loss of contrast is higher for thicker media with larger single-scattering albedos and smaller particles.

Analytical formulae for the OTF result from approximations for local optical characteristics, which are presented in Chapter 2. For instance, the phase function of large particles can be calculated in the framework of ray optics. Therefore, for the value of the phase function we obtain:

$$p(\theta) = \frac{4\pi\sigma_{sca}(\theta)}{\sigma_{sca}}$$

where the volume scattering coefficient is given by the following equation:

$$\sigma_{sca}(\theta) = N\pi\sum_{i=1}^{3}\int_0^\infty \left\{\sigma_{sca}^{(i)}\right\} f(a)\,da$$

and $\sigma_{sca}^{(1)} = (4J_1^2(\theta x)/\theta^2)a^2$, $\sigma_{sca}^{(2)} = R(\theta)a^2$, $\sigma_{sca}^{(3)} = T(\theta)\exp(-c\Delta(\theta))a^2$. Here $c = 4\chi x$, $x = 2\pi a/\lambda$, σ_{sca} is the scattering coefficient (see Chapter 2), and N is the number concentration of particles. The contribution of internally reflected rays was neglected. The formulae for $R(\theta)$, $T(\theta)$, $\Delta(\theta)$ are presented in Chapter 2. They can be approximated by the following equations as $\theta \to 0$:

$$R(\theta) = e^{-\alpha\theta}, \qquad T(\theta) = T(0)\,e^{-\beta\theta^2 - c}$$

and $\Delta(\theta) = 1 - \gamma\theta^2$, where

$$T(0) = \left(\frac{2n}{n+1}\right)^4 (n-1)^{-2}, \qquad \gamma = 1 - \frac{2n-1}{4(n-1)}$$

Note that it follows for water clouds ($n = 1.33$) and dust particles ($n = 1.53$) that $\alpha = 3$, $\beta = 4.7$ and $\alpha = \beta = 2.4$, respectively, with error $< 15\%$ at $\theta < 35^\circ$. These formulae allow us to find the Fourier–Bessel spectrum of the phase function analytically. For instance, it follows for the diffraction part (see Table 2.5) that:

$$p_1(\upsilon^*) = \begin{cases} 2(\arccos\xi - \xi\sqrt{1-\xi^2})/\pi, & \xi \le 1 \\ 0, & \xi > 1 \end{cases}$$

and (see equation (5.79)):

$$\eta_1 = \frac{4}{3\pi b} - \left\{\frac{2}{\pi}\arccos b + \frac{2(2+b^2)\sqrt{1-b^2}}{3\pi b}\right\} U_+(1-b)$$

where $\xi = v^* y/2x$, $b = v^*/2x$. The result for the OTF of polydispersed media is:

$$S(v^*) = \exp(-\tau + \eta(v^*)L) \tag{5.88}$$

where L is the geometrical thickness of a layer, $v^* = vL$, and

$$\eta(v^*) = \sum_{i=1}^{3} \eta_i(v^*) \tag{5.89}$$

$$\eta_i(v^*) = N \int_0^\infty \pi a^2 f(a) \eta_i(v^*, a)\, da \tag{5.90}$$

$$\eta_1(v^*, a) = \left\{\frac{2}{\pi}\arccos\left(\frac{v^*}{2x}\right) - \frac{4x\left(2+\left(\frac{v^*}{2x}\right)^2\right)\left(1-\left(\frac{v^*}{2x}\right)^2\right)^{1/2}}{3\pi v^*}\right\} \times U_+\left(1-\left(\frac{v^*}{2x}\right)\right) + \frac{8x}{3\pi v^*} \tag{5.91}$$

$$\eta_2(v^*, a) = \frac{1}{2\alpha^2\left\{1+\left(\frac{v^*}{\alpha}\right)^2\right\}^{1/2}}$$

$$\eta_3(v^*, a) = \frac{T(0)\,e^{-c}}{4v^*}\sqrt{\frac{\pi}{\beta}}\,\mathrm{erf}\left(\frac{v^*}{2\sqrt{\beta}}\right)$$

Here,

$$U_+(x) = \begin{cases} 1, & x \geq 0 \\ 0, & x < 0 \end{cases}, \qquad \mathrm{erf}(x) = \frac{2}{\sqrt{\pi}}\int_0^x e^{-c^2}\, dc$$

is the probability integral. Functions $\eta_1(v^*)$, $\eta_2(v^*)$, and $\eta_3(v^*)$ describe the influence of diffraction, reflection, and transmission processes, respectively, on the loss of contrast of the image. We can see that functions $\eta_i(v^*)$ decrease with angular frequency and so does the OTF.

At large absorption ($c \to \infty$), function $\eta_3(v^*)$ vanishes and the OTF decreases. For nonabsorbing medium the components $\eta_2(v^*)$ and $\eta_3(v^*)$ do not depend on the size of particles and function $\eta_1(v^*)$ increases with size of particle. Note that function $\eta_1(v^*, a)$ can be rewritten as follows (the error will be $<2\%$):

$$\eta_1(v^*, a) = 1 - \frac{v^*}{\pi}\left[1 - \left(\frac{v^*}{24x}\right)^2\right], \qquad v^* < 2x \tag{5.92}$$

$$\eta_1(v^*, a) = \frac{8x}{3\pi v^*}, \qquad v^* \geq 2x \tag{5.93}$$

Thus, we can obtain as $v^* \to \infty$:

$$\eta_1(v^*) = \frac{8k}{3v^*} NM_3, \qquad M_3 = \int_0^\infty a^3 f(a)\, da$$

where $k = 2\pi/\lambda$. The increase in size of particles in a disperse medium (e.g. due to condensation processes) causes contrast in the image to rise.

Note that the OTF at $v^* = 0$ gives the transmission coefficient of a layer illuminated by a normally incident plane wave $T(\mu_0 = 1)$:

$$T(\mu_0 = 1) = e^{-\tau(1-\omega_0^*)}$$

where

$$\omega_0^* = \frac{1}{2}\left\{1 + \frac{1}{2\alpha^2} + \frac{T(0)}{4\beta}\varphi\right\}, \qquad \varphi = \frac{\int_0^\infty a^2 e^{-ca} f(a)\, da}{\int_0^\infty a^2 f(a)\, da}$$

Let us now consider a scattering medium with a gamma PSD. It follows from equations (5.89)–(5.93) that:

$$\eta(v^*) = \left\{1 + \frac{8x_{ef}}{3\pi v^*} P(\mu+4, \Delta) - P(\mu+3, \Delta) - \frac{2\Delta}{\pi(\mu+2)}[1 - P + 2, \Delta)] \right.$$
$$+ \frac{\Delta^3}{6\pi\mu(\mu+1)(\mu+2}[1 - P(\mu, \Delta)] + \frac{1}{2\alpha^2[1 + (v^*/\alpha)^2]^{1/2}}$$
$$\left. + \frac{T(0)\gamma}{4v^*}(\pi/\beta)^{1/2} \operatorname{erf}(v^*/2\sqrt{\beta})\right\}\Sigma \tag{5.94}$$

where $\Sigma = N\pi a_0^2(\mu+1)(\mu+2)\mu^2$, $\Delta = \mu\omega^*/2x_0$, $\gamma = (1 + 4\chi x_0/\mu)^{-(\mu+3)}$, $x_{ef} = x_0(1 + 3/\mu)$, $x_0 = ka_0$, a_0 and μ are parameters of the gamma PSD, and $P(n, y)$ is the incomplete gamma function.

The proposed solution simplifies the study of the effect of medium microstructure on image degradation during image transfer through fogs and clouds. Formula (5.94) for the gamma PSD is especially simple when μ is an integer, because in this case the incomplete gamma function $P(n, y)$ is determined by the following equation:

$$P(n, y) = 1 - \exp(-y) \sum_{j=0}^{n-1} \frac{y^j}{j!}$$

Comparisons between calculations of the OTF using equation (5.94) and Monte-Carlo calculations (Drofa and Usachev, 1980) were performed by Zege and Kokhanovsky (1994). It was found that the accuracy of approximate equations (5.36), (5.94) for water clouds with effective radius 6 µm and $\tau = 1$ is better than 5%. The accuracy decreases with optical thickness τ, which should be less than 5–7 for application of equations (5.88), (5.94).

5.3 REMOTE SENSING AND INVERSE PROBLEMS

5.3.1 Introduction

All problems of light-scattering media optics can be classified into two broad sections. The first section presents what is called the direct problem. In this case we know the microstructure of a medium (particle size distributions, shapes and chemical composition of particles, the concentrations) and determine the intensity or components of the Stokes vector of scattered light. So far, in this book only this problem has been considered. However, in the majority of cases the microstructure of a medium is not known a priori.

The problem of the determination of the microstructure of light-scattering media from measurements of characteristics of light fields is called the inverse problem (IP). This problem is more important and more complex than the direct problem (Hirleman and Bohren, 1991; Shifrin and Tonna, 1993). Moreover, the IP cannot be solved in principle in some cases. For instance, if we measure the spectral transmission of direct light through a scattering medium with large particles ($a \gg \lambda$), we are unable to find either particle size distribution or the refractive index of particles, because the extinction coefficient of such media depends only on the effective radius and concentration of particles. There are two other major problems of the IP solution, namely uniqueness and stability. The solution can be very sensitive to small measurement errors in initial data, and this may produce physically absurd results.

Thus, we can see how different direct and inverse problems are. The methods for solving them are fundamentally different as well. The investigator, solving the IP, needs to evaluate in advance the information content in a given experimental setup in order to warrant that the measured data contain sufficient information about the quantity to be retrieved.

In this section, we will consider some methods for the IP solution (mostly for multiply light-scattering media). The recent review of methods for IP solution in single scattering experiments was presented by Shifrin and Tonna (1993). Jones (1999) discussed different experimental set-ups used for the solution of inverse problems.

There are two main methods for the solution of the IP for multiply light-scattering media. The first one is called the fitting method (FM). Within the framework of this method we calculate the radiative characteristics of a light-scattering medium for a number of particle size distributions, concentrations, shapes, and refractive indices of particles. Selection of the correct particle size distribution is made by comparing experimental optical data with the corresponding theoretical results computed within the considered family of microstructure parameters. This method was successfully applied to many problems, including the determination of the effective radius of cloud droplets and the optical thickness of clouds, effective radius of particles in Venus's atmosphere, etc. (Hansen and Hovenier, 1974; Nakajima and King, 1990; King *et al.*, 1992).

The second method involves approximate analytical solutions of the RTE. This technique, which we will call the analytical method (AM), has some advantages and disadvantages compared with the FM. The main disadvantage is that the AM is not universal and can be applied only to a restricted number of problems. The advantage of the AM lies in its relative simplicity. In the following sections we will consider application of the AM to the solution of some inverse problems for light-scattering media with large particles.

5.3.2 Method of small angles

Let us consider a plane-parallel light-scattering layer with a forward-peaked phase function, illuminated by a plane wave of intensity I_0 along the normal to the layer. In this case the normalized intensity of transmitted light can be found with the following equation (Ishimaru, 1978; Zege *et al.*, 1991; Alexandrov *et al.*, 1993):

$$I(\tau, \vartheta) = \int_0^\infty \Gamma(s) J_0(s\vartheta) s\, ds \tag{5.95}$$

where

$$\Gamma(s) = I_0 \exp\{-\tau + \sigma_{sca}(s) z\} \tag{5.96}$$

$$\sigma_{sca}(s) = \frac{1}{2} \int_0^\infty \sigma_{sca}(\theta) J_0(s\theta) \theta\, d\theta \tag{5.97}$$

$\sigma_{sca}(\theta)$ is the volume scattering coefficient (VSC), τ is optical thickness, z is geometrical depth, $J_0(s\theta)$ is the Bessel function, ϑ is the observation angle (see equation (3.102)). For simplification we will consider the case of unit incident intensity I_0.

Information on the size of particles is contained in the function $\Gamma(s)$ (Angelsky and Maksimyak, 1993), which is the transverse coherence function of a plane wave transmitted through a scattering layer (Ishimaru, 1978). Note that equation (5.96) provides a simple method to study coherence loss due to light scattering and propagation processes (Zege and Kokhanovsky, 1994). This function can be obtained from equation (5.95):

$$\Gamma(s) = 2\pi \int_0^\infty I(\vartheta) J_0(s\vartheta) \vartheta\, d\vartheta \tag{5.98}$$

from the intensity correlation function $K(s)$ (Borovoi, 1982; Borovoi *et al.*, 1986):

$$\Gamma(s) = I_0 \sqrt{e^{-2\tau} + K(s)} \tag{5.99}$$

or can directly be measured. As follows from equation (5.96), there is a simple relationship between the coherence function $\Gamma(s)$ and the spectrum of the VSC $\sigma_{sca}(s)$:

$$\sigma_{sca}(s) = \frac{\tau + \ln \Gamma(s)/I_0}{z} \tag{5.100}$$

The VSC can be obtained from equations (5.97), (5.100):

$$\sigma_{sca}(\theta) = \frac{1}{2} \int_0^\infty \sigma_{sca}(s) J_0(s\theta) s\, ds \tag{5.101}$$

Thus, the problem of PSD determination from the intensity of multiply scattered light is reduced to the same problem for singly scattered light (from the function $\sigma_{sca}(\theta)$). Analytical methods for the determination of PSD from $\sigma_{sca}(\theta)$ data were developed in many papers (e.g. Shifrin and Tonna, 1993) and will not be considered in any detail here.

Note that it follows for monodispersed large particles that:

$$\sigma_{sca}(\theta) = \frac{4J_1^2(\theta x)}{\theta^2}\pi a^2 \tag{5.102}$$

and (Goodman, 1968):

$$\sigma_{sca}(s) = \begin{cases} 2a^2\{\arccos t - t\sqrt{1-t^2}\}, & t \leq 1 \\ 0, & t > 1 \end{cases} \tag{5.103}$$

where $t = s/2x$, x is the size parameter. Equations (5.102), (5.103) can be used to estimate the size of large monodispersed particles (Momota *et al.*, 1994). We can see that the inversion procedure in Fourier space is very simple. The same simple equations can be obtained if we measure not $\Gamma(s)$ but the OTF $S(v^*)$. This question was considered by Zege and Kokhanovsky (1992). For instance, it follows at $v^* \to \infty$ (see equation (5.93)) that:

$$S(v^*) = \exp\left(-\tau\left(1 - \frac{\gamma k a_{ef}}{v^*}\right)\right) \tag{5.104}$$

where $\gamma \sim 1$, $k = 2\pi/\lambda$, $v^* = vL$, $\tau = \sigma_{ext}L$, λ is the wavelength, L is the geometrical thickness of a layer, v is the angular frequency, σ_{ext} is the extinction coefficient. Thus, we can obtain at large frequencies:

$$a_{ef} = \frac{v^*}{k}\left\{1 + \frac{\ln S(v^*)}{\tau}\right\} \tag{5.105}$$

The influence of multiple light scattering on spectral extinction methods (Shifrin and Tonna, 1993) of the inverse problem solution was studied by Khlebtsov (1984).

5.3.3 Determination of the optical thickness of aerosol layers

Atmospheric aerosols are of considerable importance in different fields of modern science and technology. Their physical and chemical properties have been studied by many authors (Junge, 1963; Twomey, 1977; Whitby, 1978; d'Almeida *et al.*, 1991). In particular, it was recognized that there are three distinctive modes of aerosol matter with different radii: nucleation mode (0.001–0.1 μm), accumulation mode (0.1–1 μm), and coarse mode (> 1 μm).

Nucleation mode is produced by gas-to-particle conversion; accumulation mode is produced by coagulation and heterogeneous condensation; and coarse mode has an origin in mechanical processes. The light-scattering efficiency of nucleation mode is very low, so phase function $p(\theta)$ of the atmospheric aerosols depends mostly on accumulation and coarse modes (Shifrin and Zolotov, 1996).

In the single-scattering approximation (excluding multiple light scattering and surface reflection effects), the reflection function $R(\mu, \mu_0, \psi)$ of the atmosphere over ocean is (Sobolev, 1956; Stephens, 1994):

$$R(\mu, \mu_0, \psi) = \frac{\omega_0 p_a(\theta)}{4(\mu + \mu_0)} \left\{ 1 - \exp\left[-\tau \left(\frac{1}{\mu} + \frac{1}{\mu_0} \right) \right] \right\} \tag{5.106}$$

where ω_0 is single-scattering albedo, $p_a(\theta)$ is atmospheric phase function, $\mu = \cos\vartheta$, $\mu_0 = \cos\vartheta_0$, ϑ_0 is solar angle, ϑ is observation angle, $\theta = \arccos(-\mu\mu_0 + \sqrt{(1-\mu^2)(1-\mu_0^2)}\cos\psi)$, ψ is azimuth angle, and τ is optical thickness (see equation (3.16)). Note that reflection from a Lambertian surface can be accounted by equation (3.18). It follows from equation (5.106) that:

$$\tau = \frac{\mu_0 \mu}{\mu_0 + \mu} \ln \left[1 - \frac{4(\mu + \mu_0)}{\omega p_a(\theta)} R(\mu, \mu_0, \psi) \right]^{-1} \tag{5.107}$$

This equation can be used to find approximate estimates of the atmospheric aerosol optical thickness over ocean from measurements of the reflection function $R(\mu, \mu_0, \psi)$ (e.g. from space). To find the value of τ we need to know the single scattering albedo ω_0 and the phase function $p_a(\theta)$. They can be retrieved from the sky radiance over ocean as well (Wang and Gordon, 1993, 1994). Note that absorption of light by oceanic aerosols is weak (d'Almeida *et al.*, 1991) and $\omega_0 \sim 1$.

The phase function at small scattering angles depends mostly on large particles ($a \geq \lambda$, where a is the radius of a particle and λ is the wavelength). Accumulation mode (along with molecular scattering) is responsible for phase function behaviour at large scattering angles θ. This is due to the fact that the number concentration of large particles N_l is very small in comparison with the number concentration of small particles N_s. According to the International Radiation Commission (WCP-112, 1986), the ratio $\gamma = N_l/N_s$ is about 10^{-6}. In addition, it should be pointed out that phase functions of small particles are greater than or nearly equal to values of phase functions of large particles at scattering angles exceeding 90° (see Table A5.1 in Appendix 5). The phase functon of atmospheric aerosols at scattering angles exceeding 90° is of a particular interest for application of equation (5.107). This function was studied by Kokhanovsky (1998) in detail using the statistical approach. It was found that the coefficient of the variance of the phase function has a minimum at the scattering angle $\theta = 150°$ if the effective radius of $a_{ef} = 0.1$–$1.5\,\mu$m, the coefficient of the variance of the PSD of $\Delta = 0.2$–1.1, real part of the refractive index of $n = 1.45 \div 1.6$, and imaginary part of the refractive index of $k = 0.001$–0.01. The value of $p(150°) \approx 0.22$ in this case. This fact can be used for estimation of the optical thickness of aerosol from space for a fixed observation geometry ($\theta = 150°$). It could be done, for example, with POLDER multidirectional measurements (Deschamps *et al.*, 1994). Further details on aerosol remote sensing from space are given in Kaufman *et al.* (1997), Taure *et al.* (1999), von Hoyningen–Huene *et al.* (2003), and Kokhanovsky *et al.* (2004).

Equation (5.107) can be used to investigate the error in determining optical thickness related to uncertainties in values of the single-scattering albedo and the phase function (at least, at a first coarse approximation). It follows from equation (5.107) that:

$$\frac{\Delta\tau_0}{\tau_0} = K\sqrt{\frac{(\Delta\omega_0)^2}{\omega_0^2} + \frac{(\Delta p_a)^2}{p_a^2}} \tag{5.108}$$

where

$$K = \frac{X}{(X-1)\ln(1-X)} \tag{5.109}$$

$$X = \frac{4(\mu+\mu_0)}{\omega_0 p_a(\theta)} R(\mu, \mu_0, \psi) \tag{5.110}$$

As has been mentioned, the absorption of light by oceanic aerosols is low ($\omega_0 \approx 1$). It follows that $(\Delta\omega_0)^2/\omega_0^2 \ll (\Delta p_a)^2/p_a^2$ and $\Delta\tau_0/\tau_0 = K\Delta p_a/p_a$. We can see that the accuracy of the optical thickness determination over ocean depends mostly on uncertainty in the value of the atmospheric phase function.

It is interesting that coefficient K depends only on parameter $X = 1 - \exp[-\tau_0(1/\mu + 1/\mu_0)]$ (see equations (5.107) and (5.110)). For atmospheric aerosols over the ocean, τ_0 is less than 0.3 at $\lambda = 0.7\,\mu$m in most cases. The value of $M = 1/\mu + 1/\mu_0$ changes from 2 to 4 for solar and observation angles less than 60°. This leads to values of $X < 0.7$ and $K < 2$. We can see that the error in optical thickness determination can be twice larger than uncertainty in the value of the atmospheric phase function.

Let us now consider the case of nadir measurements and a scattering angle of 150°. This angle will occur at the solar zenith angle of $\vartheta_0 = 30^\circ$. It follows in this case that $X \leq 0.5$ (at $\tau \leq 0.3$) and $K \leq 1.5$.

The atmospheric phase function can be presented in the following from (Liou, 1992):

$$p_a(\theta) = \frac{p(\theta) + \xi p_m(\theta)}{1+\xi} \tag{5.111}$$

where $p(\theta)$ is the aerosol phase function, $p_m(\theta) = 0.75(1+\cos^2\theta)$ is the molecular scattering function, and ξ is the ratio of molecular optical thickness to the aerosol scattering optical depth.

It follows from equation (5.51) that:

$$\frac{dp_a(\theta)}{p_a(\theta)} = C\frac{dp(\theta)}{p(\theta)}, \qquad C = \left[1 + \xi\frac{p_m(\theta)}{p(\theta)}\right]^{-1} \tag{5.112}$$

Note that ξ decreases with the wavelength and is not smaller than 0.1 at $\lambda = 0.7$ for most cases (McCartney, 1977; Bucholtz, 1995). The value of C becomes less than 0.6 at $\theta = 150^\circ$, and the error in optical thickness determination is about equal to uncertainty in aerosol phase function (20–30% at the scattering angle of 150°). Therefore, the error in optical thickness determination can be as large as 20–30% in the case under consideration.

Additional information on aerosol microphysical and optical properties can be obtained from polarized sky radiance (Zhang and Gordon, 1997; Zhao *et al.*, 1997).

5.3.4 Determination of the microstructure of cloudy media

There have been a number of studies about the determination of cloud optical thickness τ, the liquid water path w and the effective drop radius a_{ef} from measurements of the spectral reflectance in the visible and near infrared (Hansen and Pollack, 1970; Curran and Wu, 1982; Twomay and Cocks, 1982, 1989; Arking and Childs, 1985; King, 1987; Foot, 1988, Rossow *et al.*, 1989; Nakajima and King, 1990). The underlying principle on which these techniques are based is the fact that the reflection function of clouds in the visible wavelength is primarily a function of the cloud optical thickness, but the reflection function in the near infrared is primarily a function of the droplet size (Kokhanovsky *et al.*, 2003).

To find values of a_{ef} and w from reflection measurements the simple equations derived in Chapter 2 can be used. In the case of nadir measurements and totally absorbing underlying surface they can be written in the following form (see Section 5.1.1.1):

$$R(1,\mu_0) = R^0_\infty(1,\mu_0)\exp\left(-y\frac{K_0(1)K_0(\mu_0)}{R^0_\infty(1,\mu_0)}\right) - tK_0(1)K_0(\mu_0)\exp(-y(1+z)) \tag{5.113}$$

$$t = \frac{\sinh(y)}{\sinh((1+z)y)}, \qquad K_0(\mu_0) = \tfrac{3}{7}(1+2\mu_0), \qquad R^0_\infty(1,\mu_0) = 0.49\frac{1+4\mu_0}{1+\mu_0} \tag{5.114}$$

$$y = 6\sqrt{\alpha a_{ef}}, \qquad z = \frac{0.12w}{\rho a_{ef}}\left(1+\frac{6}{x_{ef}^{2/3}}\right) \tag{5.115}$$

The absorption of radiation by drops is weak at visible ($\lambda = \lambda_v$), and it follows from equation (5.113) that:

$$R(1,\mu_0,\lambda=\lambda_v) = R^0_\infty(1,\mu_0) - \frac{K_0(1)K_0(\mu_0)}{1+z_v} \tag{5.116}$$

where

$$z_v = \frac{0.12w}{\rho a_{ef}}\left(1+\frac{6}{(k_v a_{ef})^{2/3}}\right), \qquad k_v = \frac{2\pi}{\lambda_v} \tag{5.117}$$

Thus, we can obtain from equation (5.117):

$$z_v = \frac{K_0(1)K_0(\mu_0)}{R_\infty(1,\mu_0) - R(1,\mu_0,\lambda=\lambda_v)} - 1 \tag{5.118}$$

It follows from equation (5.113) at near infrared ($\lambda = \lambda_i$) that:

$$R(1,\mu_0,\lambda=\lambda_i) = R^0_\infty(1,\mu_0)\exp\left(-6\sqrt{\alpha_i a_{ef}}\frac{K_0(1)K_0(\mu_0)}{R^0_\infty(1,\mu_0)}\right) - \frac{\sinh\left(6\sqrt{\alpha_i a_{ef}}\right)K_0(1)K_0(\mu_0)}{\sinh\left(6(1+z_i)\sqrt{\alpha_i a_{ef}}\right)} \tag{5.119}$$

where

$$z_i = \frac{0.12w}{\rho a_{ef}}\left(1 + \frac{6}{(k_i a_{ef})^{2/3}}\right), \qquad \alpha_i = \frac{4\pi\chi}{\lambda_i}, \qquad k_i = \frac{2\pi}{\lambda} \tag{5.120}$$

The transcendent equation for determination of a_{ef} can be obtained from equations (5.117), (5.119), (5.120):

$$R(1,\mu_0,\lambda=\lambda_i) - R_\infty^0(1,\mu_0)\exp\left(-6\sqrt{\alpha_i a_{ef}}\frac{K_0(1)K_0(\mu_0)}{R_\infty^0(1,\mu_0)}\right) + \frac{\sinh(6\sqrt{\alpha_i a_{ef}})K_0(1)K_0(\mu_0)}{\sinh(6(1+\gamma z_v) + \sqrt{\alpha_i a_{ef}})} = 0 \tag{5.121}$$

where the value of z_v is determined by equation (5.118) and $\gamma = z_i/z_v$.

It follows from equations (5.117), (5.120) for the value of γ that:

$$\gamma = \frac{1 + 6(k_i a_{ef})^{-2/3}}{1 + 6(k_v a_{ef})^{-2/3}} \tag{5.122}$$

We can use equations (5.117), (5.118) to find the liquid water path w if the value of the effective radius of droplets is known (e.g. from equation (5.121)). The accuracy of this simplified algorithm was studied by Kokhanovsky and Zege (1996).

It should be pointed out that the interest in retrieving the liquid water path and effective radius of drops derives not only from the fact that such retrieval seems to be possible, but from the fact that cloud radiative properties, especially plane and spherical albedos, fractional absorption, total transmittance, depend almost exclusively on these parameters. This forms the basis of cloud radiative parameterization schemes (Slingo, 1989), which are important for climate studies and require that a global database on the effective radius and liquid water path of clouds is available. Global measurements of values of a_{ef} and τ are derivable from spaceborne remote sensing observations only (Han *et al.*, 1994).

Inverse problems for mixed-phase and ice clouds was studied by Oshchepkov and Isaka (1997), Goloub *et al.* (2000), and Rolland *et al.* (2000).

Note that applications of exact solutions of the RTE for plane-parallel media to cloud microstructure and cloud optical thickness retrievals does not necessarily mean better accuracy (Loeb and Davies, 1996; Loeb and Coakley, Jr, 1998) than the method presented here. This is related to complexity of the structure of real-world clouds.

5.3.5 The snow grain size determination

The equations given in Section 5.1.2 can be used for rapid and yet accurate estimations of spectral snow reflectance for various observation geometries and snow grain sizes. This has several applications including aerosol and cloud property retrievals over snow fields.

However, the most promising area for application of our semi-analytical theory is the snow grain size determination using data from airborne and satellite-based radiometers and spectrometers (Filey *et al.*, 1997; Nolin and Dozier, 2000).

The snow grain size is an important parameter, which can be used, in particular, as an indicator of snow age. Changes in the grain size help identify snow dunes, melt areas, and blue ice regions. They often indicate the change in the snowpack energy balance.

The rate of the grain growth is exponentially proportional to temperature. This means that changes in the grain size are indicators of thermodynamic processes in the snowpack. Its estimation is of importance for calculating the absorption of radiation in the snowpack. Note that an increased absorption of solar light by snow leads to increased probabilities of snow melting and avalanches.

To start with, let us find the analytical relationship between the snow bi-directional reflectance and the snow grain size. For this we substitute equations (5.14), (5.35), and (5.47) in equation (5.11):

$$R = R_0 \exp(-bf\sqrt{\gamma d}) \tag{5.123}$$

where

$$b = \frac{4}{3}\sqrt{\frac{2B}{(1 - \langle g^G \rangle)}} \tag{5.124}$$

We conclude from equation (5.123) that the spectral dependence of $\ln R$ can be presented as a linear function of $X(\lambda) = \sqrt{\gamma(\lambda)}$. In particular, it follows: $\ln R(\lambda) = \zeta + \nu X(\lambda)$, where $\zeta = \ln R_0$ and $\nu = -bf\sqrt{d}$. Our estimations show that the spectral variability of parameters ζ and ν can be neglected.

Note that it follows (see equation (5.11)):

$$\alpha = b\sqrt{\gamma d} \tag{5.125}$$

This means that the spectral reflectance by snow is determined by the light absorption over the length equal to the grain diameter d (e.g. by the parameter $c = \gamma d$). This is similar in some respects to the conventional transmittance spectroscopy. Then the value of c is directly measured and γ is obtained as the ratio c/d, where d is the length of the cell used in measurements.

Substituting values $B \approx 1.84$ and $\langle g^G \rangle \approx 0.5$ in equation (5.124), we obtain for media with fractal particles: $b_n \approx 3.62$. It follows for spheres with $B \approx 1.27$ and $\langle g^G \rangle \approx 0.78$ that $b_s \approx 4.53$. Figure 5.13 confirms this finding for spheres. We have, therefore, $b_s/b_n \approx 1.25$. Therefore, we conclude that the value of b for spheres is approximately 25% too high compared with fractal particles. Therefore, smaller grains are needed to produce the same value of R (see equation (5.123)) for snow made of ice spheres compared with fractal snow.

It follows from equation (5.123):

$$d = \frac{1}{\gamma b^2 f^2} \ln^2\left(\frac{R}{R_0}\right) \tag{5.126}$$

This physically-based equation can be used as an alternative to snow grain size retrieval techniques based on fitting procedures. Equation (5.126) can also be used for estimates of the influence of various factors on the retrieved value of d. In particular, we obtain from equation (5.126) that the uncertainty ε_γ in the value of γ translates into the same value of uncertainty for the value of d.

It follows from equation (5.126), under the assumption that $R_0 \approx 1.0$ for both spheres and fractals that: $d_n/d_s = (b_s/b_n)^2 \approx 1.6$. Therefore, we have $d_n \approx 1.6\, d_s$ and, therefore, currently used remote sensing techniques may substantially underestimate the size of snow grains. We also have: $\varepsilon_d = 1 - d_s/d_n \approx 0.4$ as it was outlined above. Note that the necessity to account for atmospheric scattering above the snow complicates the estimation of ε_d given above.

It should be emphasized that all parameters in equation (5.126) (except R_0) are given by the simple analytical formulae specified above. This allows for the determination of d from the measured value of R if R_0 is obtained from pre-calculated look-up tables for fractal particles. Note that R_0 is close to 1 for fractal particles at nadir observation and non-grazing solar angles.

The information on R_0 is not needed if the snow plane albedo is measured. Then we have (see equation (5.19)):

$$d = \frac{\Lambda \ln^2(r_p)}{\gamma(1 + 2\cos\vartheta_0)^2} \tag{5.127}$$

where $\Lambda = 49/9b^2 \approx 0.42$ and we accounted for the fact that $b \approx 3.6$ for fractals. This allows for the immediate determination of d from r_p if the solar zenith angle ϑ_0 is known. It follows from equation (5.127) that uncertainty ε_γ in the value of γ leads to the same value of uncertainty in the retrieved grain size d. This has already been shown above.

An even simpler equation can be derived for d if the spherical albedo r of snow is measured. Indeed, it follows:

$$r = \exp(-b\sqrt{\gamma d}) \tag{5.128}$$

and, therefore,

$$d = \frac{1}{\gamma b^2} \ln^2 r \tag{5.129}$$

Interestingly, equation (5.128) transforms into a well-known formula (Bohren, 1983):

$$r = 1 - b\sqrt{\gamma d} \tag{5.130}$$

for small values of light absorption in snow. Note that Bohren (1983) proposed the value $b = 6.0$, which is somewhat larger than the corresponding value for fractal particles (see above). It is also larger than values of b for asymptotically large ice spheres (see Figure 5.13). Clearly, only experiments can give answers to the possible range of changes of this parameter in natural snow. Note that we have from equation (5.130): $b = (1 - r)/\sqrt{\gamma d}$ for weakly absorbing media. This allows us to find the value of b from experimental measurements.

The accuracy of equation (5.128) is studied in Figure 5.21 using experimental data for snow reflectance obtained in Antarctica (Grenfell *et al.*, 1994), where snow pollution is only of minor importance. The data for the ice refractive index were

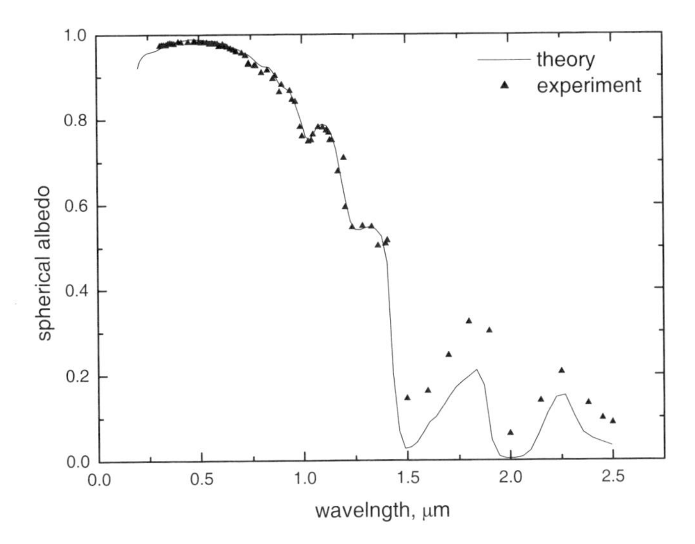

Figure 5.21. The dependence of the spherical albedo on the wavelength according to measurements performed by Grenfell *et al.* (1994) and calculations using equation (5.128) at $b = 3.6$ and $d = 0.22$ mm. The ice refractive index data were taken from Warren (1984).

taken from Warren (1984). There is a problem with the choice of an appropriate value of d, which is representative for measurements. Note that Grenfell *et al.* (1994) give the grain size vertical profiles measured visually in the field at the South Pole (1985–1986) and Vostok (1990–1991) stations. The average values of d were 0.15 mm for the South Pole station and 0.22 mm for the Vostok station (in the upper 10 cm of snow, see Grenfell *et al.* (1994; table 3)). They also give the averaged snow size profiles at the South Pole station (1986), obtained by photographic analysis ($d = 0.28$ mm in the upper 10 cm of snow) in Grenfell *et al.* (1994; table 4). The resulted averaged grain diameter is equal to 0.22 mm. So we used this diameter in our calculations in Figure 5.21.

It follows from Figure 5.21 that experimental data can be explained by the theory developed here for a wide spectral range starting from UV to approximately 1.4 μm. For larger wavelengths, the weakly absorbing media assumption is not valid, so errors are increased. Then it is preferable to use the exact radiative transfer equation with numerical calculations of the local optical characteristics of snow fractal grains. Such numerical results for a semi-infinite homogeneous snow layer have already been reported (Mishchenko *et al.*, 1999). However, the simplicity of the retrieval scheme as underlined above is then lost.

Clearly, such good fits as in Figure 5.21 at $\lambda \leq 1.4$ μm (and also for larger wavelengths) can be obtained with Mie theory as well. However, then the value of d is merely a fitting parameter. This parameter can differ from the snow grain size defined by equation (5.36) substantially, as it was emphasized above. This underlines the necessity of a careful analysis of the spectral snow reflectance with respect to the snow grain size determination.

The analytical solution of the inverse problem proposed here can be applied to the case of uniform snow only. However, aged snow is characterized by the vertical inhomogeneity of its physical properties. This means that the value of d derived from equation (5.126) will change with the wavelength for vertically inhomogeneous snow. However, this could be of advantage. Indeed, in this case the reflection function is influenced by different parts of the layer depending on the wavelengths. Clearly, highly absorbing wavelengths will give information on the top snow layer microstructure only. More deep layers of snow can be sensed by weakly absorbing wavelengths. This peculiarity was first studied by Li *et al.* (2001) using Airborne Visible/Infrared Imaging Spectrometer data at wavelengths 0.86, 1.05, 1.24, and 1.73 μm. Clearly, this issue deserves further exploration. For this, however, our technique should be extended to the case of vertically inhomogeneous media. This could be done (e.g. using the Chamberlian's effective reflective layer technique (Yanovitskij, 1997)).

Another important question, which can be addressed in the framework of our approach is snow pollution identification (e.g. by soot (Hansen and Nazarenko, 2004)). In this case, assuming that the diameter of grains is obtained from IR channels (Zege *et al.*, 1998), where γ is determined almost exclusively by ice absorption, it follows from equation (5.126) in the visible spectrum, where ice is almost transparent:

$$\gamma = \frac{1}{b^2 f^2 d} \ln^2\left(\frac{R}{R_0}\right) \tag{5.131}$$

The value of γ can substantially deviate from zero due to snow pollution. Clearly, the derived value of γ can be used for the estimation of the concentration of pollutants (Zege *et al.*, 1998). Then the information on the type of pollutants (e.g. snow, dust, or both) should be known a priori or assessed from spectral curves $R(\lambda)$ (and related spectra $\gamma(\lambda)$).

Equation (5.131) offers a simple way for the estimation of the relative absorption coefficient of ice in snow $\gamma_r(\lambda) = \gamma(\lambda)/\gamma(\lambda_0)$, where λ_0 is the selected reference wavelength. Namely, it follows from equation (5.131):

$$\gamma_r(\lambda) = \frac{\ln^2(\Xi R(\lambda))}{\ln^2(\Xi R(\lambda_0))} \tag{5.132}$$

where we accounted for the fact that $\Xi \equiv 1/R_0(\lambda_0)$ does not vary considerably with wavelength.

An even more simple result follows from equations (5.127) and (5.129):

$$\gamma_r(\lambda) = \frac{\ln^2 r_p(\lambda)}{\ln^2 r_p(\lambda_0)} = \frac{\ln^2 r(\lambda)}{\ln^2 r(\lambda_0)} \tag{5.133}$$

This allows for the direct determination of the relative ice absorption coefficient from the spectral albedo measurements.

Issues related to snow cover determination can be explored using the combination of our approach and the linear or nonlinear multiple endmember spectral mixtures analysis as proposed by Painter *et al.* (2003).

5.4 BIO-OPTICS

5.4.1 Circular dichroism and optical rotation spectra of light-scattering layers

Most light-scattering biological media (Chang and Yu, 1990; Rosen and Pendleton, 1995) are characterized by circular dichroism θ (different absorption of right and left circular polarized light beams) and optical rotation φ (different real part of refractive index of particles for right and left circular polarized light beams, which causes rotation of the plane of polarization of linear polarized waves).

Following Van de Hulst (1957), we shall assume that a plane-parallel layer of a disperse medium is equivalent, insofar as transmission is concerned, to a homogeneous layer with some relative effective index M_j (with respect to the surrounding medium):

$$M_j = 1 - 2\pi N k^{-3} i S_j(0) \tag{5.134}$$

where N is the number of particles per unit volume, $S_j(0)$ is the amplitude scattering matrix in the forward direction in a circular polarization representation, and the values of the index $j = 1, 2$ correspond to left and right circular polarization for the incident wave. This relation is applicable for optically thin ($\tau_0 = \sigma_{ext} L \ll 1$, σ_{ext} is the extinction coefficient and L is the geometrical thickness of layer) electrodynamically passive ($|M_j - 1| \ll 1$) media with a small concentration of inclusions. Under these assumptions, the polarization angle φ and ellipticity θ per unit length in the light-scattering layer is given by (Bohren and Huffman, 1983):

$$\varphi = \frac{\pi}{\lambda} \operatorname{Re}(M_1 - M_2), \qquad \theta = \frac{\pi}{\lambda} \operatorname{Im}(M_2 - M_1) \tag{5.135}$$

or, using (5.134):

$$\varphi = \pi N k^{-2} \operatorname{Re}(i\Delta S), \qquad \theta = -\pi N k^{-2} \operatorname{Im}(i\Delta S) \tag{5.136}$$

where

$$\Delta S = S_2(0) - S_1(0) \tag{5.137}$$

Equations (5.136) are general. They can be applied for studies of optical rotation $\varphi(\lambda)$ and circular dichroism $\theta(\lambda)$ spectra of optically thin light-scattering layers with optically active inclusions.

The function $\Delta S(\lambda)$ can be determined by use of the exact solution of the problem of light diffraction by an optically active sphere (see Appendix 2). However, the latter is a complex problem and requires numerical calculations. We will instead use the Van de Hulst (anomalous diffraction) approximation to find the function $S_j(0)$. Note that the conditions for applicability of the Van de Hulst approximation were discussed in Chapter 2.

The Van de Hulst approximation for an optically active sphere may be obtained by means of asymptotic analysis of the exact scattering series for optically active particles (Bohren and Huffman, 1983):

$$S_j(0) = x^2 K(i\rho_j) \tag{5.138}$$

where $\rho_j = 2x(m_j - 1)$; m_1 and m_2 are the relative refractive indices for the left and right circularly polarized light, respectively, and $K(\nu) = \frac{1}{2} + e^{-\nu}/\nu + (e^{-\nu} - 1)/\nu^2$ is the Van de Hulst function.

For a polydisperse system, instead of (5.138) we obtain:

$$\tilde{S}_j(0) = \int_0^\infty S_j(0) f(a)\, da = \int_0^\infty x^2 K(ip_j) f(a)\, da \tag{5.139}$$

where $f(a)$ is the particle size distribution.

We now take into account the weakness of the spatial dispersion effects (Bohren and Huffman, 1983):

$$|m_1 - m_2| \ll 1 \tag{5.140}$$

Thus, from equation (5.139) we obtain:

$$\Delta\tilde{S}(0) = \int_0^\infty \left(\frac{\partial S}{\partial \rho}\right)_{\rho=\bar{\rho}} \Delta\rho f(a)\, da \tag{5.141}$$

where $\bar{\rho} = 2x(\bar{m} - 1)$; $\Delta\rho = 2x(m_2 - m_1)$; $\bar{m} = (m_1 + m_2)/2$. After substitution of (5.141) into (5.136), it follows that:

$$\varphi = \frac{\pi C_v}{\lambda} \operatorname{Im}(D\Delta m), \qquad \theta = \frac{\pi C_v}{\lambda} \operatorname{Re}(D\Delta m) \tag{5.142}$$

where

$$C_v = \frac{4\pi N}{3} \int_0^\infty a^3 f(a)\, da, \qquad \Delta m = m_1 - m_2 \tag{5.143}$$

and

$$D = \frac{\displaystyle\int_0^\infty a^3 f(a)\, d(\rho)\, da}{\displaystyle\int_0^\infty a^3 f(a)\, da} \tag{5.144}$$

Here C_v is the volume concentration of particles and $d(\bar{\rho}) = 3(\partial K/\partial\rho)_{\rho=\bar{\rho}}$.

Since $D(\rho)$ and Δm are, in general, complex functions, we write:

$$D(\rho) = D'(\rho) + iD''(\rho), \qquad \Delta m = \Delta n - i\Delta\chi \tag{5.145}$$

Now, from (5.142) and (5.145), it follows that:

$$\varphi = (D''\varphi_0 - D'\theta_0), \qquad \theta = (D'\varphi_0 + D''\theta_0) \tag{5.146}$$

where $\varphi_0 = \pi\Delta n C_v/\lambda$ and $\theta_0 = \pi\Delta\chi C_v/\lambda$ are, respectively, optical rotation and circular dichroism for a homogeneous medium of the scatter material.

Thus, the circular dichroism spectrum $\theta(\lambda)$ and the optical rotation dispersion spectrum $\varphi(\lambda)$ of a disperse layer are linear combinations of the corresponding spectra $\varphi_0(\lambda)$ and $\theta_0(\lambda)$ for the homogeneous material inside small particles. The linear transformation operator has the form:

$$\hat{D} = \begin{pmatrix} D'' & -D' \\ D' & D'' \end{pmatrix} \tag{5.147}$$

and

$$\Phi = \hat{D}\Phi_0 \tag{5.148}$$

where

$$\Phi = \begin{pmatrix} \varphi \\ \theta \end{pmatrix}, \qquad \Phi_0 = \begin{pmatrix} \varphi_0 \\ \theta_0 \end{pmatrix} \tag{5.149}$$

These expressions make simple analysis of the influence of the dimension of the inclusions on the observed spectra $\varphi(\lambda)$ and $\theta(\lambda)$ possible (the red shift of the Cotton effect, inversion of the spectra (Gordon, 1972)). Moreover, on this basis it is also possible to solve the inverse problem of obtaining the circular dichroism and optical rotation spectra for disperse systems in suspensions, which is important, for example, in biochemical studies:

$$\Phi_0 = \hat{D}^{-1}\Phi \tag{5.150}$$

where

$$\hat{D}^{-1} = \frac{1}{\Delta}\begin{pmatrix} D'' & D' \\ -D' & D'' \end{pmatrix}, \qquad \Delta = |D|^2 \neq 0 \tag{5.151}$$

It is interesting to see that, if $D'' \ll D'$ and D' is only weakly dependent on λ in the spectral interval under study, then it follows from equations (5.150) and (5.151) that:

$$\varphi_0(\lambda) \sim \theta(\lambda), \qquad \theta_0(\lambda) \sim -\varphi(\lambda) \tag{5.152}$$

Such a transformation of the spectra has been observed experimentally (Gordon, 1972). For the small values of the phase shifts in (5.151), such a situation is impossible. Indeed, when $\rho \to 0$, we find from (5.144) that $D' \to 0$, $D'' \to 1$, and (see equation (5.151)):

$$\varphi = \varphi_0, \qquad \theta = \theta_0 \tag{5.153}$$

which corresponds to the well-known result (Bohren and Huffman, 1983), obtained in the framework of the Rayleigh–Gans approximation. We can see (compare equations (5.152), (5.153)) how different results for Van de Hulst and Rayleigh–Gans particles can be.

We shall now assume for the sake of simplicity that, in the spectral region under consideration, particles do not absorb. Then, we obtain from equation (5.146):

$$\varphi = D''\varphi_0, \qquad \theta = D'\varphi_0 \tag{5.154}$$

Thus, even in the absence of absorption in the medium, transmitted radiation is characterized by an elliptical polarization. The circular dichroism in this case is caused by different values of the scattering cross-sections C^l_{sca} and C^r_{sca} for left (l) and right (r) circular polarizations. According to the optical theorem (Bohren and Huffman, 1983), we have:

$$C^j_{sca} = C^j_{ext} = 4\pi k^{-2}\,\mathrm{Re}(S_j(0)) = 4\pi a^2\,\mathrm{Re}(K(i\rho_j)) \tag{5.155}$$

where we have assumed that the medium is transparent, and C^j_{ext} is the extinction cross-section, $j = l, r$. It is evident from equation (5.155) that one of the circular polarizations will experience stronger scattering to the side, which will lead to ellipticity of transmitted light.

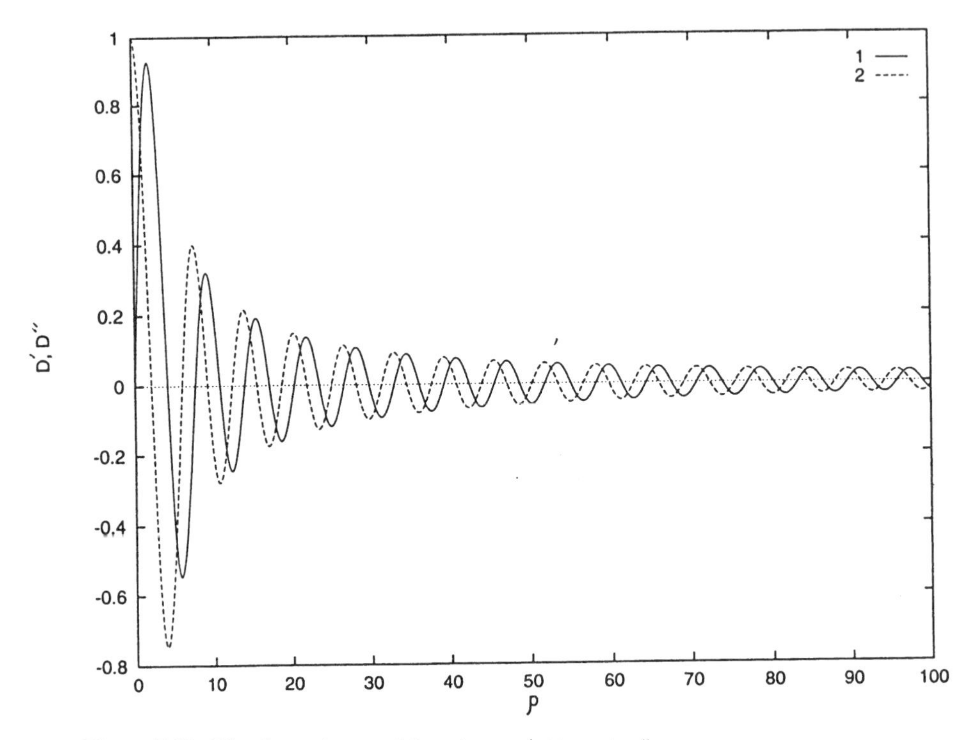

Figure 5.22. The dependence of functions D' (1) and D'' (2) on the phase shift ρ.

The dependence of the quantities φ and θ on particle dimensions is contained in functions $D'(\rho)$ and $D''(\rho)$ (see equations (5.144) and (5.148)), which for a number of particular particle size distributions may be calculated analytically. This is related to the simple form of the function $d(\rho)$ (see equation (5.144)):

$$d(\rho) = \frac{6(e^{-i\rho} - 1)}{\rho^3} + \frac{6i\,e^{-i\rho}}{\rho^2} - \frac{3e^{-i\rho}}{\rho} \tag{5.156}$$

where $\rho = 2x(\bar{m} - 1)$.

Let us assume now that the suspension is monodisperse. Then $D \equiv d(\rho)$ and (e.g. for nonabsorbing inclusions) we find from equation (5.144) (see Figure 5.22) that:

$$D'(\rho) = 3\left(\frac{2(\cos\rho - 1)}{\rho^3} + \frac{2\sin\rho}{\rho^2} - \frac{\cos\rho}{\rho}\right) \tag{5.157}$$

$$D''(\rho) = 3\left(-\frac{2\sin\rho}{\rho^3} + \frac{2\cos\rho}{\rho^2} + \frac{\sin\rho}{\rho}\right) \tag{5.158}$$

The presence of weak absorption has almost no influence on these expressions, and increase in particle size leads to smearing of the oscillations of functions D' and D''.

We shall now study functions (5.157), (5.158) in more detail. To this end, we calculate their derivatives:

$$\frac{\partial D'}{\partial \rho} = 3\left(\frac{\sin\rho}{\rho} + \frac{3\cos\rho}{\rho^2} - \frac{6\sin\rho}{\rho^3} - \frac{6(\cos\rho - 1)}{\rho^4}\right) \tag{5.159}$$

$$\frac{\partial D''}{\partial \rho} = 3\left(\frac{\cos\rho}{\rho} - \frac{3\sin\rho}{\rho^2} - \frac{6\cos\rho}{\rho^3} + \frac{6\sin\rho}{\rho^4}\right) \tag{5.160}$$

For large ρ we find from equation (5.159), (5.160) that:

$$\frac{\partial D'}{\partial \rho} = 0 \quad \text{for } \operatorname{ctg}\rho = \rho/3, \qquad \frac{\partial D''}{\partial \rho} = 0 \quad \text{for } \operatorname{tg}\rho = \rho/3 \tag{5.161}$$

while from equations (5.157), (5.158) it follows that:

$$\frac{\partial D''}{\partial \rho} = 0 \quad \text{for } \operatorname{ctg}\rho = \rho/2, \qquad D' = 0 \quad \text{for } \operatorname{tg}\rho = \rho/2 \tag{5.162}$$

We underline the fact that equation $\operatorname{tg}\rho = \rho/2$ ($\operatorname{ctg}\rho = \rho/2$) has the approximate solution $\rho = \pi(l + \frac{1}{2})$ ($\rho = \pi l$), where $l = 1, 2, 3 \ldots$. Thus, it is clear that the maxima and minima of the functions are shifted with respect to each other by a quarter of period $\pi/2$. The zeros of function $D'(\rho)$ ($D''(\rho)$) approximately coincide with the maxima of function $|D''(\rho)|$ ($|D'(\rho)|$). This has an important consequence: the smallest values of rotation of polarization plane φ correspond to maximal values of ellipticity θ and vice versa. There is an analogous effect for circular dichroism and optical rotation dispersion spectra of chiral molecules in a solution close to the isolated absorption band, but it is brought about in that case by other factors. We note that the previously mentioned feature has already been studied in the case of disperse media with anisotropic particles, for which linear dichroism and birefringence are common (Meeten, 1980).

The results presented here may be used for stereochemical analysis of disperse systems.

5.4.2 Blood optics

The investigation of microphysical and optical properties of blood is a vital issue in bio-optics (Ishimaru, 1978) and relates to important human health problems. Scattered light can be used to determine the content of oxygen in human blood, optical constants of erythrocytes, their shape and size distribution. The concentration of erythrocytes in blood (haematocrit) is 30–60% in most cases. Another constituent of blood is the transparent solution of different salts. Thus, blood is a close-packed system of nonspherical scaterers. Erythrocytes have the shape of a disc with thickness 1 μm in the centre and 2 μm near the edges. The diameters of discs are about 7 μm. The real part of the refractive index n is less than 1.1 and the imaginary part χ is less than 10^{-3} in the visible and near infrared. The value of χ is less than 10^{-6} at $0.68\ \mu\text{m} \leq \lambda \leq 1\ \mu\text{m}$ (Ishimaru, 1978). Thus, blood (as many other biological media, including chloroplasts and other cells) is a weakly absorbing light-scattering media. The optical density of blood is high. Thus, the simple equations obtained in Chapter 3, can be used to calculate diffuse reflection and transmission of a blood sample contained in a plane-parallel layer having the optical thickness $\tau = \sigma_{\text{ext}} L \gg 1$ (L is the geometrical thickness of the sample):

$$r = \frac{\sinh x}{\sinh(x+y)}, \qquad t = \frac{\sinh y}{\sinh(x+y)} \tag{5.163}$$

where $x = k\tau$, $y = 4qk$, $k = \sqrt{3(1-\omega_0)(1-g)}$. It is possible to propose different methods for retrieval of parameters x and y from equations (5.163) (Dubova *et al.*, 1981). For instance, if we measure the values of $r(\lambda)$ and $t(\lambda)$, it follows that:

$$x = \operatorname{arcsin} h\left(\frac{Q}{2r}\right), \qquad y = \operatorname{arccosh}\left(\frac{Q}{2t}\right), \qquad Q = \sqrt{(1+r^2+t^2)^2 - 4r^2} \tag{5.164}$$

The obtained values of x and y can be used to calculate the transport extinction coefficient

$$\sigma_{tr}(\lambda) = \sigma_{ext}(1-g) = \frac{4x}{3yL} \tag{5.165}$$

and the absorption coefficient

$$\sigma_{abs}(\lambda) = \frac{xy}{4L} \tag{5.166}$$

It was shown in Chapter 2 that for weakly absorbing particles of any shape:

$$\sigma_{abs}(\lambda) = \frac{4\pi\chi}{\lambda} f(n) C_v \tag{5.167}$$

where $f(n) \to 1$ at $n \to 1$. Equations (5.163)–(5.167) allow us to obtain dependence $\chi(\lambda)$ for pigments in erythrocytes *in vivo*.

This is the attractive point of diffuse spectroscopy methods in comparison with the usual transmittance methods (Ishimaru, 1978). The content of the oxygen is determined by the ratio of oxyhaemoglobin molecules HbO_2 to haemoglobin molecules Hb inside erythrocytes. The spectral behaviour of $\chi_{\text{Hb}}(\lambda)$ for Hb and χ_{HbO_2} for HbO_2 is different and well known (see figure 3.18 in the book by Ishimaru (1978)). Thus, measuring the spectrum of absorption coefficient

$$\chi(\lambda) = c_{\text{Hb}} \chi_{\text{Hb}}(\lambda) + c_{\text{HbO}_2} \chi_{\text{HbO}_2}(\lambda) \tag{5.168}$$

of pigments in erythrocytes at two wavelengths we can obtain values of concentrations of haemoglobin c_{Hb} and oxyhaemoglobin c_{HbO_2}. The ratio $\xi = c_{\text{HbO}_2}/(c_{\text{HbO}_2} + c_{\text{Hb}})$ is an important parameter in medicine practice.

Note that the use of red and near-infrared laser radiation for non-invasive imaging in medical diagnostics was discussed by Rinneberg (1995). Tuchin (1997) discussed modern light-scattering methods in tissue optics. Note that tissues are optically thick media. Thus, the diffusion approximation (Zege and Katsev, 1978) can be applied in this case.

5.5 PLANETARY OPTICS

The main developments in radiative transfer theory were made by astrophysicists (Chandrasekhar, 1950; Sobolev, 1956, 1972; Ambarzumian, 1961; Van de Hulst, 1980). A huge number of papers and books are devoted to the radiative characteristics of stellar and planetary atmospheres. In this section we will briefly consider just two examples: the atmospheres of Venus and Mars – the planets closest to the Earth.

At the moment, Venus is the only planet with a dense atmosphere in which spacecraft has descended and performed measurements of internal radiation (Wauben, 1992). The interpretation of the ground-based polarimetry of Venus by Hansen and Hovenier (1974) and the above-mentioned *in situ* measurements showed that the atmosphere of Venus can be represented as a plane-parallel layer consisting of two homogeneous sublayers: a pure CO_2 gas layer below a cloud layer containing a mixture of concentrated sulphuric acid droplets and CO_2 molecules. The effective radius of droplets is about 1 μm, the effective variance is 0.07, and the real part of the refractive index is 1.44 (Hansen and Hovenier, 1974). The Venus atmosphere probes found that the optical thickness of the upper layer is between 20 and 53 at visible wavelengths (Wauben, 1992). The Venus surface albedo is small (0.02–0.15) and in many radiative transport problems can be ignored. The single scattering albedo of particles in Venus's atmosphere in the visible ($\lambda = 0.63\,\mu$m) is about 0.9985 and the assymmetry parameter $g \approx 0.7$. Note that the reflection function almost does not depend on optical thickness ($(1-g)\tau = 9 \div 15$) at these conditions. Thus, from the point of view of radiative transport we can consider the atmosphere of Venus as a semi-infinite weakly absorbing medium and the spherical albedo can be found from the following equation:

$$r = \exp\left(-\frac{4s}{\sqrt{3}}\right) \tag{5.169}$$

where $s = \sqrt{(1-\omega_0)/(1-g)}$. The spectral measurements of $r(\lambda)$ allow us to obtain the spectral behaviour of $s(\lambda)$ in the atmosphere of Venus. Other radiative characteristics of Venus can be studied using the equations for weakly absorbing layers presented in Table 3.3.

Contrary to the case for Venus, it is possible to see details of the Martian surface from our planet. Thus, the optical thickness of the Martian atmosphere, which consists of particles with the refractive index $n \sim 1.5$ (dust aerosols) and CO_2 molecules is low and the approximation of thin layers can be used to find the reflection function (Sobolev, 1972):

$$R(\mu, \mu_0, \phi) = \frac{\omega_0 p(\theta)}{4(\mu + \mu_0)}\left[1 - \exp\left(-\tau\left(\frac{1}{\mu} + \frac{1}{\mu_0}\right)\right)\right] + A_s\, e^{-\tau(1/\mu + 1/\mu_0)} \tag{5.170}$$

where the second term (see equations (3.18)–(3.22)) accounts approximately for the reflection of light from the Martian surface, assuming that the surface is Lambertian with albedo A_s. We can obtain from equation (5.70) as $\tau \to 0$:

$$R(\mu, \mu_0, \phi) = A_s - \frac{[4(\mu + \mu_0)A_s - \omega_0 p(\theta)]\tau}{4\mu\mu_0} \tag{5.171}$$

This equation can be used for the calculation of the reflection functions of the Martian atmosphere. The problem is that values of A_s, ω_0, τ, and $p(\theta)$ are not known a priori and can vary considerably. It is common to assume that $\omega_0 \sim 1$, $\tau \leq 0.05$, and $A_s = 0.1$–0.3 for the Martian atmosphere. Equation (5.171) can be used for estimation of the values of A_s and τ from measurements $R(\mu, \mu_0, \phi)$ under certain assumptions about the product $\omega_0 p(\theta)$. For instance, we can use the fact that the variability of the phase function $p(\theta)$ for aerosol particles is smallest at $\theta = 150°$ ($p(150°) \sim 0.2$). Thus, this angle is a favourable scattering angle for remote sensing of the Martian atmosphere. Note that the scattering angle θ can vary from 133° to 180° if we measure the reflection function of the Martian atmosphere from the Earth.

Appendix 1

Refractive indices

Table A1.1. Water at the temperature 25°C (Hale and Querry, 1973).

λ (μm)	n	χ	λ (μm)	n	χ
0.2	1.396	1.1E−7	2.0	1.306	1.1E−3
0.225	1.373	4.9E−8	2.2	1.296	2.89E−4
0.250	1.362	3.35E−8	2.4	1.279	9.56E−4
0.275	1.354	2.35E−8	2.6	1.242	3.17E−3
0.300	1.349	1.6E−8	2.65	1.219	6.7E−3
0.325	1.346	1.08E−8	2.70	1.188	0.019
0.350	1.343	6.5E−9	2.75	1.157	0.059
0.375	1.341	3.5E−9	2.80	1.142	0.115
0.400	1.339	1.86E−9	2.85	1.149	0.185
0.425	1.338	1.3E−9	2.90	1.201	0.268
0.450	1.337	1.02E−9	2.95	1.292	0.298
0.475	1.336	9.35E−10	3.00	1.371	0.272
0.500	1.335	1.00E−9	3.05	1.426	0.240
0.525	1.334	1.32E−9	3.10	1.467	0.192
0.550	1.333	1.96E−9	3.15	1.483	0.135
0.575	1.333	3.60E−9	3.20	1.478	0.0924
0.600	1.332	1.09E−8	3.25	1.467	0.0610
0.625	1.332	1.39E−8	3.30	1.450	0.0368
0.650	1.331	1.64E−8	3.35	1.432	0.0261
0.675	1.331	2.23E−8	3.40	1.420	0.0195
0.700	1.331	3.35E−8	3.45	1.410	0.0132
0.725	1.330	9.15E−8	3.50	1.400	0.0094
0.750	1.330	1.56E−7	3.6	1.385	0.00515
0.775	1.330	1.48E−7	3.7	1.374	0.00360
0.800	1.329	1.25E−7	3.8	1.364	0.00340
0.825	1.329	1.82E−7	3.9	1.357	0.00380
0.850	1.329	2.93E−7	4.0	1.351	0.00460
0.875	1.328	3.91E−7	4.1	1.346	0.00562
0.900	1.328	4.86E−7	4.2	1.342	0.00688
0.925	1.328	1.06E−6	4.3	1.338	0.00845
0.950	1.327	2.93E−6	4.4	1.334	0.0103
0.975	1.327	3.48E−6	4.5	1.332	0.0134
1.0	1.327	2.89E−6	4.6	1.330	0.0147
1.2	1.324	9.89E−6	4.7	1.330	0.0157
1.4	1.321	1.38E−4	4.8	1.330	0.0150
1.6	1.317	8.55E−5	4.9	1.328	0.0137
1.8	1.312	1.15E−4	5.0	1.325	0.0124

Table A1.2. Ice (Warren, 1984).

λ (μm)	n	χ	λ (μm)	n	χ
0.21	1.3800	1.325E−8	0.81	1.3047	1.400E−7
0.25	1.3509	8.623E−9	0.82	1.3045	1.430E−7
0.30	1.3339	5.504E−9	0.83	1.3044	1.450E−7
0.35	1.3249	3.765E−9	0.84	1.3042	1.510E−7
0.40	1.3194	2.710E−9	0.85	1.3040	1.830E−7
0.41	1.3185	2.510E−9	0.86	1.3038	2.150E−7
0.42	1.3177	2.260E−9	0.87	1.3035	3.350E−7
0.43	1.3170	2.080E−9	0.88	1.3035	3.350E−7
0.44	1.3163	1.910E−9	0.89	1.3033	3.920E−7
0.45	1.3157	1.540E−9	0.90	1.3032	4.200E−7
0.46	1.3151	1.530E−9	0.91	1.3030	4.440E−7
0.47	1.3145	1.550E−9	0.92	1.3028	4.740E−7
0.48	1.3140	1.640E−9	0.93	1.3027	5.110E−7
0.49	1.3135	1.780E−9	0.94	1.3025	5.530E−7
0.50	1.3130	1.910E−9	0.95	1.3023	6.020E−7
0.51	1.3126	2.140E−9	0.96	1.3022	7.550E−7
0.52	1.3122	2.160E−9	0.97	1.3020	9.260E−7
0.53	1.3118	2.540E−9	0.98	1.3018	1.120E−6
0.54	1.3114	2.930E−9	0.99	1.3017	1.330E−6
0.55	1.3110	3.110E−9	1.00	1.3015	1.620E−6
0.56	1.3106	3.290E−9	1.01	1.3013	2.000E−6
0.57	1.3103	3.520E−9	1.02	1.3012	2.250E−6
0.58	1.3100	4.040E−9	1.03	1.3010	2.330E−6
0.59	1.3097	4.880E−9	1.04	1.3008	2.330E−6
0.60	1.3094	5.730E−9	1.05	1.3006	2.170E−6
0.61	1.3091	6.890E−9	1.06	1.3005	1.960E−6
0.62	1.3088	8.580E−9	1.07	1.3003	1.810E−6
0.63	1.3085	1.040E−8	1.08	1.3001	1.740E−6
0.64	1.3083	1.220E−8	1.09	1.3000	1.730E−6
0.65	1.3080	1.430E−8	1.10	1.2998	1.700E−6
0.66	1.3078	1.660E−8	1.11	1.2996	1.760E−6
0.67	1.3076	1.890E−8	1.12	1.2995	1.820E−6
0.68	1.3073	2.090E−8	1.13	1.2993	2.040E−6
0.69	1.3071	2.400E−8	1.14	1.2991	2.250E−6
0.70	1.3069	2.900E−8	1.15	1.2989	2.290E−6
0.71	1.3067	3.440E−8	1.16	1.2987	3.040E−6
0.72	1.3065	4.030E−8	1.17	1.2985	3.840E−6
0.73	1.3062	4.300E−8	1.18	1.2984	4.770E−6
0.74	1.3060	4.920E−8	1.19	1.2982	5.760E−6
0.75	1.3058	5.870E−8	1.20	1.2980	6.710E−6
0.76	1.3057	7.080E−8	1.21	1.2978	8.660E−6
0.77	1.3055	8.580E−8	1.22	1.2976	1.020E−5
0.78	1.3053	1.020E−7	1.23	1.2974	1.130E−5
0.79	1.3051	1.180E−7	1.24	1.2972	1.220E−5
0.80	1.3049	1.340E−7	1.25	1.2970	1.290E−5

Table A1.2 (*cont.*).

λ (μm)	n	χ	λ (μm)	n	χ
1.26	1.2969	1.320E−5	1.905	1.2777	4.687E−4
1.27	1.2967	1.350E−5	1.923	1.2769	7.615E−4
1.28	1.2965	1.330E−5	1.942	1.2761	1.010E−3
1.29	1.2963	1.320E−5	1.961	1.2754	1.313E−3
1.30	1.2931	1.320E−5	1.980	1.2747	1.539E−3
1.31	1.2958	1.310E−5	2.00	1.2740	1.588E−3
1.32	1.2956	1.320E−5	2.020	1.2733	1.540E−3
1.33	1.2954	1.320E−5	2.041	1.2724	1.412E−3
1.34	1.2952	1.340E−5	2.062	1.2714	1.244E−3
1.35	1.2950	1.390E−5	2.083	1.2703	1.068E−3
1.36	1.2948	1.420E−5	2.105	1.2690	8.414E−4
1.37	1.2945	1.480E−5	2.130	1.2674	5.650E−4
1.38	1.2943	1.580E−5	2.150	1.2659	4.320E−4
1.39	1.2941	1.740E−5	2.170	1.2644	3.500E−4
1.40	1.2938	1.980E−5	2.190	1.2628	2.870E−4
1.41	1.2936	2.500E−5	2.220	1.2604	2.210E−4
1.42	1.2933	5.400E−5	2.240	1.2586	2.030E−4
1.43	1.2930	1.040E−4	2.245	1.2582	2.010E−4
1.44	1.2927	2.030E−4	2.250	1.2577	2.030E−4
1.449	1.2925	2.708E−4	2.260	1.2567	2.140E−4
1.46	1.2923	3.511E−4	2.270	1.2558	2.320E−4
1.471	1.2921	4.299E−4	2.290	1.2538	2.890E−4
1.481	1.2919	5.181E−4	2.310	1.2518	3.810E−4
1.493	1.2917	5.855E−4	2.330	1.2497	4.620E−4
1.504	1.2915	5.899E−4	2.350	1.2475	5.480E−4
1.515	1.2913	5.635E−4	2.370	1.2451	6.180E−4
1.527	1.2911	5.480E−4	2.390	1.2427	6.800E−4
1.538	1.2908	5.266E−4	2.410	1.2400	7.300E−4
1.563	1.2903	4.394E−4	2.430	1.2373	7.820E−4
1.587	1.2896	3.701E−4	2.460	1.2327	8.480E−4
1.613	1.2889	3.372E−4	2.500	1.2258	9.250E−4
1.650	1.2878	2.410E−4	2.520	1.2220	9.200E−4
1.680	1.2869	1.890E−4	2.550	1.2155	8.920E−4
1.700	1.2862	1.660E−4	2.565	1.2118	8.700E−4
1.730	1.2852	1.450E−4	2.580	1.2079	8.900E−4
1.760	1.2841	1.280E−4	2.590	1.2051	9.300E−4
1.800	1.2826	1.030E−4	2.600	1.2021	1.010E−3
1.830	1.2814	8.600E−5	2.620	1.1957	1.350E−3
1.840	1.2809	8.220E−5	2.675	1.1741	3.420E−3
1.850	1.2805	8.030E−5	2.725	1.4173	7.920E−3
1.855	1.2802	8.500E−5	2.778	1.1077	2.000E−2
1.860	1.2800	9.900E−5	2.817	1.0674	3.800E−2
1.870	1.2795	1.500E−4	2.833	1.0476	5.200E−2
1.890	1.2785	2.950E−4	2.849	1.0265	6.800E−2

(*continued*)

Table A1.2 (*cont.*).

λ (μm)	n	χ	λ (μm)	n	χ
2.865	1.0036	9.230E−2	3.367	1.5265	5.400E−2
2.882	0.9820	1.270E−1	3.390	1.5114	4.220E−2
2.8999	0.9650	1.690E−1	3.413	1.4973	3.420E−2
2.915	0.9596	2.210E−1	3.436	1.4845	2.740E−2
2.933	0.9727	2.760E−1	3.460	1.4721	2.200E−2
2.950	0.9917	3.120E−1	3.484	1.4612	1.860E−2
2.967	1.0067	3.470E−1	3.509	1.4513	1.520E−2
2.985	1.0219	3.880E−1	3.534	1.4421	1.260E−2
3.003	1.0427	4.380E−1	3.559	1.4337	1.060E−2
3.021	1.0760	4.930E−1	3.624	1.4155	8.020E−3
3.040	1.1295	5.540E−1	3.732	1.3942	6.850E−3
3.058	1.2127	6.120E−1	3.775	1.3873	6.600E−3
3.077	1.3251	6.250E−1	3.847	1.3773	6.960E−3
3.096	1.4260	5.930E 1	3.969	1.3645	9.160E−3
3.115	1.4966	5.390E−1	4.099	1.3541	1.110E−2
3.135	1.5510	4.910E−1	4.239	1.3446	1.450E−2
3.155	1.5999	4.380E−1	4.348	1.3388	2.000E−2
3.175	1.6363	3.720E−1	4.387	1.3381	2.300E−2
3.195	1.6502	3.000E−1	4.444	1.3385	2.600E−2
3.215	1.6428	2.380E−1	4.505	1.3405	2.900E−2
3.236	1.6269	1.930E−1	4.547	1.3429	2.930E−2
3.257	1.6128	1.580E−1	4.560	1.3442	3.000E−2
3.279	1.5924	1.210E−1	4.580	1.3463	2.850E−2
3.300	1.5733	1.030E−1	4.719	1.3442	1.730E−2
3.322	1.5577	8.360E−2	4.904	1.3345	1.290E−2
3.345	1.5413	6.680E−2	5.000	1.3290	1.200E−2

Table A1.3. Dust, water soluble, oceanic and soot components of the atmospheric aerosol, respectively (WCP-112, 1986).

λ (μm)	*n*	χ	*n*	χ	*n*	χ	*n*	χ
0.200	1.530	7.00E−02	1.530	7.00E−02	1.429	2.87E−05	1.500	0.350
0.250	1.530	3.00E−02	1.530	3.00E−02	1.404	1.45E−06	1.620	0.450
0.300	1.530	8.00E−03	1.530	8.00E−03	1.395	5.83E−07	1.740	0.470
0.337	1.530	8.00E−03	1.530	5.00E−03	1.392	1.20E−07	1.750	0.470
0.400	1.530	8.00E−03	1.530	5.00E−03	1.385	9.90E−09	1.750	0.460
0.488	1.530	8.00E−03	1.530	5.00E−03	1.382	6.41E−09	1.750	0.450
0.515	1.530	8.00E−03	1.530	5.00E−03	1.381	3.70E−09	1.750	0.450
0.550	1.530	8.00E−03	1.530	6.00E−03	1.381	4.26E−09	1.750	0.440
0.633	1.530	8.00E−03	1.530	6.00E−03	1.377	1.62E−08	1.750	0.430
0.694	1.530	8.00E−03	1.530	7.00E−03	1.376	5.04E−08	1.750	0.430
0.860	1.520	8.00E−03	1.520	1.20E−02	1.372	1.09E−06	1.750	0.430
1.060	1.520	8.00E−03	1.520	1.70E−02	1.367	6.01E−05	1.750	0.440
1.300	1.460	8.00E−03	1.510	2.00E−02	1.365	1.41E−04	1.760	0.450
1.536	1.400	8.00E−03	1.510	2.30E−02	1.359	2.43E−04	1.770	0.460
1.800	1.330	8.00E−03	1.460	1.70E−02	1.351	3.11E−04	1.790	0.480
2.000	1.260	8.00E−03	1.420	8.00E−03	1.347	1.07E−03	1.800	0.490
2.250	1.220	9.00E−03	1.420	1.00E−02	1.334	8.50E−04	1.810	0.500
2.500	1.180	9.00E−03	1.420	1.20E−02	1.309	2.39E−03	1.820	0.510
2.700	1.180	1.30E−02	1.400	5.50E−02	1.249	1.56E−02	1.830	0.520
3.000	1.160	1.20E−02	1.420	2.20E−02	1.439	0.197	1.840	0.540
3.200	1.220	1.00E−02	1.430	8.00E−03	1.481	6.69E−02	1.860	0.540
3.392	1.260	1.30E−02	1.430	7.00E−03	1.439	1.51E−02	1.870	0.550
3.500	1.280	1.10E−02	1.450	5.00E−03	1.423	7.17E−03	1.880	0.560
3.750	1.270	1.10E−02	1.452	4.00E−03	1.398	2.90E−03	1.900	0.570
4.000	1.260	1.20E−02	1.455	5.00E−03	1.388	3.69E−03	1.920	0.580
4.500	1.260	1.40E−02	1.460	1.30E−02	1.377	9.97E−03	1.940	0.590
5.000	1.250	1.60E−02	1.450	1.20E−02	1.366	9.57E−03	1.970	0.600

Appendix 2

Exact solutions of light-scattering problems for uniform, two-layered, and optically active spherical particles

Exact expressions for the elements of the amplitude scattering matrix (1.65) for an isotropic spherical particle illuminated by a linearly polarized plane electromagnetic wave are (Van de Hulst, 1980):

$$S_{11}(\theta) = \sum_{n=1}^{\infty} \frac{2n+1}{n(n+1)} \{a_n \tau_n(\cos\theta) + b_n \pi_n(\cos\theta)\} \tag{A2.1}$$

$$S_{22}(\theta) = \sum_{n=1}^{\infty} \frac{2n+1}{n(n+1)} \{a_n \pi_n(\cos\theta) + b_n \tau_n(\cos\theta)\} \tag{A2.2}$$

$$S_{12}(\theta) = S_{21}(\theta) = 0 \tag{A2.3}$$

where

$$\left.\begin{aligned} a_n &= \frac{\psi'_n(y)\psi_n(x) - m\psi_n(y)\psi'_n(x)}{\psi'_n(y)\xi_n(x) - m\psi_n(y)\xi'_n(x)} \\ b_n &= \frac{m\psi'_n(y)\psi_n(x) - \psi_n(y)\psi'_n(x)}{m\psi'_n(y)\xi_n(x) - \psi_n(y)\xi'(x)} \end{aligned}\right\} \tag{A2.4}$$

$m = n - i\chi$ is the relative refractive index of a particle ($m = m_p/m_h$, m_p and m_h are refractive indices of a particle and a host medium, respectively), $y = mx$, $x = 2\pi a/\lambda$, a is the radius of a particle, λ is the incident wavelength in a host nonabsorbing medium, $\psi_n(x) = \sqrt{(\pi x/2)}J_{n+1/2}(x)$, $\xi_n(x) = \sqrt{(\pi x/2)}H^{(2)}_{n+1/2}(x)$, $J_{n+1/2}$ and $H^{(2)}_{n+1/2}$ are Bessel and Hankel functions (see Appendix 3). The angular functions $\pi_n(\cos\theta)$, $\tau_n(\cos\theta)$ are determined by the following formulae (see Appendix 3):

$$\pi_n(\cos\theta) = \frac{P_n^{(1)}(\cos\theta)}{\sin\theta}, \qquad \tau_n(\cos\theta) = \frac{dP_n^{(1)}(\cos\theta)}{d\theta} \tag{A2.5}$$

where $P_n^{(1)}(\cos\theta)$ is the associated Legendre polynomial (see Appendix 3), θ is the scattering angle.

Equations (A2.1)–(A2.3) hold for multilayered spherical particles as well, but the values of a_n, b_n should be changed (Kerker, 1969). For instance, it follows for two-layered spheres (Bohren and Huffman, 1983) that:

$$a_n = \frac{\psi_n(y)[\psi_n'(m_2 y) - A_n\chi_n'(m_2 y)] - m_2\psi_n'(y)[\psi_n(m_2 y) - A_n\chi_n(m_2 y)]}{\xi_n(y)[\psi_n'(m_2 y) - A_n\chi_n'(m_2 y)] - m_2\xi_n'(y)[\psi_n(m_2 y) - A_n\chi_n(m_2 y)]} \quad \text{(A2.6)}$$

$$b_n = \frac{m_2\psi_n(y)[\psi_n'(m_2 y) - B_n\chi_n'(m_2 y)] - \psi_n'(y)[\psi_n(m_2 y) - B_n\chi_n(m_2 y)]}{m_2\xi_n(y)[\psi_n'(m_2 y) - B_n\chi_n'(m_2 y)] - \xi_n'(y)[\psi_n(m_2 y) - B_n\chi_n(m_2 y)]} \quad \text{(A2.7)}$$

where

$$A_n = \frac{m_2\psi_n(m_2 x)\psi_n'(m_1 x) - m_1\psi_n'(m_2 x)\psi_n(m_1 x)}{m_2\chi_n(m_2 x)\psi_n'(m_1 x) - m_1\chi_n'(m_2 x)\psi_n(m_1 x)}$$

$$B_n = \frac{m_2\psi_n(m_1 x)\psi_n'(m_2 x) - m_1\psi_n(m_2 x)\psi_n'(m_1 x)}{m_2\chi_n'(m_2 x)\psi_n(m_1 x) - m_1\psi_n'(m_1 x)\chi_n(m_2 x)}$$

and m_1, m_2 are relative to a host medium refractive indices of a core and a shell, respectively, $x = ka$, $y = kb$, $k = 2\pi/\lambda$, a is the radius of a core, b is the radius of a particle.

Values of S_1 and S_2 for chiral spheres can be calculated using equations (A2.1), (A2.2), and

$$S_{12}(\theta) = -S_{21}(\theta) = \sum_{n=1}^{\infty} \frac{2n+1}{n(n+1)} c_n(\pi_n + \tau_n) \quad \text{(A2.8)}$$

where

$$c_n = i\frac{W_n(R)A_n(L) - W_n(L)A_n(R)}{W_n(L)V_n(R) + V_n(L)W_n(R)} \quad \text{(A2.9)}$$

It follows for coefficients a_n, b_n in equations (A2.1) and (A2.2) in the case of chiral spheres that:

$$a_n = \frac{V_n(R)A_n(L) + V_n(L)A_n(R)}{W_n(L)V_n(R) + V_n(L)W_n(R)} \quad \text{(A2.10)}$$

$$b_n = \frac{W_n(L)B_n(R) + W_n(R)B_n(L)}{W_n(L)V_n(R) + V_n(L)W_n(R)} \quad \text{(A2.11)}$$

where

$$W_n(J) = m\psi_n(m_J x)\xi_n'(x) - \xi_n(x)\psi_n'(m_J x)$$

$$V_n(J) = \psi_n(m_J x)\xi_n'(x) - m\xi_n(x)\psi_n'(m_J x)$$

$$A_n(J) = m\psi_n(m_J x)\psi_n'(x) - \psi_n(x)\psi_n'(m_J x)$$

$$B_n(J) = \psi_n(m_J x)\psi_n'(x) - m\psi_n(x)\psi_n'(m_J x)$$

Values of J are equal to L or R, $m_L = N_L/N$, $m_R = N_R/N$, $m = (2m_L m_R)/(m_L + m_R)$, N is the refractive index of a host medium, N_L and N_R are refractive

Table A2.1. Light scattering characteristics for isotropic spheres.

Value	Formula
Q_{ext}	$\frac{2}{x^2}\sum_{n=1}^{\infty}(2n+1)\,\mathrm{Re}(a_n+b_n)$
Q_{sca}	$\frac{2}{x^2}\sum_{n=1}^{\infty}(2n+1)[\lvert a_n\rvert^2+\lvert b_n\rvert^2]$
$p(\theta)$	$\frac{2\pi(i_1+i_2)}{k^2C_{sca}}$, $i_1=\lvert S_1\rvert^2$, $\lvert i_2\rvert=\lvert S_2\rvert^2$, $C_{sca}=\pi a^2Q_{sca}$
g	$\frac{4}{x^2Q_{sca}}\sum_{n=1}^{\infty}\left[\frac{n(n+2)}{n+1}\mathrm{Re}(a_na_{n+1}^*+b_nb_{n+1}^*)+\frac{2n+1}{n(n+1)}\mathrm{Re}(a_nb_n^*)\right]$

indices of particles for left and right handed circularly polarized waves, $x=2\pi a/\lambda$ is the size parameter.

Expressions for the values of the extinction efficiency $Q_{ext}=C_{ext}/(\pi a^2)$, the scattering efficiency $Q_{sca}=C_{sca}/(\pi a^2)$, the phase function $p(\theta)$ at natural light illumination, and the asymmetry parameter $g=\frac{1}{2}\int_0^{\pi}p(\theta)\sin\theta\cos\theta\,d\theta$ are presented in Table A2.1. It follows for polydispersed media that:

$$\sigma_{ext}=N\int_0^{\infty}\pi a^2Q_{ext}f(a)\,da,\qquad \sigma_{sca}=N\int_0^{\infty}\pi a^2Q_{sca}f(a)\,da$$

$$p(\theta)=\frac{2\pi N\int_0^{\infty}(i_1+i_2)f(a)\,da}{k^2\sigma_{sca}},\qquad g=\frac{\int_0^{\infty}a^2Q_{sca}g(a)f(a)\,da}{\int_0^{\infty}a^2Q_{sca}f(a)\,da}$$

Appendix 3

Special functions

The main special functions in the Mie problem are defined as follows:

$$\psi_l(Z) = \sqrt{\frac{\pi Z}{2}} J_{l+1/2}(Z)$$

$$\xi_l(Z) = \sqrt{\frac{\pi Z}{2}} H^{(2)}_{l+1/2}(Z)$$

$$\pi_l(\cos\theta) = \frac{dP_l(\cos\theta)}{d(\cos\theta)}$$

$$\tau_l(\cos\theta) = \frac{dP^1_l(\cos\theta)}{d(\cos\theta)}$$

where $J_{l+1/2}$, $H^{(2)}_{l+1/2}$ are Bessel and Hankel functions, $P_l(\cos\theta)$ and $P^1_l(\cos\theta)$ are Legendre and associated Legendre polynomials, respectively. The same special functions are used in solving the problem of light scattering by multilayered and optically active spheres.

The following formulae are often useful:

$$(-1)^m \int_{-1}^{1} P^m_i(Z) P^m_k(Z)\, dZ = \begin{cases} 0, & i \neq k \\ \dfrac{2}{2i+1} \dfrac{(i+m)!}{(i-m)!}, & i = k \end{cases}$$

$$P_0(Z) = 1, \qquad P_1(Z) = Z, \qquad P_2(Z) = \frac{3Z^2}{2} - \frac{1}{2}, \qquad P^1_1(Z) = \sqrt{1-Z^2}$$

$$J_l(Z) = \left(\frac{Z}{2}\right)^l \sum_{k=0}^{\infty} \frac{(-1)^k \left(\dfrac{Z}{2}\right)^{2k}}{k!\Gamma(l+k+1)}$$

$$\xi_l(Z) = \psi_l(Z) + i\chi_l(Z)$$

$$\psi_l'(Z)\xi_l(Z) - \psi_l(Z)\xi_l'(Z) = 1$$

$$\psi_{l+1}(Z) = \frac{2l+1}{Z}\psi_l(Z) - \psi_{l-1}(Z)$$

$$\chi_{l+1}(Z) = \frac{2l+1}{Z}\chi_l(Z) - \chi_{l-1}(Z)$$

$$\psi_l'(Z) = \frac{l}{2l+1}\psi_{l-1}(Z) - \frac{l+1}{2l+1}\psi_{l+1}(Z)$$

$$\chi_l'(Z) = -\frac{l}{Z}\chi_l(Z) - \chi_{l-1}(Z)$$

$$\psi_0(Z) = \sin Z, \qquad \psi_1(Z) = \frac{\sin Z}{Z} - \cos Z$$

$$\chi_0(Z) = \cos Z, \qquad \chi_1(Z) = \frac{\cos Z}{Z} + \sin Z$$

$$\psi_0'(Z) = \cos Z, \qquad \psi_1'(Z) = \left(1 - \frac{1}{Z^2}\right)\sin Z + \frac{\cos Z}{Z}$$

$$\chi_0'(Z) = -\sin Z, \qquad \chi_1'(Z) = \left(1 - \frac{1}{Z^2}\right)\cos Z - \frac{\sin Z}{Z}$$

$$\pi_1(\cos\theta) = \frac{(2l-1)\cos\theta}{l-1}\pi_{l-1}(\cos\theta) - \frac{l}{l-1}\pi_{l-2}(\cos\theta)$$

$$\tau_l(\cos\theta) = l\pi_l(\cos\theta)\cos\theta - (l+1)\pi_{l-1}(\cos\theta)$$

$$\pi_1(\cos\theta) = 1, \qquad \pi_2(\cos\theta) = 3\cos\theta$$

$$\tau_1(\cos\theta) = \cos\theta, \qquad \tau_2(\cos\theta) = 3\cos 2\theta$$

$$\pi_l(-\cos\theta) = (-1)^{l-1}\pi_l(\cos\theta)$$

$$\tau_l(-\cos\theta) = (-1)^l\tau_l(\cos\theta)$$

$$\pi_l(1) = \tau_l(1) = \frac{l(l+1)}{2}$$

$$\pi_l(-1) = -\tau_l(-1) = (-1)^{l+1}\frac{l(l+1)}{2}$$

Asymptotics are given by the following equations:

1. $Z \to 0$:

$$\psi_l(Z) = \frac{Z^{n+1}}{1 \cdot 3 \cdot 5 \cdots (2l+1)} \left\{ 1 - \frac{\frac{Z^2}{2}}{1!(2l+3)} + \frac{\left(\frac{Z^2}{2}\right)^2}{2!(2l+3)(2l+5)} - \cdots \right\}$$

$$\chi_l(Z) = -\frac{1 \cdot 3 \cdot 5 \cdots (2l-1)}{Z^l} \left\{ 1 - \frac{\frac{Z^2}{2}}{1!(1-2l)} + \frac{\left(\frac{Z^2}{2}\right)^2}{2!(1-2l)(3-2l)} - \cdots \right\}$$

2. $Z \to \infty$:

$$\psi_l(Z) = \frac{\cos\left(Zf - \frac{\pi}{4}\right)}{\sqrt{\sin\tau}}$$

$$\chi_l(Z) = -\frac{\sin\left(Zf - \frac{\pi}{4}\right)}{\sqrt{\sin\tau}}$$

$$\psi_l'(Z) = -\sqrt{\sin\tau}\sin\left(Zf - \frac{\pi}{4}\right)$$

$$\chi_l'(Z) = -\sqrt{\sin\tau}\cos\left(Zf - \frac{\pi}{4}\right)$$

where

$$f = \sin\tau - \tau\cos\tau, \qquad \tau = \arccos\left(\frac{l+\frac{1}{2}}{Z}\right)$$

3. $l \to \infty$, $(l+\frac{1}{2})\theta = \text{const}$:

$$\pi_l(\cos\theta) = \frac{l(l+1)}{2}\{J_0[(l+\tfrac{1}{2})\theta] + J_2[(l+\tfrac{1}{2})\theta]\}$$

$$\tau_l(\cos\theta) = \frac{l(l+1)}{2}\{J_0[(l+\tfrac{1}{2})\theta] - J_2[(l+\tfrac{1}{2})\theta]\}$$

4. $l \to \infty$, $(l+\frac{1}{2})(\pi-\theta) = \text{const}$:

$$\pi_l(\cos\theta) = (-1)^{l-1}\frac{l(l+1)}{2}\{J_0[(l+\tfrac{1}{2})(\pi-\theta)] + J_2[(l+\tfrac{1}{2})(\pi-\theta]\}$$

$$\tau_l(\cos\theta) = (-1)^{1}\frac{l(l+1)}{2}\{J_0[(l+\tfrac{1}{2})(\pi-\theta)] - J_2[(l+\tfrac{1}{2})(\pi-\theta)]\}$$

Appendix 4

Light scattering on the Internet

A lot of radiative transfer and light-scattering codes are available via the Internet. Some Internet sites are presented below. The author of this book can be reached at the following e-mail address: akokhanovsky@hotmail.com. You may also check the website `www.iup.physik.uni-bremen.de/~alexk`

Author	e-mail	Internet address	Programs
B. Draine	draine@astro.princeton.edu	http://www.astro.princeton.edu/~draine	DDSCAT (version 5a10) discrete dipole approximatin code
P. J. Flatau	pflatau@ucsd.edu	http://atol.ucsd.edu/~pflatau/scatlib	SCATTERLIB, library of single light-scattering codes
V. Ilin	vi2087@vi2087.spb.edu	http://www.astro.spbu.ru/staff/ilin2/ilin.html	Database of optical constants, single light-scattering codes
A. Macke	amacke@ifm.geomar.de	http://www.ifm.uni-kiel.de/fb/fb1/me/research/Projekte/RemSens/SourceCodes/codes.html	Ray-tracing codes for nonspherical particles, a Monte Carlo radiative transfer code
B. Michel	info@lightscattering.de	http://www.lightscattering.de	Optics online, light-scattering, effective medium theory
M. Mishchenko	crmim@giss.nasa.gov	http://www.giss.nasa.gov/~crmim/	T-matrix codes for nonspherical particles, a radiative transfer code for semi-infinite turbid media
T. Rother	tom.rother@dlr.de	http://vl.nz.dlr.de	Scattering codes online
T. Wriedt	thw@iwt.uni-bremen.de	http://www.t-matrix.de	Electromagnetic scattering codes; radiative transfer codes

Appendix 5

Phase functions

Table A5.1. Phase functions of selected clouds and aerosols at the wavelength 0.5 μm (g and ω_0 are the asymmetry parameter and single scattering albedo, respectively).

θ (deg)	Water cloud C1 $g = 0.8552$ $\omega_0 = 1$	Ice cloud (Liou, 1992) $g = 0.75$ $\omega_0 = 1$	Oceanic aerosol $g = 0.7884$ $\omega_0 = 1$	Dust aerosol $g = 0.8803$ $\omega_0 = 0.6491$	Water soluble aerosol $g = 0.7259$ $\omega_0 = 0.8666$	Soot aerosol $g = 0.3616$ $\omega_0 = 0.2311$
0.0	3.386E+03	1.083E+05	1.008E+03	9.522E+03	2.348E+02	3.840
0.1	3.361E+03	6.037E+04	9.931E+02	9.211E+03	2.341E+02	3.840
0.2	3.286E+03	3.231E+04	9.532E+02	8.368E+03	2.320E+02	3.840
0.3	3.166E+03	1.809E+04	8.934E+02	7.161E+03	2.286E+02	3.840
0.4	3.005E+03	9.985E+03	8.227E+02	5.833E+03	2.239E+02	3.839
0.5	2.813E+03	5.477E+03	7.495E+02	4.590E+03	2.182E+02	3.839
0.6	2.597E+03	3.210E+03	6.798E+02	3.564E+03	2.116E+02	3.839
0.7	2.366E+03	2.106E+03	6.166E+02	2.789E+03	2.042E+02	3.839
0.8	2.130E+03	1.502E+03	5.610E+02	2.232E+03	1.964E+02	3.838
0.9	1.894E+03	1.095E+03	5.119E+02	1.832E+03	1.882E+02	3.838
1.0	1.667E+03	7.875E+02	4.686E+02	1.530E+03	1.798E+02	3.838
1.1	1.451E+03	5.550E+02	4.299E+02	1.289E+03	1.715E+02	3.837
1.2	1.253E+03	3.855E+02	3.952E+02	1.093E+03	1.633E+02	3.837
1.3	1.074E+03	2.758E+02	3.643E+02	9.328E+02	1.554E+02	3.836
1.4	9.133E+02	2.027E+02	3.365E+02	8.042E+02	1.479E+02	3.836
1.5	7.728E+02	1.563E+02	3.116E+02	7.004E+02	1.407E+02	3.835
1.6	6.511E+02	1.267E+02	2.891E+02	6.143E+02	1.340E+02	3.835
1.7	5.470E+02	1.069E+02	2.687E+02	5.409E+02	1.278E+02	3.834
1.8	4.587E+02	9.266E+01	2.502E+02	4.780E+02	1.220E+02	3.833
1.9	3.846E+02	8.151E+01	2.334E+02	4.246E+02	1.166E+02	3.832
2.0	3.229E+02	7.210E+01	2.181E+02	3.795E+02	1.116E+02	3.832

(*continued*)

Table 5.1 (*cont.*)

θ (deg)	Water cloud C1 $g = 0.8552$ $\omega_0 = 1$	Ice cloud (Liou, 1992) $g = 0.75$ $\omega_0 = 1$	Oceanic aerosol $g = 0.7884$ $\omega_0 = 1$	Dust aerosol $g = 0.8803$ $\omega_0 = 0.6491$	Water soluble aerosol $g = 0.7259$ $\omega_0 = 0.8666$	Soot aerosol $g = 0.3616$ $\omega_0 = 0.2311$
3.0	7.524E+01	2.223E+01	1.196E+02	1.479E+02	7.572E+01	3.822
4.0	3.437E+01	9.666	7.321E+01	7.243E+01	5.472E+01	3.708
5.0	2.146E+01	5.198	4.863E+01	4.075E+01	4.142E+01	3.790
6.0	1.538E+01	3.208	3.441E+01	2.525E+01	3.247E+01	3.769
7.0	1.212E+01	2.182	2.562E+01	1.706E+01	2.617E+01	3.744
8.0	9.934	1.598	1.995E+01	1.217E+01	2.155E+01	3.716
9.0	8.871	1.236	1.617E+01	9.058	1.805E+01	3.685
10.0	8.316	1.013	1.354E+01	7.083	1.535E+01	3.650
11.0	7.378	8.296E−01	1.157E+01	5.611	1.320E+01	3.613
12.0	6.786	7.494E 01	9.983	4.577	1.147E+01	3.573
13.0	6.371	6.753E−01	8.735	3.866	1.008E+01	3.531
14.0	5.912	6.220E−01	7.752	3.318	8.939	3.486
15.0	5.714	5.681E−01	7.006	2.924	8.0004	3.440
16.0	5.319	5.248E−01	6.385	2.602	7.0291	3.391
17.0	5.018	4.883E−01	5.850	2.325	6.555	3.341
18.0	4.755	4.598E−01	5.418	2.099	5.991	3.290
19.0	4.493	4.409E−01	5.056	1.947	5.512	3.237
20.0	4.183	4.227E−01	4.713	1.804	5.093	3.183
21.0	3.982	1.935	4.376	1.683	4.718	3.128
22.0	3.738	5.333	4.024	1.580	4.380	3.072
23.0	3.471	6.137	3.720	1.482	4.074	3.015
24.0	3.322	6.043	3.478	1.394	3.796	2.958
25.0	3.092	4.660	3.266	1.313	3.543	2.900
26.0	2.915	3.665	3.061	1.243	3.314	2.843
27.0	2.744	2.955	2.869	1.180	3.108	2.785
28.0	2.586	2.404	2.711	1.125	2.922	2.727
29.0	2.422	1.982	2.577	1.071	2.754	2.669
30.0	2.275	1.638	2.443	1.018	2.601	2.611
31.0	2.138	1.342	2.318	9.733E−01	2.462	2.553
32.0	2.013	1.130	2.184	9.366E−01	2.335	2.496
33.0	1.864	9.85E−01	2.041	9.009E−01	2.215	2.439
34.0	1.768	8.54E−01	1.910	8.594E−01	2.102	2.383
35.0	1.643	7.24E−01	1.795	8.270E−01	1.994	2.327
36.0	1.548	6.26E−01	1.696	7.972E−01	1.890	2.272
37.0	1.443	5.55E−01	1.602	7.610E−01	1.787	2.217
38.0	1.347	5.55E−01	1.509	7.221E−01	1.689	2.163
39.0	1.261	4.45E−01	1.426	6.884E−01	1.598	2.110
40.0	1.179	4.12E−01	1.352	6.587E−01	1.514	2.058
41.0	1.104	4.03E−01	1.281	6.351E−01	1.439	2.006
42.0	1.026	3.74E−01	1.209	6.108E−01	1.370	1.955
43.0	9.604E−01	3.57E−01	1.141	5.860E−01	1.305	1.906

Table 5.1 (*cont.*)

θ (deg)	Water cloud C1 $g = 0.8552$ $\omega_0 = 1$	Ice cloud (Liou, 1992) $g = 0.75$ $\omega_0 = 1$	Oceanic aerosol $g = 0.7884$ $\omega_0 = 1$	Dust aerosol $g = 0.8803$ $\omega_0 = 0.6491$	Water soluble aerosol $g = 0.7259$ $\omega_0 = 0.8666$	Soot aerosol $g = 0.3616$ $\omega_0 = 0.2311$
44.0	8.976E−01	4.09E−01	1.077	5.598E−01	1.244	1.857
45.0	8.312E−01	5.44E−01	1.011	5.360E−01	1.188	1.809
46.0	7.795E−01	7.44E−01	9.467E−01	5.139E−01	1.137	1.762
47.0	7.260E−01	8.88E−01	8.903E−01	4.948E−01	1.091	1.716
48.0	6.746E−01	8.78E−01	8.396E−01	4.783E−01	1.046	1.670
49.0	6.313E−01	8.10E−01	7.938E−01	4.595E−01	1.003	1.626
50.0	5.839E−01	7.15E−01	7.493E−01	4.396E−01	9.607E−01	1.583
51.0	5.467E−01	6.56E−01	7.070E−01	4.243E−01	9.199E−01	1.541
52.0	5.055E−01	5.91E−01	6.664E−01	4.082E−01	8.798E−01	1.500
53.0	4.731E−01	5.09E−01	6.309E−01	3.914E−01	8.401E−01	1.460
54.0	4.370E−01	4.52E−01	5.989E−01	3.738E−01	8.016E−01	1.421
55.0	4.098E−01	4.16E−01	5.699E−01	3.582E−01	7.651E−01	1.383
56.0	3.773E−01	3.89E−01	5.416E−01	3.435E−01	7.309E−01	1.346
57.0	3.504E−01	3.72E−01	5.124E−01	3.291E−01	6.988E−01	1.310
58.0	3.287E−01	3.58E−01	4.862E−01	3.156E−01	6.687E−01	1.275
59.0	3.022E−01	3.47E−01	4.611E−01	3.032E−01	6.404E−01	1.240
60.0	2.808E−01	3.46E−01	4.347E−01	2.916E−01	6.136E−01	1.207
61.0	2.606E−01	3.44E−01	4.088E−01	2.801E−01	5.880E−01	1.175
62.0	2.423E−01	3.43E−01	3.856E−01	2.675E−01	5.637E−01	1.144
63.0	2.239E−01	3.44E−01	3.647E−01	2.571E−01	5.411E−01	1.114
64.0	2.076E−01	3.43E−01	3.454E−01	2.469E−01	5.199E−01	1.084
65.0	1.931E−01	3.40E−01	3.276E−01	2.373E−01	4.999E−01	1.056
66.0	1.785E−01	3.38E−01	3.109E−01	2.285E−01	4.810E−01	1.028
67.0	1.648E−01	3.37E−01	2.956E−01	2.200E−01	4.626E−01	1.002
68.0	1.532E−01	3.36E−01	2.811E−01	2.109E−01	4.446E−01	9.758E−01
69.0	1.419E−01	3.36E−01	2.667E−01	2.023E−01	4.273E−01	9.508E−01
70.0	1.322E−01	3.36E−01	2.536E−01	1.950E−01	4.107E−01	9.268E−01
71.0	1.214E−01	3.35E−01	2.413E−01	1.878E−01	3.946E−01	9.035E−01
72.0	1.134E−01	3.34E−01	2.300E−01	1.803E−01	3.787E−01	8.811E−01
73.0	1.049E−01	3.33E−01	2.193E−01	1.730E−01	3.635E−01	8.594E−01
74.0	9.713E−02	3.32E−01	2.092E−01	1.656E−01	3.492E−01	8.356E−01
75.0	9.033E−02	3.31E−01	1.993E−01	1.592E−01	3.361E−01	8.185E−01
76.0	8.429E−02	3.29E−01	1.899E−01	1.531E−01	3.240E−01	7.991E−01
77.0	7.781E−02	3.27E−01	1.815E−01	1.474E−01	3.130E−01	7.805E−01
78.0	7.263E−02	3.25E−01	1.733E−01	1.424E−01	3.028E−01	7.626E−01
79.0	6.780E−02	3.24E−01	1.647E−01	1.375E−01	2.930E−02	7.454E−01
80.0	6.780E−02	3.22E−01	1.562E−01	1.326E−01	2.836E−01	7.290E−01
81.0	6.285E−02	3.21E−01	1.486E−01	1.277E−01	2.742E−01	7.131E−01
82.0	5.887E−02	3.18E−01	1.425E−01	1.228E−01	2.650E−01	6.980E−01
83.0	5.510E−02	3.14E−01	1.366E−01	1.182E−01	2.560E−01	6.835E−01

(*continued*)

Table 5.1 (*cont.*)

θ (deg)	Water cloud C1 $g = 0.8552$ $\omega_0 = 1$	Ice cloud (Liou, 1992) $g = 0.75$ $\omega_0 = 1$	Oceanic aerosol $g = 0.7884$ $\omega_0 = 1$	Dust aerosol $g = 0.8803$ $\omega_0 = 0.6491$	Water soluble aerosol $g = 0.7259$ $\omega_0 = 0.8666$	Soot aerosol $g = 0.3616$ $\omega_0 = 0.2311$
84.0	5.155E−02	3.11E−01	1.312E−01	1.138E−01	2.474E−01	6.696E−01
85.0	4.834E−02	3.11E−01	1.262E−01	1.098E−01	2.392E−01	6.563E−01
86.0	4.564E−02	3.08E−01	1.214E−01	1.059E−01	2.313E−01	6.436E−01
87.0	4.056E−02	3.03E−01	1.168E−01	1.022E−01	2.237E−01	6.315E−01
88.0	3.822E−02	2.98E−01	1.127E−01	9.863E−02	2.165E−01	6.200E−01
89.0	3.640E−02	2.95E−01	1.090E−01	9.532E−02	2.095E−01	6.090E−01
90.0	3.442E−02	2.91E−01	1.055E−01	9.224E−02	2.028E−01	5.986E−01
91.0	3.293E−02	2.89E−01	1.024E−01	8.925E−02	1.965E−01	5.886E−01
92.0	3.153E−02	2.87E−01	9.985E−02	8.637E−02	1.906E−01	5.792E−01
93.0	3.021E−02	2.85E−01	9.762E−02	8.377E−02	1.851E−01	5.703E−01
94.0	2.926E−02	2.83E−01	9.549E−02	8.139E−02	1.800E−01	5.619E−01
95.0	2.851E−02	2.82E−01	9.339E−02	7.907E−02	1.751E−01	5.539E−01
96.0	2.768E−02	2.81E−01	9.099E−02	7.776E−02	1.706E−01	5.464E−01
97.0	2.699E−02	2.78E−01	8.867E−02	7.458E−02	1.663E−01	5.393E−01
98.0	2.659E−02	2.75E−01	8.673E−02	7.252E−02	1.622E−01	5.327E−01
99.0	2.612E−02	2.72E−01	8.492E−02	7.057E−02	1.585E−01	5.264E−01
100.0	2.560E−02	2.68E−01	8.327E−02	6.874E−02	1.549E−01	5.206E−01
101.0	2.521E−02	2.62E−01	8.198E−02	6.698E−02	1.514E−01	5.152E−01
102.0	2.506E−02	2.56E−01	8.084E−02	6.530E−02	1.481E−01	5.101E−01
103.0	2.459E−02	2.53E−01	7.966E−02	6.368E−02	1.449E−01	5.054E−01
104.0	2.460E−02	2.48E−01	7.842E−02	6.217E−02	1.418E−01	5.011E−01
105.0	2.453E−02	2.42E−01	7.727E−02	6.078E−02	1.388E−01	5.971E−01
106.0	2.467E−02	2.35E−01	7.624E−02	5.948E−02	1.359E−01	5.934E−01
107.0	2.463E−02	2.27E−01	7.543E−02	5.825E−02	1.331E−01	5.901E−01
108.0	2.482E−02	2.21E−01	7.484E−02	5.706E−02	1.303E−01	5.870E−01
109.0	2.524E−02	2.16E−01	7.409E−02	5.593E−02	1.276E−01	5.843E−01
110.0	2.571E−02	2.11E−01	7.328E−02	5.483E−02	1.251E−01	4.812E−01
111.0	2.659E−02	2.05E−01	7.269E−02	5.379E−02	1.227E−01	4.797E−01
112.0	2.788E−02	1.99E−01	7.231E−02	5.291E−02	1.206E−01	4.777E−01
113.0	2.942E−02	1.96E−01	7.182E−02	5.217E−02	1.189E−01	4.761E−01
114.0	3.175E−02	1.89E−01	7.117E−02	5.147E−02	1.174E−01	4.747E−01
115.0	3.386E−02	1.79E−01	7.004E−02	5.079E−02	1.161E−01	4.735E−01
116.0	3.654E−02	1.64E−01	6.836E−02	5.017E−02	1.151E−01	4.726E−01
117.0	3.916E−02	1.42E−01	6.661E−02	4.960E−02	1.142E−01	4.718E−01
118.0	4.138E−02	1.27E−01	6.544E−02	4.905E−02	1.134E−01	4.713E−01
119.0	4.363E−02	1.18E−01	6.471E−02	4.849E−02	1.127E−01	4.710E−01
120.0	4.548E−02	1.11E−01	6.397E−02	4.794E−02	1.121E−01	4.709E−01
121.0	4.701E−02	1.04E−01	6.314E−02	4.747E−02	1.117E−01	4.710E−01
122.0	4.785E−02	9.93E−02	6.280E−02	4.709E−02	1.114E−01	4.712E−01
123.0	4.835E−02	9.86E−02	6.342E−02	4.678E−02	1.112E−01	4.716E−01
124.0	4.873E−02	9.79E−02	6.483E−02	4.652E−02	1.112E−01	4.721E−01

Table 5.1 *(cont.)*

θ (deg)	Water cloud C1 $g = 0.8552$ $\omega_0 = 1$	Ice cloud (Liou, 1992) $g = 0.75$ $\omega_0 = 1$	Oceanic aerosol $g = 0.7884$ $\omega_0 = 1$	Dust aerosol $g = 0.8803$ $\omega_0 = 0.6491$	Water soluble aerosol $g = 0.7259$ $\omega_0 = 0.8666$	Soot aerosol $g = 0.3616$ $\omega_0 = 0.2311$
125.0	4.792E−02	9.70E−02	6.632E−02	4.626E−02	1.113E−01	4.729E−01
126.0	4.757E−02	9.67E−02	6.750E−02	4.603E−02	1.116E−01	4.737E−01
127.0	4.717E−02	9.68E−02	6.842E−02	4.584E−02	1.121E−01	4.747E−01
128.0	4.714E−02	9.76E−02	6.938E−02	4.571E−02	1.130E−01	4.758E−01
129.0	4.757E−02	9.97E−02	7.066E−02	4.570E−02	1.141E−01	4.770E−01
130.0	4.933E−02	1.01E−01	7.267E−02	4.579E−02	1.156E−01	4.784E−01
131.0	5.242E−02	1.03E−01	7.517E−02	4.593E−02	1.174E−01	4.798E−01
132.0	5.760E−02	1.04E−01	7.760E−02	4.610E−02	1.193E−01	4.810E−01
133.0	6.597E−02	1.02E−01	8.005E−02	4.629E−02	1.214E−01	4.830E−01
134.0	7.796E−02	1.02E−01	8.321E−02	4.645E−02	1.235E−01	4.847E−01
135.0	9.477E−02	9.96E−02	8.768E−02	4.662E−02	1.257E−01	4.865E−01
136.0	1.150E−01	1.02E−01	9.384E−02	4.684E−02	1.279E−01	4.884E−01
137.0	1.435E−01	1.12E−01	1.016E−01	4.712E−02	1.302E−01	4.903E−01
138.0	1.756E−01	1.21E−01	1.104E−01	4.747E−02	1.327E−01	4.923E−01
139.0	2.109E−01	1.27E−01	1.201E−01	4.789E−02	1.356E−01	4.943E−01
140.0	2.467E−01	1.35E−01	1.310E−01	4.842E−02	1.389E−01	4.964E−01
141.0	2.773E−01	1.44E−01	1.439E−01	4.918E−02	1.429E−01	4.985E−01
142.0	2.959E−01	1.54E−01	1.594E−01	5.022E−02	1.476E−01	5.007E−01
143.0	2.988E−01	1.66E−01	1.778E−01	5.152E−02	1.532E−01	5.029E−01
144.0	2.886E−01	1.78E−01	1.986E−01	5.299E−02	1.595E−01	5.051E−01
145.0	2.622E−01	1.91E−01	2.213E−01	5.458E−02	1.664E−01	5.073E−01
146.0	2.289E−01	2.04E−01	2.447E−01	5.627E−02	1.738E−01	5.096E−01
147.0	1.980E−01	2.13E−01	2.664E−01	5.816E−02	1.816E−01	5.118E−01
148.0	1.734E−01	2.25E−01	2.827E−01	6.044E−02	1.900E−01	5.140E−01
149.0	1.613E−01	2.43E−01	2.934E−01	6.322E−02	1.990E−01	5.163E−01
150.0	1.546E−01	2.63E−01	2.996E−01	6.654E−02	2.088E−01	5.185E−01
151.0	1.554E−01	2.87E−01	3.028E−01	7.036E−02	2.192E−01	5.207E−01
152.0	1.564E−01	3.04E−01	3.037E−01	7.468E−02	2.302E−01	5.229E−01
153.0	1.565E−01	3.08E−01	3.015E−01	7.975E−02	2.419E−01	5.251E−01
154.0	1.561E−01	3.09E−01	2.972E−01	8.587E−02	2.542E−01	5.272E−01
155.0	1.534E−01	3.07E−01	2.933E−01	9.325E−02	2.673E−01	5.293E−01
156.0	1.507E−01	2.99E−01	2.916E−01	1.020E−01	2.809E−01	5.314E−01
157.0	1.473E−01	2.80E−01	2.934E−01	1.116E−01	2.946E−01	5.334E−01
158.0	1.472E−01	2.63E−01	2.968E−01	1.219E−01	3.079E−01	5.354E−01
159.0	1.425E−01	2.52E−01	2.984E−01	1.335E−01	3.207E−01	5.373E−01
160.0	1.449E−01	2.36E−01	3.009E−01	1.467E−01	3.331E−01	5.392E−01
161.0	1.435E−01	2.13E−01	3.067E−01	1.608E−01	3.447E−01	5.410E−01
162.0	1.398E−01	1.95E−01	3.156E−01	1.736E−01	3.548E−01	5.427E−01
163.0	1.421E−01	1.77E−01	3.223E−01	1.842E−01	3.635E−01	5.444E−01
164.0	1.394E−01	1.66E−01	3.286E−013	1.937E−01	3.715E−01	5.459E−01

(continued)

Table 5.1 (*cont.*)

θ (deg)	Water cloud C1 $g = 0.8552$ $\omega_0 = 1$	Ice cloud (Liou, 1992) $g = 0.75$ $\omega_0 = 1$	Oceanic aerosol $g = 0.7884$ $\omega_0 = 1$	Dust aerosol $g = 0.8803$ $\omega_0 = 0.6491$	Water soluble aerosol $g = 0.7259$ $\omega_0 = 0.8666$	Soot aerosol $g = 0.3616$ $\omega_0 = 0.2311$
165.0	1.399E−01	1.57E−01	3.399E−01	2.028E−01	3.800E−01	5.475E−01
166.0	1.444E−01	1.58E−01	3.544E−01	2.102E−01	3.896E−01	5.489E−01
167.0	1.440E−01	1.69E−01	3.652E−01	2.173E−01	4.010E−01	5.502E−01
168.0	1.488E−01	1.90E−01	3.776E−01	2.255E−01	4.152E−01	5.515E−01
169.0	1.491E−01	2.35E−01	3.868E−01	2.333E−01	4.321E−01	5.127E−01
170.0	1.511E−01	2.71E−01	3.914E−01	2.428E−01	4.517E−01	5.538E−01
171.0	1.700E−01	2.84E−01	4.000E−01	2.538E−01	4.739E−01	5.547E−01
172.0	1.696E−01	2.88E−01	4.065E−01	2.654E−01	4.979E−01	5.556E−01
173.0	1.795E−01	2.63E−01	4.210E−01	2.781E−01	5.236E−01	5.564E−01
174.0	2.040E−01	2.67E−01	4.433E−01	2.910E−01	5.503E−01	5.571E−01
175.0	2.400E−01	3.04E−01	4.699E−01	3.014E−01	5.774E−01	5.577E−01
176.0	2.984E−01	3.81E−01	4.969E−01	3.048E−01	6.043E−01	5.582E−01
177.0	4.177E−01	5.76E−01	5.258E−01	2.955E−01	6.307E−01	5.586E−01
178.0	4.992E−01	7.98E−01	5.604E−01	2.706E−01	6.566E−01	5.588E−01
179.0	3.968E−01	1.01	6.173E−01	2.514E−01	6.801E−01	5.590E−01
180.0	6.779E−01	1.18	7.098E−01	2.588E−01	6.912E−01	5.590E−01

Table A5.2. Phase functions of large transparent spheres in the framework of the geometrical optics approximation at different refractive indices of particles $n = 1.1, 1.2, 1.3, 1.333, 1.4, 1.5, 1.6, 1.7, 1.8, 1.9, 2.0$, calculated with my tracing Monte Carlo code (Kokhanovsky and Nakajima, 1998).

	1.1	1.2	1.3	1.333	1.4	1.5	1.6	1.7	1.8	1.9	2.0
0.0	1.201E+02	3.631E+01	1.911E+01	1.635E+01	1.264E+01	9.232E+00	7.319E+00	6.073E+00	5.263E+00	4.624E+00	4.324E+00
1.0	1.184E+02	3.610E+01	1.905E+01	1.634E+01	1.260E+01	9.239E+00	7.316E+00	6.068E+00	5.224E+00	4.603E+00	4.249E+00
1.5	1.160E+02	3.581E+01	1.897E+01	1.624E+01	1.253E+01	9.234E+00	7.299E+00	6.059E+00	5.204E+00	4.582E+00	4.206E+00
2.0	1.128E+02	3.547E+01	1.883E+01	1.612E+01	1.244E+01	9.189E+00	7.263E+00	6.039E+00	5.182E+00	4.562E+00	4.175E+00
2.5	1.088E+02	3.519E+01	1.868E+01	1.604E+01	1.238E+01	9.123E+00	7.242E+00	6.020E+00	5.161E+00	4.534E+00	4.163E+00
3.0	1.042E+02	3.477E+01	1.857E+01	1.593E+01	1.231E+01	9.074E+00	7.201E+00	5.991E+00	5.146E+00	4.511E+00	4.152E+00
3.5	9.913E+01	3.421E+01	1.845E+01	1.581E+01	1.220E+01	9.031E+00	7.163E+00	5.948E+00	5.113E+00	4.489E+00	4.134E+00
4.0	9.378E+01	3.365E+01	1.830E+01	1.572E+01	1.211E+01	8.976E+00	7.130E+00	5.910E+00	5.069E+00	4.451E+00	4.106E+00
4.5	8.819E+01	3.301E+01	1.811E+01	1.560E+01	1.208E+01	8.917E+00	7.088E+00	5.884E+00	5.043E+00	4.423E+00	4.066E+00
5.0	8.252E+01	3.235E+01	1.791E+01	1.545E+01	1.200E+01	8.877E+00	7.050E+00	5.859E+00	5.020E+00	4.395E+00	4.048E+00
5.5	7.686E+01	3.162E+01	1.770E+01	1.529E+01	1.188E+01	8.833E+00	7.020E+00	5.815E+00	4.985E+00	4.362E+00	4.032E+00
6.0	7.132E+01	3.082E+01	1.749E+01	1.512E+01	1.177E+01	8.766E+00	6.987E+00	5.780E+00	4.950E+00	4.344E+00	4.004E+00
6.5	6.593E+01	3.001E+01	1.724E+01	1.495E+01	1.168E+01	8.698E+00	6.944E+00	5.765E+00	4.923E+00	4.317E+00	3.986E+00
7.0	6.076E+01	2.922E+01	1.699E+01	1.476E+01	1.156E+01	8.633E+00	6.901E+00	5.738E+00	4.903E+00	4.293E+00	3.967E+00
7.5	5.586E+01	2.838E+01	1.673E+01	1.456E+01	1.145E+01	8.569E+00	6.851E+00	5.700E+00	4.885E+00	4.275E+00	3.947E+00
8.0	5.127E+01	2.747E+01	1.645E+01	1.436E+01	1.134E+01	8.506E+00	6.813E+00	5.666E+00	4.853E+00	4.258E+00	3.940E+00
8.5	4.697E+01	2.659E+01	1.619E+01	1.414E+01	1.120E+01	8.422E+00	6.767E+00	5.633E+00	4.814E+00	4.224E+00	3.918E+00
9.0	4.297E+01	2.572E+01	1.589E+01	1.393E+01	1.107E+01	8.704E+00	6.709E+00	5.582E+00	4.786E+00	4.188E+00	3.873E+00
9.5	3.926E+01	2.482E+01	1.559E+01	1.370E+01	1.094E+01	8.793E+00	6.657E+00	5.537E+00	4.762E+00	4.169E+00	3.854E+00
10.0	1.583E+01	2.394E+01	1.532E+01	1.346E+01	1.079E+01	8.483E+00	6.610E+00	5.510E+00	4.735E+00	4.150E+00	3.847E+00
10.5	3.267E+01	2.305E+01	1.499E+01	1.324E+01	1.066E+01	8.327E+00	6.567E+00	5.465E+00	4.708E+00	4.127E+00	3.845E+00
11.0	2.980E+01	2.217E+01	1.464E+01	1.302E+01	1.051E+01	8.205E+00	6.516E+00	5.421E+00	4.674E+00	4.097E+00	3.827E+00
11.5	2.720E+01	2.131E+01	1.431E+01	1.276E+01	1.036E+01	8.085E+00	6.489E+00	5.387E+00	4.625E+00	4.073E+00	3.777E+00
12.0	2.481E+01	2.045E+01	1.400E+01	1.250E+01	1.022E+01	7.974E+00	6.413E+00	5.339E+00	4.593E+00	4.044E+00	3.744E+00
12.5	2.263E+01	1.961E+01	1.369E+01	1.226E+01	1.008E+01	7.872E+00	6.307E+00	5.297E+00	4.578E+00	4.018E+00	3.720E+00
13.0	2.066E+01	1.878E+01	1.337E+01	1.201E+01	9.923E+00	7.765E+00	6.242E+00	5.260E+00	4.545E+00	3.998E+00	3.693E+00
13.5	1.886E+01	1.797E+01	1.304E+01	1.176E+01	9.750E+01	7.658E+00	6.184E+00	5.214E+00	4.504E+00	3.973E+00	3.681E+00

(continued)

Table A5.2 *(cont.)*

	1.1	1.2	1.3	1.333	1.4	1.5	1.6	1.7	1.8	1.9	2.0
14.0	1.724E+01	1.720E+01	1.272E+01	1.151E+01	9.597E+00	7.561E+00	6.126E+00	5.166E+00	4.475E+00	3.943E+00	3.668E+00
14.5	1.576E+01	1.646E+01	1.240E+01	1.126E+01	9.440E+00	7.462E+00	6.054E+00	5.121E+00	4.443E+00	3.916E+00	3.646E+00
15.0	1.441E+01	1.574E+01	1.207E+01	1.100E+01	9.288E+00	7.358E+00	5.991E+00	5.078E+00	4.407E+00	3.895E+00	3.630E+00
15.5	1.318E+01	1.504E+01	1.175E+01	1.075E+01	9.132E+00	7.250E+00	5.933E+00	5.035E+00	4.369E+00	3.871E+00	3.616E+00
16.0	1.207E+01	1.435E+01	1.142E+01	1.050E+01	8.979E+00	7.143E+00	5.873E+00	4.986E+00	4.335E+00	3.845E+00	3.613E+00
16.5	1.106E+01	1.370E+01	1.110E+01	1.025E+01	8.819E+00	7.042E+00	5.815E+00	4.944E+00	4.308E+00	3.816E+00	3.631E+00
17.0	1.013E+01	1.307E+01	1.079E+01	9.993E+00	8.646E+00	6.939E+00	5.750E+00	4.901E+00	4.269E+00	3.788E+00	3.558E+00
17.5	9.290E+00	1.245E+01	1.048E+01	9.739E+00	8.491E+00	6.840E+00	5.676E+00	4.854E+00	4.232E+00	3.761E+00	3.463E+00
18.0	8.534E+00	1.186E+01	1.017E+01	9.489E+00	8.347E+00	6.734E+00	5.603E+00	4.810E+00	4.198E+00	3.736E+00	3.439E+00
18.5	7.845E+00	1.130E+01	9.867E+00	9.239E+00	8.276E+00	6.626E+00	5.537E+00	4.759E+00	4.161E+00	3.713E+00	3.411E+00
19.0	7.207E+00	1.077E+01	9.584E+00	8.999E+00	8.055E+00	6.519E+00	5.470E+00	4.706E+00	4.128E+00	3.687E+00	3.391E+00
19.5	6.624E+00	1.026E+01	9.299E+00	8.767E+00	7.737E+00	6.416E+00	5.409E+00	4.657E+00	4.091E+00	3.657E+00	3.374E+00
20.0	6.085E+00	9.765E+00	9.012E+00	8.531E+00	7.548E+00	6.318E+00	5.343E+00	4.608E+00	4.054E+00	3.630E+00	3.350E+00
20.5	5.594E+00	9.293E+00	8.741E+00	8.303E+00	7.388E+00	6.209E+00	5.268E+00	4.554E+00	4.014E+00	3.606E+00	3.324E+00
21.0	5.151E+00	8.849E+00	8.467E+00	8.075E+00	7.229E+00	6.108E+00	5.198E+00	4.505E+00	3.972E+00	3.576E+00	3.306E+00
21.5	4.748E+00	8.421E+00	8.200E+00	7.845E+00	7.063E+00	6.011E+00	5.132E+00	4.463E+00	3.936E+00	3.551E+00	3.290E+00
22.0	4.373E+00	8.017E+00	7.938E+00	7.630E+00	6.904E+00	5.898E+00	5.065E+00	4.414E+00	3.897E+00	3.521E+00	3.264E+00
22.5	4.033E+00	7.628E+00	7.682E+00	7.413E+00	6.746E+00	5.792E+00	4.997E+00	4.362E+00	3.858E+00	3.489E+00	3.241E+00
23.0	3.718E+00	7.256E+00	7.432E+00	7.199E+00	6.592E+00	5.694E+00	4.929E+00	4.311E+00	3.827E+00	3.464E+00	3.222E+00
23.5	3.425E+00	6.903E+00	7.187E+00	6.992E+00	6.440E+00	5.594E+00	4.859E+00	4.261E+00	3.792E+00	3.434E+00	3.197E+00
24.0	3.160E+00	6.564E+00	6.955E+00	6.785E+00	6.289E+00	5.499E+00	4.790E+00	4.235E+00	3.748E+00	3.405E+00	3.174E+00
24.5	2.916E+00	4.244E+00	6.725E+00	6.591E+00	6.140E+00	5.400E+00	4.719E+00	4.196E+00	3.711E+00	3.382E+00	3.156E+00
25.0	2.691E+00	5.945E+00	6.499E+00	6.392E+00	5.996E+00	5.301E+00	4.646E+00	4.133E+00	3.675E+00	3.351E+00	3.133E+00
25.5	2.483E+00	5.658E+00	6.283E+00	6.195E+00	5.853E+00	5.206E+00	4.581E+00	4.077E+00	3.638E+00	3.322E+00	3.112E+00
26.0	2.287E+00	5.387E+00	6.073E+00	6.015E+00	5.708E+00	5.106E+00	4.514E+00	4.024E+00	3.605E+00	3.294E+00	3.095E+00
26.5	2.107E+00	5.119E+00	5.867E+00	5.830E+00	5.569E+00	5.002E+00	4.445E+00	3.975E+00	3.566E+00	3.260E+00	3.075E+00
27.0	1.944E+00	4.859E+00	5.662E+00	5.648E+00	5.431E+00	4.905E+00	4.377E+00	3.919E+00	3.526E+00	3.232E+00	3.052E+00
27.5	1.791E+00	4.623E+00	5.508E+00	5.478E+00	5.293E+00	4.812E+00	4.314E+00	3.867E+00	3.489E+00	3.211E+00	3.033E+00
28.0	1.651E+00	4.403E+00	5.344E+00	5.311E+00	5.157E+00	4.724E+00	4.247E+00	3.816E+00	3.453E+00	3.183E+00	3.013E+00
28.5	1.519E+00	4.188E+00	5.131E+00	5.145E+00	5.023E+00	4.632E+00	4.177E+00	3.766E+00	3.416E+00	3.152E+00	2.994E+00
29.0	1.398E+00	3.977E+00	4.944E+00	4.979E+00	4.894E+00	4.536E+00	4.110E+00	3.719E+00	3.377E+00	3.122E+00	2.981E+00

29.5	1.287E+00	3.778E+00	4.769E+00	4.817E+00	4.767E+00	4.442E+00	4.042E+00	3.668E+00	3.340E+00	3.093E+00	2.963E+00
30.0	1.183E+00	3.595E+00	4.597E+00	4.671E+00	4.644E+00	4.349E+00	3.974E+00	3.619E+00	3.303E+00	3.069E+00	2.942E+00
30.5	1.089E+00	3.425E+00	4.433E+00	4.522E+00	4.521E+00	4.265E+00	3.909E+00	3.569E+00	3.264E+00	3.040E+00	2.928E+00
31.0	1.000E+00	3.259E+00	4.276E+00	4.374E+00	4.402E+00	4.183E+00	3.846E+00	3.516E+00	3.229E+00	3.008E+00	2.914E+00
31.5	9.122E−01	3.107E+00	4.120E+00	4.238E+00	4.286E+00	4.092E+00	3.782E+00	3.471E+00	3.199E+00	2.981E+00	2.901E+00
32.0	8.349E−01	2.958E+00	3.971E+00	4.102E+00	4.168E+00	4.006E+00	3.828E+00	3.425E+00	3.163E+00	2.955E+00	2.896E+00
32.5	7.642E−01	2.790E+00	3.830E+00	3.968E+00	4.057E+00	3.920E+00	3.830E+00	3.372E+00	3.123E+00	2.929E+00	2.890E+00
33.0	6.958E−01	2.635E+00	3.690E+00	3.843E+00	3.944E+00	3.834E+00	3.703E+00	3.323E+00	3.085E+00	2.901E+00	2.892E+00
33.5	6.327E−01	2.504E+00	3.557E+00	3.717E+00	3.836E+00	3.759E+00	3.611E+00	3.274E+00	3.050E+00	2.879E+00	2.911E+00
34.0	5.747E−01	2.376E+00	3.430E+00	3.596E+00	3.734E+00	3.676E+00	3.537E+00	3.224E+00	3.015E+00	2.854E+00	2.950E+00
34.5	5.220E−01	2.255E+00	3.306E+00	3.479E+00	3.628E+00	3.594E+00	3.467E+00	3.176E+00	2.078E+00	2.829E+00	3.068E+00
35.0	4.742E−01	2.138E+00	3.183E+00	3.365E+00	3.527E+00	3.511E+00	3.394E+00	3.132E+00	2.941E+00	2.806E+00	3.225E+00
35.5	4.289E−01	2.030E+00	3.061E+00	3.256E+00	3.431E+00	3.434E+00	3.329E+00	3.082E+00	2.905E+00	2.782E+00	2.879E+00
36.0	3.873E−01	1.937E+00	2.948E+00	3.151E+00	3.333E+00	3.363E+00	3.269E+00	3.035E+00	2.872E+00	2.768E+00	2.458E+00
36.5	3.495E−01	1.842E+00	2.840E+00	3.049E+00	3.240E+00	3.284E+00	3.205E+00	2.992E+00	2.840E+00	2.757E+00	2.430E+00
37.0	3.143E−01	1.741E+00	2.738E+00	2.949E+00	3.150E+00	3.214E+00	3.143E+00	2.950E+00	2.806E+00	2.711E+00	2.405E+00
37.5	2.823E−01	1.645E+00	2.637E+00	2.848E+00	3.060E+00	3.140E+00	3.086E+00	2.905E+00	2.768E+00	2.661E+00	2.381E+00
38.0	2.522E−01	1.557E+00	2.534E+00	2.751E+00	2.969E+00	3.065E+00	3.024E+00	2.855E+00	2.729E+00	2.641E+00	2.356E+00
38.5	2.243E−01	1.475E+00	2.437E+00	2.662E+00	2.881E+00	2.998E+00	2.965E+00	2.808E+00	2.694E+00	2.618E+00	2.332E+00
39.0	1.995E−01	1.395E+00	2.344E+00	2.576E+00	2.800E+00	2.931E+00	2.906E+00	2.771E+00	2.661E+00	2.600E+00	2.309E+00
39.5	1.777E−01	1.319E+00	2.256E+00	2.494E+00	2.718E+00	2.861E+00	2.851E+00	2.726E+00	2.629E+00	2.583E+00	2.285E+00
40.0	1.572E−01	1.245E+00	2.170E+00	2.416E+00	2.640E+00	2.794E+00	2.796E+00	2.679E+00	2.597E+00	2.566E+00	2.261E+00
40.5	1.379E−01	1.777E+00	2.086E+00	2.339E+00	2.567E+00	2.726E+00	2.737E+00	2.641E+00	2.561E+00	2.554E+00	2.235E+00
41.0	1.216E−01	1.112E+00	2.005E+00	2.275E+00	2.491E+00	2.664E+00	2.686E+00	2.600E+00	2.525E+00	2.548E+00	2.214E+00
41.5	1.071E−01	1.048E+00	1.928E+00	2.220E+00	2.416E+00	2.603E+00	2.632E+00	2.559E+00	2.495E+00	2.557E+00	2.191E+00
42.0	9.391E−02	9.902E+01	1.853E+00	2.103E+00	2.342E+00	2.536E+00	2.576E+00	2.514E+00	2.463E+00	2.574E+00	2.165E+00
42.5	8.234E−02	9.347E+01	1.781E+00	1.980E+00	2.269E+00	2.477E+00	2.531E+00	2.473E+00	2.428E+00	2.646E+00	2.141E+00
43.0	7.216E−02	8.806E+01	1.710E+00	1.908E+00	2.200E+00	2.424E+00	2.484E+00	2.435E+00	2.399E+00	2.769E+00	2.117E+00
43.5	6.361E−02	8.280E+01	1.643E+00	1.865E+00	2.132E+00	2.368E+00	2.431E+00	2.396E+00	2.370E+00	2.500E+00	2.093E+00
44.0	5.579E−02	7.779E−01	1.578E+00	1.814E+00	2.070E+00	2.305E+00	2.379E+00	2.357E+00	2.339E+00	2.164E+00	2.072E+00
44.5	4.916E−02	7.319E−01	1.514E+00	1.732E+00	2.007E+00	2.246E+00	2.327E+00	2.317E+00	2.308E+00	2.136E+00	2.050E+00
45.0	4.387E−02	6.884E−01	1.450E+00	1.663E+00	1.941E+00	2.192E+00	2.277E+00	2.280E+00	2.274E+00	2.105E+00	2.026E+00
45.5	3.921E−02	6.471E−01	1.391E+00	1.599E+00	1.879E+00	2.139E+00	2.234E+00	2.242E+00	2.244E+00	2.079E+00	2.003E+00

(continued)

Table A5.2 *(cont.)*

	1.1	1.2	1.3	1.333	1.4	1.5	1.6	1.7	1.8	1.9	2.0
46.5	3.254E−02	5.685E−01	1.284E+00	1.477E+00	1.767E+00	2.033E+00	2.138E+00	2.168E+00	2.192E+00	2.028E+00	1.962E+00
47.0	2.989E−02	5.331E−01	1.229E+00	1.420E+00	1.710E+00	1.980E+00	2.094E+00	2.134E+00	2.165E+00	2.000E+00	1.937E+00
47.5	2.822E−02	4.972E−01	1.177E+00	1.365E+00	1.653E+00	1.929E+00	2.050E+00	2.098E+00	2.138E+00	1.974E+00	1.911E+00
48.0	2.674E−02	4.629E−01	1.128E+00	1.309E+00	1.600E+00	1.878E+00	2.005E+00	2.061E+00	2.113E+00	1.949E+00	1.891E+00
48.5	2.541E−02	4.318E−01	1.080E+00	1.258E+00	1.547E+00	1.827E+00	1.962E+00	2.024E+00	2.090E+00	1.923E+00	1.870E+00
49.0	2.481E−02	4.025E−01	1.038E+00	1.210E+00	1.498E+00	1.780E+00	1.918E+00	1.990E+00	2.069E+00	1.896E+00	1.845E+00
49.5	2.396E−02	3.744E−01	9.963E−01	1.161E+00	1.449E+00	1.735E+00	1.878E+00	1.957E+00	2.051E+00	1.871E+00	1.822E+00
50.0	2.303E−02	3.474E−01	9.528E−01	1.112E+00	1.400E+00	1.692E+00	1.838E+00	1.924E+00	2.035E+00	1.845E+00	1.803E+00
50.5	2.251E−02	3.215E−01	9.114E−01	1.066E+00	1.354E+00	1.648E+00	1.799E+00	1.891E+00	2.026E+00	1.817E+00	1.781E+00
51.0	2.182E−02	2.979E−01	8.737E−01	1.021E+00	1.309E+00	1.602E+00	1.762E+00	1.858E+00	2.022E+00	1.790E+00	1.759E+00
51.5	2.109E−02	2.753E−01	8.379E−01	9.789E−01	1.265E+00	1.561E+00	1.722E+00	1.933E+00	2.035E+00	1.765E+00	1.739E+00
52.0	2.059E−02	2.555E−01	8.027E−01	9.406E−01	1.222E+00	1.519E+00	1.683E+00	1.946E+00	2.112E+00	1.742E+00	1.718E+00
52.5	2.000E−02	2.367E−01	7.696E−01	9.007E−01	1.178E+00	1.478E+00	1.643E+00	1.839E+00	2.147E+00	1.718E+00	1.696E+00
53.0	1.941E−02	2.166E−01	7.405E−01	8.615E−01	1.136E+00	1.438E+00	1.605E+00	1.789E+00	1.910E+00	1.695E+00	1.675E+00
53.5	1.892E−02	1.986E−01	7.175E−01	8.245E−01	1.096E+00	1.399E+00	1.571E+00	1.750E+00	1.686E+00	1.672E+00	1.651E+00
54.0	1.843E−02	1.828E−01	7.181E−01	7.885E−01	1.060E+00	1.364E+00	1.533E+00	1.712E+00	1.658E+00	1.647E+00	1.629E+00
54.5	1.792E−02	1.675E−01	6.586E−01	7.545E−01	1.021E+00	1.327E+00	1.499E+00	1.676E+00	1.632E+00	1.620E+00	1.609E+00
55.0	1.737E−02	1.529E−01	5.742E−01	7.203E−01	9.826E−01	1.283E+00	1.469E+00	1.646E+00	1.617E+00	1.596E+00	1.589E+00
55.5	1.663E−02	1.386E−01	5.444E−01	6.866E−01	9.462E−01	1.242E+00	1.434E+00	1.617E+00	1.599E+00	1.574E+00	1.571E+00
56.0	1.608E−02	1.265E−01	5.162E−01	6.566E−01	9.111E−01	1.205E+00	1.400E+00	1.586E+00	1.562E+00	1.552E+00	1.552E+00
56.5	1.586E−02	1.158E−01	4.903E−01	6.265E−01	8.773E−01	1.169E+00	1.369E+00	1.556E+00	1.533E+00	1.528E+00	1.530E+00
57.0	1.551E−02	1.053E−01	4.650E−01	5.961E−01	8.438E−01	1.136E+00	1.338E+00	1.525E+00	1.509E+00	1.506E+00	1.508E+00
57.5	1.500E−02	9.539E−02	4.393E−01	5.684E−01	8.122E−01	1.104E+00	1.306E+00	1.497E+00	1.485E+00	1.485E+00	1.489E+00
58.0	1.464E−02	8.597E−02	4.151E−01	5.411E−01	7.814E−01	1.071E+00	1.276E+00	1.470E+00	1.461E+00	1.464E+00	1.472E+00
58.5	1.422E−02	7.790E−02	3.925E−01	5.138E−01	7.512E−01	1.038E+00	1.248E+00	1.444E+00	1.441E+00	1.443E+00	1.452E+00
59.0	1.387E−02	7.050E−02	3.697E−01	4.884E−01	7.214E−01	1.006E+00	1.217E+00	1.422E+00	1.433E+00	1.419E+00	1.430E+00
59.5	1.365E−02	6.381E−02	3.476E−01	4.651E−01	6.936E−01	9.755E−01	1.188E+00	1.400E+00	1.397E+00	1.395E+00	1.411E+00
60.0	1.321E−02	5.804E−02	3.273E−01	4.423E−01	6.664E−01	9.435E−01	1.160E+00	1.379E+00	1.348E+00	1.375E+00	1.393E+00
60.5	1.271E−02	5.294E−02	3.079E−01	4.191E−01	6.367E−01	9.161E−01	1.133E+00	1.358E+00	1.323E+00	1.355E+00	1.374E+00
61.0	1.236E−02	4.814E−02	2.893E−01	3.968E−01	6.094E−01	8.883E−01	1.105E+00	1.341E+00	1.295E+00	1.333E+00	1.354E+00
61.5	1.201E−02	4.399E−02	2.717E−01	3.765E−01	5.847E−01	8.590E−01	1.075E−00	1.329E+00	1.271E+00	1.311E+00	1.337E+00

62.0	1.176E−02	4.029E−02	2.537E−01	3.561E−01	5.591E−01	8.298E−01	1.050E−00	1.318E+00	1.247E+00	1.290E+00	1.318E+00
62.5	1.140E−02	3.699E−02	2.375E−01	3.364E−01	5.355E−01	8.027E−01	1.025E+00	1.322E+00	1.225E+00	1.270E+00	1.299E+00
63.0	1.103E−02	3.442E−02	2.225E−01	3.193E−01	5.136E−01	7.764E−01	1.001E+00	1.361E+00	1.204E+00	1.249E+00	1.283E+00
63.5	1.090E−02	3.214E−02	2.073E−01	3.036E−01	4.896E−01	7.499E−01	9.764E−01	1.446E+00	1.182E+00	1.230E+00	1.265E+00
64.0	1.068E−02	3.038E−02	1.933E−01	2.855E−01	4.668E−01	7.262E−01	9.509E−01	1.281E+00	1.159E+00	1.209E+00	1.247E+00
64.5	1.054E−02	2.935E−02	1.803E−01	2.656E−01	4.460E−01	7.004E−01	9.277E−01	1.048E+00	1.136E+00	1.187E+00	1.230E+00
65.0	1.043E−02	2.809E−02	1.673E−01	2.487E−01	4.261E−01	6.761E−01	9.051E−01	1.024E+00	1.113E+00	1.168E+00	1.211E+00
65.5	1.011E−02	2.663E−02	1.557E−01	2.327E−01	4.066E−01	6.534E−01	8.816E−01	1.000E+00	1.091E+00	1.148E+00	1.193E+00
66.0	1.029E−02	2.596E−02	1.449E−01	2.176E−01	3.863E−01	6.295E−01	8.602E−01	9.757E−01	1.069E+00	1.132E+00	1.178E+00
66.5	1.061E−02	2.556E−02	1.340E−01	2.041E−01	3.671E−01	6.079E−01	8.384E−01	9.521E−01	1.048E+00	1.114E+00	1.160E+00
67.0	1.065E−02	2.490E−02	1.240E−01	1.911E−01	3.496E−01	5.872E−01	8.170E−01	9.295E−01	1.029E+00	1.095E+00	1.140E+00
67.5	1.135E−02	2.473E−02	1.150E−01	1.792E−01	3.331E−01	5.651E−01	7.960E−01	9.077E−01	1.042E+00	1.076E+00	1.123E+00
68.0	1.272E−02	2.452E−02	1.063E−01	1.673E−01	3.163E−01	5.441E−01	7.749E−01	8.844E−01	1.096E+00	1.056E+00	1.106E+00
68.5	1.733E−02	2.353E−02	9.779E−02	1.556E−01	2.990E−01	5.238E−01	7.543E−01	8.610E−01	1.081E+00	1.038E+00	1.089E+00
69.0	1.475E−02	2.293E−02	9.034E−02	1.452E−01	2.830E−01	5.041E−01	7.352E−01	8.409E−01	1.017E+00	1.017E+00	1.074E+00
69.5	8.189E−03	2.291E−02	8.548E−02	1.351E−01	2.685E−01	4.849E−01	7.176E−01	8.213E−01	9.839E−01	9.997E−01	1.060E+00
70.0	8.006E−03	2.234E−02	7.959E−02	1.259E−01	2.541E−01	4.654E−01	6.994E−01	7.994E−01	9.556E−01	9.835E−01	1.042E+00
70.5	7.856E−03	2.169E−02	7.184E−02	1.167E−01	2.404E−01	4.457E−01	6.814E−01	7.796E−01	9.303E−01	9.660E−01	1.024E+00
71.0	7.587E−03	2.134E−02	6.578E−02	1.077E−01	2.274E−01	4.283E−01	6.651E−01	7.600E−01	9.056E−01	9.473E−01	1.007E+00
71.5	7.290E−03	2.126E−02	6.092E−02	1.003E−01	2.139E−01	4.125E−01	6.485E−01	7.397E−01	8.829E−01	9.289E−01	9.908E−01
72.0	7.180E−03	2.121E−02	5.641E−02	9.325E−02	2.013E−01	3.957E−01	6.336E−01	7.208E−01	8.631E−01	9.111E−01	9.739E−01
72.5	7.035E−03	2.073E−02	5.192E−02	8.623E−02	2.070E−01	3.799E−01	6.208E−01	7.016E−01	8.432E−01	8.955E−01	9.598E−01
73.0	6.877E−03	2.022E−02	4.848E−02	8.007E−02	2.119E−01	3.643E−01	6.072E−01	6.825E−01	8.230E−01	8.798E−01	9.458E−01
73.5	6.811E−03	2.005E−02	4.572E−02	7.447E−02	1.944E−01	3.478E−01	5.946E−01	6.636E−01	8.023E−01	8.615E−01	9.294E−01
74.0	6.594E−03	1.964E−02	4.245E−02	6.890E−02	1.764E−01	3.325E−01	5.844E−01	6.457E−01	7.828E−01	8.448E−01	9.136E−01
74.5	6.369E−03	1.914E−02	3.982E−02	6.399E−02	1.627E−01	3.196E−01	5.774E−01	6.282E−01	7.654E−01	8.282E−01	8.995E−01
75.0	6.280E−03	1.902E−02	3.804E−02	5.989E−02	1.511E−01	3.062E−01	5.722E−01	6.100E−01	7.488E−01	8.125E−01	8.838E−01
75.5	6.152E−03	1.907E−02	3.619E−02	5.589E−02	1.405E−01	2.926E−01	5.735E−01	5.940E−01	7.305E−01	8.011E−01	8.690E−01
76.0	6.067E−03	1.849E−02	3.453E−02	5.203E−02	1.308E−01	2.804E−01	5.905E−01	5.787E−01	7.121E−01	7.868E−01	8.560E−01
76.5	6.016E−03	1.780E−02	3.349E−02	4.883E−02	1.222E−01	2.684E−01	6.666E−01	5.615E−01	6.970E−01	7.678E−01	8.405E−01
77.0	5.830E−03	1.743E−02	3.250E−02	4.612E−02	1.136E−01	2.565E−01	6.399E−01	5.455E−01	6.796E−01	7.510E−01	8.261E−01
77.5	5.630E−03	1.720E−02	3.161E−02	4.351E−02	1.059E−01	2.462E−01	4.661E−01	5.318E−01	6.584E−01	7.362E−01	8.119E−01
78.0	5.471E−03	1.710E−02	3.116E−02	4.140E−02	9.872E−02	2.365E−01	3.733E−01	5.165E−01	6.404E−01	7.221E−01	7.968E−01
78.5	5.477E−03	1.677E−02	3.060E−02	3.943E−02	9.160E−02	2.256E−01	3.584E−01	4.997E−01	6.235E−01	7.072E−01	7.832E−01

(continued)

Table A5.2 (*cont.*)

	1.1	1.2	1.3	1.333	1.4	1.5	1.6	1.7	1.8	1.9	2.0
79.0	5.532E−03	1.659E−02	3.003E−02	3.765E−02	8.505E−02	2.159E−01	3.441E−01	4.833E−01	6.068E−01	6.920E−01	7.684E−01
79.5	5.295E−03	1.643E−02	2.970E−02	3.649E−02	7.916E−02	2.074E−01	3.297E−01	4.692E−01	5.920E−01	6.767E−01	7.543E−01
80.0	6.203E−03	1.638E−02	2.939E−02	3.543E−02	7.428E−02	1.992E−01	3.160E−01	4.551E−01	5.763E−01	6.603E−01	7.408E−01
80.5	6.926E−03	1.646E−02	2.935E−02	3.474E−02	6.978E−02	1.914E−01	3.042E−01	4.411E−01	5.615E−01	6.462E−01	7.270E−01
81.0	6.179E−03	1.584E−02	2.896E−02	3.398E−02	6.514E−01	1.838E−01	2.926E−01	4.285E−01	5.461E−01	6.326E−01	7.147E−01
81.5	5.877E−03	1.528E−02	2.832E−02	3.326E−02	6.056E−02	1.759E−01	2.794E−01	4.132E−01	5.306E−01	6.162E−01	6.998E−01
82.0	5.335E−03	1.582E−02	2.782E−02	3.253E−02	5.679E−02	1.688E−01	2.677E−01	3.988E−01	5.170E−01	7.152E−01	6.861E−01
82.5	4.838E−03	1.602E−02	2.757E−02	3.211E−02	5.416E−02	1.631E−01	2.570E−01	3.883E−01	5.031E−01	7.548E−01	6.734E−01
83.0	4.863E−03	1.562E−02	2.780E−02	3.224E−02	5.189E−02	1.572E−01	2.455E−01	3.769E−01	4.882E−01	6.666E−01	6.598E−01
83.5	4.796E−03	1.542E−02	2.751E−02	3.190E−02	4.934E−02	1.520E−01	2.349E−01	3.649E−01	4.734E−01	6.333E−01	6.463E−01
84.0	2.743E−01	1.504E−02	2.710E−02	3.154E−02	4.758E−02	1.475E−01	2.253E−01	3.539E−01	4.595E−01	6.088E−01	6.325E−01
84.5	5.363E−01	1.461E−02	2.687E−02	3.129E−02	4.596E−02	1.428E−01	2.136E−01	3.429E−01	4.454E−01	5.864E−01	6.186E−01
85.0	4.297E−01	1.449E−02	2.656E−02	3.097E−02	4.436E−02	1.389E−01	2.035E−01	3.361E−01	4.324E−01	5.665E−01	6.053E−01
85.5	2.927E−01	1.430E−02	2.620E−02	3.064E−02	4.323E−02	1.356E−01	1.948E−01	3.299E−01	4.204E−01	5.497E−01	5.929E−01
86.0	2.333E−01	1.396E−02	2.596E−02	3.036E−02	4.215E−02	1.326E−01	1.855E−01	3.086E−01	4.070E−01	5.339E−01	5.824E−01
86.5	1.949E−01	1.391E−02	2.600E−02	3.011E−02	4.113E−02	1.298E−01	1.764E−01	2.855E−01	3.932E−01	5.185E−01	5.712E−01
87.0	1.670E−01	1.393E−02	2.597E−02	2.985E−02	4.049E−02	1.278E−01	1.679E−01	2.737E−01	3.810E−01	5.024E−01	5.581E−01
87.5	1.454E−01	1.390E−02	2.613E−02	2.986E−02	4.039E−02	1.270E−01	1.604E−01	2.634E−01	3.692E−01	4.879E−01	5.469E−01
88.0	1.272E−01	1.377E−02	2.638E−02	2.987E−02	4.008E−02	1.268E−01	1.526E−01	2.527E−01	3.582E−01	4.753E−01	5.351E−01
88.5	1.129E−01	1.351E−02	2.663E−02	2.995E−02	4.001E−02	1.268E−01	1.456E−01	2.433E−01	3.469E−01	4.612E−01	5.226E−01
89.0	1.001E−01	1.337E−02	2.681E−02	2.962E−02	3.982E−02	1.273E−01	1.384E−01	2.344E−01	3.347E−01	4.482E−01	5.108E−01
89.5	8.908E−02	1.320E−02	2.705E−02	2.916E−02	3.939E−02	1.289E−01	1.308E−01	2.243E−01	3.229E−01	4.353E−01	4.994E−01
90.0	7.941E−02	1.308E−02	2.893E−02	2.953E−02	3.937E−02	1.325E−01	1.246E−01	2.151E−01	3.126E−01	4.211E−01	4.882E−01
90.5	7.029E−02	1.307E−02	2.839E−02	2.990E−02	3.924E−02	1.380E−01	1.187E−01	2.070E−01	3.031E−01	4.079E−01	4.763E−01
91.0	6.240E−02	1.305E−02	2.502E−02	2.895E−02	3.864E−02	1.461E−01	1.126E−01	1.971E−01	2.921E−01	3.953E−01	4.649E−01
91.5	5.514E−02	1.281E−02	2.390E−02	2.807E−02	3.843E−02	1.585E−01	1.067E−01	1.878E−01	2.810E−01	3.838E−01	4.533E−01
92.0	4.891E−02	1.269E−02	2.396E−02	2.801E−02	3.846E−02	1.805E−01	1.071E−01	1.805E−01	2.709E−01	3.729E−01	4.418E−01
92.5	4.315E−02	1.276E−02	2.356E−02	2.764E−02	3.840E−02	2.394E−01	1.054E−01	1.729E−01	2.619E−01	3.622E−01	4.317E−01
93.0	3.832E−02	1.294E−02	2.325E−02	2.752E−02	3.839E−02	2.671E−01	9.771E−02	1.655E−01	2.534E−01	3.512E−01	4.220E−01
93.5	3.404E−02	1.332E−02	2.307E−02	2.756E−02	3.824E−02	1.501E−01	9.262E−02	1.591E−01	2.442E−01	3.391E−01	4.108E−01
94.0	3.025E−02	1.371E−02	2.285E−02	2.737E−02	3.809E−02	4.932E−02	8.795E−02	1.525E−01	2.356E−01	3.293E−01	4.658E−01

94.5	2.727E−02	1.408E−02	2.289E−02	2.728E−02	3.803E−02	4.885E−02	8.440E−02	1.452E−01	2.265E−01	3.194E−01	5.255E−01
95.0	2.451E−02	1.475E−02	2.254E−02	2.695E−02	3.792E−02	4.875E−02	8.110E−02	1.384E−01	2.177E−01	3.084E−01	4.937E−01
95.5	2.191E−02	1.570E−02	2.221E−02	2.669E−02	3.799E−02	4.806E−02	7.746E−02	1.324E−01	2.097E−01	2.981E−01	4.479E−01
96.0	1.983E−02	1.676E−02	2.235E−02	2.669E−02	3.834E−02	4.737E−02	7.449E−02	1.271E−01	2.022E−01	2.886E−01	4.230E−01
96.5	1.853E−02	1.792E−02	2.245E−02	2.662E−02	3.910E−02	4.721E−02	7.208E−02	1.221E−01	1.955E−01	2.792E−01	4.032E−01
97.0	1.746E−02	1.901E−02	2.224E−02	2.645E−02	3.964E−02	4.684E−02	6.954E−02	1.170E−01	1.882E−01	2.705E−01	3.864E−01
97.5	1.645E−02	2.039E−02	2.184E−02	2.610E−02	3.985E−02	4.648E−02	6.709E−02	1.118E−01	1.797E−01	2.612E−01	3.733E−01
98.0	1.569E−02	2.225E−02	2.177E−02	2.623E−02	4.058E−02	4.684E−02	6.537E−02	1.071E−01	1.722E−01	2.514E−01	3.593E−01
98.5	1.484E−02	2.446E−02	2.154E−02	2.611E−02	4.125E−02	4.674E−02	6.408E−02	1.034E−01	1.661E−01	2.423E−01	3.462E−01
99.0	1.412E−02	2.716E−02	2.149E−02	2.585E−02	4.222E−02	4.615E−02	6.282E−02	9.983E−02	1.595E−01	2.343E−01	3.347E−01
99.5	1.398E−02	3.074E−02	2.158E−02	2.591E−02	4.369E−02	4.614E−02	6.177E−02	9.594E−02	1.529E−01	2.264E−01	3.235E−01
100.0	1.375E−02	3.591E−02	2.132E−02	2.588E−02	4.541E−02	4.620E−02	6.103E−02	9.463E−02	1.470E−01	2.177E−01	3.125E−01
100.5	1.330E−02	4.620E−02	2.139E−01	2.607E−02	4.730E−02	4.602E−02	6054E−02	9.245E−02	1.412E−01	2.096E−01	3.024E−01
101.0	1.284E−02	5.903E−02	2.149E−02	2.606E−02	4.894E−02	4.597E−02	6.027E−02	8.830E−02	1.361E−01	2.024E−01	2.925E−01
101.5	1.247E−02	3.754E−02	2.103E−02	2.556E−02	5.024E−02	4.584E−02	6.013E−02	8.577E−02	1.315E−01	1.951E−01	2.825E−01
102.0	1.218E−02	1.041E−02	2.086E−02	2.538E−02	5.224E−02	4.522E−02	5.956E−02	8.326E−02	1.266E−01	1.877E−01	2.725E−01
102.5	1.180E−02	1.039E−02	2.104E−02	2.674E−02	5.478E−02	4.503E−02	5.917E−02	8.116E−02	1.219E−01	1.816E−01	2.641E−01
103.0	1.137E−02	1.027E−02	2.094E−02	2.774E−02	5.685E−02	4.516E−02	5.957E−02	7.984E−02	1.177E−01	1.753E−01	2.561E−01
103.5	1.112E−02	1.027E−02	2.072E−02	2.688E−02	5.889E−02	4.455E−02	5.897E−02	7.785E−02	1.140E−01	1.685E−01	2.469E−01
104.0	1.082E−02	1.313E−02	2.050E−02	2.612E−02	6.107E−02	4.409E−02	5.833E−02	7.578E−02	1.101E−01	1.619E−01	2.382E−01
104.5	1.047E−02	1.615E−02	2.051E−02	2.603E−02	6.438E−02	4.445E−02	5.872E−02	7.515E−02	1.069E−01	1.560E−01	2.298E−01
105.0	1.031E−02	1.498E−02	2.052E−02	2.583E−02	6.810E−02	4.446E−02	5.856E−02	7.454E−02	1.037E−01	1.510E−01	2.223E−01
105.5	1.003E−02	1.311E−02	2.057E−02	2.562E−02	7.251E−02	4.417E−02	5.855E−02	7.358E−02	1.010E−01	1.463E−01	2.291E−01
106.0	9.874E−03	1.205E−02	2.037E−02	2.528E−02	7.804E−02	4.408E−02	5.850E−02	7.266E−02	9.850E−02	1.408E−01	2.298E−01
106.5	9.818E−03	1.134E−02	2.007E−02	2.484E−02	7.648E−02	4.375E−02	5.825E−02	7.210E−02	9.588E−02	1.362E−01	2.146E−01
107.0	9.586E−03	1.117E−02	2.015E−02	2.493E−02	7.403E−02	4.335E−02	5.829E−02	7.186E−02	9.408E−02	1.321E−01	2.042E−01
107.5	9.451E−03	1.086E−02	2.022E−02	2.509E−02	7.875E−02	4.339E−02	5.866E−02	7.168E−02	9.283E−02	1.281E−01	1.954E−01
108.0	9.344E−03	1.034E−02	2.051E−02	2.505E−02	8.215E−02	4.340E−02	5.888E−02	7.148E−02	9.118E−02	1.251E−01	1.880E−01
108.5	9.038E−03	1.006E−02	3.137E−02	2.509E−02	8.388E−02	5.689E−02	5.914E−02	7.112E−02	8.919E−02	1.212E−01	1.813E−01
109.0	8.829E−03	9.868E−03	3.649E−02	2.513E−02	8.815E−02	7.631E−02	5.989E−02	7.129E−02	8.839E−02	1.172E−01	1.749E−01
109.5	8.860E−03	9.684E−03	2.927E−02	2.499E−02	9.369E−02	7.372E−02	5.969E−02	7.114E−02	8.750E−02	1.146E−01	1.687E−01
110.5	8.410E−03	9.565E−03	2.586E−02	2.513E−02	1.076E−01	5.959E−02	6.112E−02	7.078E−02	8.537E−02	1.092E−01	1.647E−01
111.0	8.209E−03	9.428E−03	2.474E−02	2.521E−02	1.187E−01	5.680E−02	6.199E−02	7.117E−02	8.556E−02	1.077E−01	1.575E−01
111.5	8.229E−03	9.418E−03	2.385E−02	2.557E−02	1.351E−01	5.488E−02	6.171E−02	7.148E−02	8.594E−02	1.067E−01	1.531E−01

(*continued*)

Table A5.2 (*cont.*)

	1.1	1.2	1.3	1.333	1.4	1.5	1.6	1.7	1.8	1.9	2.0
112.0	8.188E−03	9.433E−03	2.310E−02	2.541E−02	1.630E−01	5.346E−02	6.115E−02	7.151E−02	8.539E−02	1.049E−01	1.492E−01
112.5	8.208E−03	9.374E−03	2.242E−02	2.546E−02	2.785E−01	5.252E−02	6.174E−02	7.147E−02	8.503E−02	1.032E−01	1.455E−01
113.0	8.173E−03	9.445E−03	2.169E−02	2.560E−02	2.032E−01	5.173E−02	6.243E−02	7.203E−02	8.538E−02	1.021E−01	1.410E−01
113.5	8.004E−03	6.096E−01	2.125E−02	2.574E−02	2.985E−02	5.093E−02	6.341E−02	7.239E−02	8.564E−02	1.018E−01	1.386E−01
114.0	7.899E−03	9.245E−01	2.140E−02	2.634E−02	3.007E−02	5.021E−02	6.419E−02	7.181E−02	8.518E−02	1.012E−01	1.357E−01
114.5	8.035E−03	5.506E−01	2.108E−02	2.675E−02	3.003E−02	4.916E−02	6.518E−02	7.149E−02	8.497E−02	1.004E−01	1.316E−01
115.0	8.134E−03	4.242E−01	2.064E−02	2.745E−02	2.991E−02	4.824E−02	6.775E−02	7.203E−02	8.541E−02	1.005E−01	1.298E−01
115.5	8.449E−03	3.578E−01	2.048E−02	2.821E−02	2.971E−02	4.759E−02	7.202E−02	7.236E−02	8.563E−02	1.005E−01	1.278E−01
116.0	9.130E−03	3.129E−01	2.046E−02	2.948E−02	2.962E−02	4.704E−02	7.865E−02	7.259E−02	8.529E−02	1.000E−01	1.254E−01
116.5	9.749E−03	2.802E−01	2.039E−02	3.073E−02	2.939E−02	4.601E−02	6.884E−02	7.324E−02	8.457E−02	9.931E−02	1.233E−01
117.0	1.087E−02	2.561E−01	1.027E−02	3.181E−02	2.925E−02	4.521E−02	5.527E−02	7.615E−02	8.467E−02	9.941E−02	1.218E−01
117.5	1.256E−02	2.346E−01	2.049E−02	3.375E−02	2.966E−02	4.515E−02	5.552E−02	7.408E−02	8.548E−02	9.983E−02	1.208E−01
118.0	1.481E−02	2.159E−01	2.061E−02	3.583E−02	2.973E−02	4.480E−02	5.559E−02	6.957E−02	8.571E−02	9.993E−02	1.199E−01
118.5	1.995E−02	2.004E−01	2.054E−02	3.775E−02	2.939E−02	4.403E−02	5.550E−02	6.939E−02	8.486E−02	9.961E−02	1.189E−01
119.0	2.180E−02	1.869E−01	2.053E−02	3.959E−02	2.945E−02	4.378E−02	5.511E−02	6.931E−02	8.401E−02	9.932E−02	1.179E−01
119.5	1.338E−02	1.752E−01	2.067E−02	4.124E−02	2.939E−02	4.373E−02	5.468E−02	6.907E−02	8.370E−02	9.909E−02	1.168E−01
120.0	6.834E−03	1.643E−01	2.069E−02	4.355E−02	2.908E−02	4.356E−02	5.480E−02	6.887E−02	8.389E−02	9.901E−02	1.169E−01
120.5	6.889E−03	1.524E−01	2.064E−02	4.651E−02	2.918E−02	4.376E−02	5.534E−02	6.926E−02	8.409E−02	9.935E−02	1.176E−01
121.0	6.682E−03	1.409E−01	2.090E−02	4.917E−02	2.912E−02	4.382E−02	5.516E−02	6.929E−02	8.401E−02	9.942E−02	1.180E−01
121.5	6.507E−03	1.307E−01	2.075E−02	5.140E−02	2.864E−02	4.367E−02	5.475E−02	6.922E−02	8.434E−02	9.942E−02	1.187E−01
122.0	6.586E−03	1.221E−01	2.081E−02	5.464E−02	2.879E−02	4.492E−02	5.501E−02	6.932E−02	8.452E−02	9.955E−02	1.197E−01
122.5	6.654E−03	1.147E−01	2.129E−02	5.820E−02	2.916E−02	4.564E−02	5.500E−02	6.922E−02	8.404E−02	9.918E−02	1.203E−01
123.0	6.513E−03	1.072E−01	2.155E−02	6.118E−02	2.898E−02	4.342E−02	5.466E−02	6.903E−02	8.330E−02	9.889E−02	1.207E−01
123.5	6.392E−03	9.980E−02	2.197E−02	6.500E−02	2.892E−02	4.183E−02	5.474E−02	6.922E−02	8.341E−02	9.877E−02	1.216E−01
124.0	6.356E−03	9.224E−02	2.245E−02	6.900E−02	2.869E−02	4.175E−02	5.474E−02	6.898E−02	8.345E−02	9.848E−02	1.226E−01
124.5	6.346E−03	8.639E−02	2.291E−02	7.337E−02	2.850E−02	4.152E−02	5.441E−02	6.851E−02	8.314E−02	9.833E−02	1.245E−01
125.0	6.464E−03	8.083E−02	2.384E−02	7.853E−02	2.850E−02	4.165E−02	5.433E−02	6.878E−02	8.370E−02	9.862E−02	1.260E−01
125.5	6.550E−03	7.553E−02	2.510E−02	8.492E−02	2.866E−02	4.185E−02	5.446E−02	6.861E−02	8.357E−02	9.865E−02	1.278E−01
126.0	6.457E−03	7.031E−02	2.625E−02	9.177E−02	2.859E−02	4.168E−02	5.450E−02	6.820E−02	8.295E−02	9.855E−02	1.299E−01
126.5	6.395E−03	6.515E−02	2.746E−02	9.974E−02	2.818E−02	4.129E−02	5.413E−02	6.846E−02	8.317E−02	9.862E−02	1.322E−01
127.0	6.347E−03	6.053E−02	2.884E−02	1.105E−01	2.812E−02	4.100E−02	5.406E−02	6.852E−02	8.322E−02	9.832E−02	1.354E−01

127.5	6.176E−03	5.611E−02	3.076E−02	1.258E−01	2.841E−02	4.102E−02	5.429E−02	6.821E−02	8.262E−02	9.746E−02	1.362E−01
128.0	6.052E−03	5.218E−02	3.328E−02	1.615E−01	2.849E−02	4.107E−02	5.415E−02	6.865E−02	8.306E−02	9.779E−02	1.386E−01
128.5	6.167E−03	4.876E−02	3.575E−02	2.339E−01	2.821E−02	4.104E−02	5.403E−02	6.875E−02	8.339E−02	9.831E−02	1.437E−01
129.0	6.330E−03	4.572E−02	3.827E−02	2.463E−01	2.827E−02	4.113E−02	5.388E−02	6.829E−02	8.483E−02	9.856E−02	1.469E−01
129.5	7.270E−03	4.259E−02	4.108E−02	1.210E−01	2.846E−02	4.122E−02	5.414E−02	6.827E−02	8.482E−02	9.960E−02	1.473E−01
130.0	7.918E−03	4.032E−02	4.435E−02	2.870E−02	2.850E−02	4.135E−02	5.450E−02	6.831E−02	8.384E−02	1.001E−01	1.488E−01
130.5	7.403E−03	3.841E−02	4.823E−02	2.755E−02	2.843E−02	4.100E−02	5.416E−02	6.856E−02	8.321E−02	1.002E−01	1.524E−01
131.0	7.289E−03	3.687E−02	5.091E−02	2.641E−02	2.830E−02	4.090E−02	5.397E−02	6.829E−02	8.315E−02	1.010E−01	1.570E−01
131.5	6.867E−03	3.545E−02	5.324E−02	2.538E−02	2.816E−02	4.095E−02	5.398E−02	6.793E−02	8.316E−02	1.020E−01	1.621E−01
132.0	6.092E−03	3.416E−02	5.717E−02	2.433E−02	2.817E−02	4.106E−02	5.401E−02	6.818E−02	8.318E−02	1.030E−01	1.733E−01
132.5	5.958E−03	3.356E−02	5.696E−01	2.340E−02	2.806E−02	4.102E−02	5.402E−02	6.837E−02	8.321E−02	1.046E−01	1.725E−01
133.0	5.911E−03	3.282E−02	1.188E+00	2.276E−02	2.788E−02	4.065E−02	5.365E−02	6.786E−02	8.292E−02	1.063E−01	1.635E−01
133.5	5.956E−03	3.194E−02	1.071E+00	2.239E−02	2.787E−02	4.046E−02	5.373E−02	7.673E−02	8.267E−02	1.077E−01	1.656E−01
134.0	6.103E−03	3.142E−02	7.746E−01	2.221E−02	2.788E−02	4.070E−02	5.399E−02	8.120E−02	6.286E−02	1.096E−01	1.692E−01
134.5	6.075E−03	3.131E−02	6.614E−01	2.203E−02	2.818E−02	4.095E−02	5.407E−02	7.532E−02	8.324E−02	1.120E−01	1.731E−01
135.0	5.992E−03	3.134E−02	5.971E−01	2.145E−02	2.836E−02	4.089E−02	5.389E−02	7.377E−02	8.316E−02	1.147E−01	1.767E−01
135.5	6.027E−03	3.097E−02	5.576E−01	2.110E−02	2.827E−02	4.099E−02	5.378E−02	7.259E−02	8.282E−02	1.170E−01	1.815E−01
136.0	5.994E−03	3.098E−02	5.343E−01	2.102E−02	2.804E−02	4.094E−02	5.375E−02	7.182E−02	8.314E−02	1.193E−01	1.868E−01
136.5	5.963E−03	1.110E−02	5.225E−01	2.097E−02	2.815E−02	4.089E−02	5.378E−02	7.131E−02	8.345E−02	1.218E−01	1.921E−01
137.0	5.999E−03	3.074E−02	5.229E−01	2.087E−02	2.802E−02	4.063E−02	5.377E−02	7.065E−02	8.304E−02	1.236E−01	1.958E−01
137.5	5.886E−03	3.049E−02	5.636E−01	4.665E−01	2.750E−02	4.029E−02	5.337E−02	7.023E−02	8.321E−02	1.262E−01	2.010E−01
138.0	5.708E−03	3.050E−02	5.900E−01	1.247E+00	2.770E−02	4.070E−02	5.379E−02	7.019E−02	8.427E−02	1.299E−01	2.096E−01
138.5	1.042E−01	3.011E−02	4.510E−01	1.261E+00	2.819E−02	4.130E−02	9.639E−02	7.044E−02	8.539E−02	1.332E−01	2.190E−01
139.0	1.644E−01	2.993E−02	3.151E−01	8.429E−01	2.815E−02	4.120E−02	1.191E−01	6.986E−02	8.623E−02	1.360E−01	2.309E−01
139.5	1.055E−01	3.008E−02	3.030E−01	6.981E−01	2.805E−02	4.095E−02	9.318E−02	6.968E−02	8.785E−02	1.386E−01	2.524E−01
140.0	7.674E−02	2.981E−02	2.913E−01	6.154E−01	2.787E−02	4.462E−02	8.389E−02	6.971E−02	8.922E−02	1.422E−01	2.886E−01
140.5	6.133E−02	2.964E−02	2.797E−01	5.596E−01	2.773E−02	4.745E−02	7.840E−02	6.885E−02	9.037E−02	1.463E−01	2.504E−01
141.0	5.132E−02	2.958E−02	2.680E−01	5.179E−01	2.788E−02	4.608E−02	7.519E−02	6.874E−02	9.259E−02	1.508E−01	1.924E−01
141.5	4.402E−02	2.957E−02	2.582E−01	4.853E−01	2.803E−02	4.460E−02	7.291E−02	6.883E−02	9.511E−02	1.551E−01	1.956E−01
142.0	3.766E−02	2.978E−02	2.488E−01	4.609E−01	2.828E−02	4.355E−02	7.106E−02	6.810E−02	9.699E−02	1.578E−01	1.972E−01
142.5	3.142E−02	2.991E−02	2.398E−01	4.387E−01	2.820E−02	4.304E−02	6.955E−02	6.778E−02	9.902E−02	1.618E−01	1.996E−01
143.0	2.652E−02	2.995E−02	2.304E−01	4.190E−01	2.813E−02	4.299E−02	6.823E−02	6.800E−02	1.020E−01	1.663E−01	2.020E−01
143.5	2.276E−02	2.918E−02	2.199E−01	4.020E−01	2.826E−02	4.325E−02	6.761E−02	6.808E−02	1.053E−01	1.702E−01	2.057E−01
144.0	1.934E−02	2.871E−02	2.118E−01	3.861E−01	2.806E−02	4.300E−02	6.690E−02	6.828E−02	1.088E−01	1.746E−01	2.100E−01

(continued)

Table A5.2 *(cont.)*

	1.1	1.2	1.3	1.333	1.4	1.5	1.6	1.7	1.8	1.9	2.0
144.5	1.666E−02	2.940E−02	2.045E−01	3.738E−01	2.766E−02	4.265E−02	6.531E−02	6.790E−02	1.115E−01	1.785E−01	2.124E−01
145.0	1.419E−02	2.954E−02	1.964E−01	3.624E−01	2.784E−02	4.272E−02	6.406E−02	6.758E−02	1.146E−01	1.835E−01	2.154E−01
145.5	1.237E−02	2.948E−02	1.882E−01	3.522E−01	2.816E−02	4.330E−02	6.594E−02	6.802E−02	1.185E−01	1.885E−01	2.185E−01
146.0	1.127E−02	2.946E−02	1.804E−01	3.429E−01	2.816E−02	4.366E−02	6.674E−02	6.893E−02	1.223E−01	1.933E−01	2.219E−01
146.5	1.010E−02	2.924E−02	1.730E−01	3.331E−01	1.104E+00	4.434E−02	6.457E−02	6.935E−02	1.263E−01	1.974E−01	2.265E−01
147.0	9.541E−03	2.916E−02	1.657E−01	3.272E−01	1.925E+00	4.567E−02	6.270E−02	6.990E−02	1.309E−01	2.008E−01	2.299E−01
147.5	9.312E−03	2.926E−02	1.588E−01	3.225E−01	1.412E+00	4.635E−02	6.126E−02	7.080E−02	1.345E−01	2.057E−01	2.339E−01
148.0	8.838E−03	2.920E−02	1.515E−01	3.053E−01	1.072E+00	4.694E−02	6.024E−02	7.232E−02	1.385E−01	2.116E−01	2.387E−01
148.5	8.531E−03	2.892E−02	1.447E−01	2.875E−01	9.176E−01	4.797E−02	5.914E−02	7.429E−02	1.440E−01	2.169E−01	2.428E−01
149.0	8.606E−03	2.890E−02	1.400E−01	2.795E−01	8.251E−01	4.920E−02	5.832E−02	7.567E−02	1.484E−01	2.219E−01	2.474E−01
149.5	8.481E−03	3.062E−02	1.351E−01	2.702E−01	7.606E−01	5.090E−02	5.772E−02	7.831E−02	1.566E−01	2.273E−01	2.527E−01
150.0	8.070E−03	3.106E−02	1.280E−01	2.600E−01	7.118E−01	5.252E−02	5.698E−02	8.143E−02	1.644E−01	2.337E−01	2.573E−01
150.5	8.254E−03	3.011E−02	1.225E−01	2.506E−01	6.739E−01	5.475E−02	5.655E−02	8.493E−02	1.679E−01	2.407E−01	2.619E−01
151.0	8.302E−03	3.008E−02	1.175E−01	2.429E−01	6.438E−01	5.774E−02	5.606E−02	8.967E−02	1.722E−01	2.478E−01	2.678E−01
151.5	7.989E−03	2.963E−02	1.111E−01	2.357E−01	6.202E−01	6.242E−02	5.523E−02	9.137E−02	1.762E−01	2.557E−01	2.737E−01
152.0	7.760E−03	2.974E−02	1.050E−01	2.271E−01	5.996E−01	8.079E−02	5.472E−02	9.262E−02	1.822E−01	2.639E−01	2.805E−01
152.5	7.821E−03	2.999E−02	1.011E−01	2.184E−01	5.812E−01	7.252E−02	5.496E−02	9.700E−02	1.894E−01	2.729E−01	2.880E−01
153.0	7.859E−03	3.002E−02	9.800E−02	2.115E−01	5.661E−01	4.433E−02	5.471E−02	1.017E−01	1.944E−01	2.831E−01	2.954E−01
153.5	7.720E−03	3.032E−02	9.365E−02	2.036E−01	5.524E−01	4.024E−02	5.427E−02	1.058E−01	1.989E−01	2.933E−01	3.041E−01
154.0	7.469E−03	30.28E−02	8.959E−02	1.964E−01	5.361E−01	4.024E−02	5.442E−02	1.102E−01	2.050E−01	3.046E−01	3.112E−01
154.5	7.494E−03	3.037E−02	8.655E−02	1.899E−01	5.221E−01	4.018E−02	5.449E−02	1.162E−01	2.115E−01	3.205E−01	3.192E−01
155.0	7.591E−03	3.071E−02	8.347E−02	1.825E−01	5.129E−01	4.013E−02	5.499E−02	1.219E−01	2.180E−01	3.418E−01	3.295E−01
155.5	7.301E−03	3.036E−02	8.194E−02	1.750E−01	5.039E−01	3.991E−02	5.481E−02	1.276E−01	2.252E−01	3.704E−01	3.398E−01
156.0	7.376E−03	3.070E−02	8.105E−02	1.676E−01	4.926E−01	3.992E−02	5.442E−02	1.341E−01	2.315E−01	4.534E−01	3.505E−01
156.5	7.561E−03	3.138E−02	7.947E−02	1.610E−01	4.816E−01	4.011E−02	5.524E−02	1.413E−01	2.385E−01	4.151E−01	3.624E−01
157.0	7.416E−03	3.134E−02	7.836E−02	1.546E−01	4.740E−01	2.204E+00	5.597E−02	1.479E−01	2.461E−01	3.122E−01	3.748E−01
157.5	7.354E−03	3.170E−02	7.752E−02	1.485E−01	4.656E−01	3.513E+00	5.678E−02	1.553E−01	2.545E−01	3.218E−01	3.866E−01
158.0	7.515E−03	3.217E−02	7.719E−02	1.431E−01	4.558E−01	2.266E+00	5.863E−02	1.636E−01	2.646E−01	3.335E−01	4.016E−01
158.5	7.507E−03	3.249E−02	7.691E−02	1.380E−01	4.487E−01	1.788E+00	6.098E−02	1.715E−01	2.731E−01	3.478E−01	4.176E−01
159.0	7.700E−03	3.304E−02	7.690E−02	1.327E−01	4.399E−01	1.554E+00	6.330E−02	1.799E−01	2.816E−01	3.650E−01	4.351E−01
159.5	7.938E−03	3.387E−02	7.690E−02	1.281E−01	4.325E−01	1.411E+00	6.676E−02	1.899E−01	2.932E−01	3.872E−01	4.546E−01

160.0	7.821E−03	3.502E−02	7.645E−02	1.230E−01	4.260E−01	1.315E+00	7.070E−02	2.000E−01	3.038E−01	3.886E−01	4.726E−01
160.5	7.767E−03	3.659E−02	7.639E−02	1.184E−01	4.169E−01	1.238E+00	7.450E−02	2.087E−01	3.149E−01	3.840E−01	4.945E−01
161.0	7.633E−03	3.821E−02	7.658E−02	1.155E−01	4.081E−01	1.179E+00	7.972E−02	2.190E−01	3.286E−01	3.990E−01	5.187E−01
161.5	7.706E−03	4.031E−02	7.547E−02	1.125E−01	3.985E−01	1.137E+00	8.643E−02	2.311E−01	3.422E−01	4.150E−01	5.400E−01
162.0	7.839E−03	4.292E−02	7.477E−02	1.090E−01	3.892E−01	1.102E+00	9.513E−02	2.431E−01	3.537E−01	4.318E−01	5.668E−01
162.5	8.219E−03	4.698E−02	7.549E−02	1.058E−01	3.862E−01	1.073E+00	1.022E−01	2.551E−01	3.686E−01	4.525E−01	5.973E−01
163.0	9.117E−03	5.231E−02	7.585E−02	1.037E−01	3.822E−01	1.051E+00	1.092E−01	2.693E−01	3.871E−01	4.719E−01	6.277E−01
163.5	9.442E−03	5.797E−02	7.597E−02	1.022E−01	4.148E−01	1.034E+00	1.208E−01	3.910E−01	4.022E−01	4.929E−01	6.610E−01
164.0	9.604E−03	6.569E−02	7.678E−02	1.016E−01	4.262E−01	1.020E+00	1.355E−01	4.708E−01	4.201E−01	5.199E−01	6.999E−01
164.5	1.130E−02	7.470E−02	7.603E−02	1.021E−01	3.858E−01	1.006E+00	1.371E−01	4.215E−01	4.404E−01	5.468E−01	7.422E−01
165.0	1.398E−02	8.456E−02	7.552E−02	1.028E−01	3.676E−01	9.956E−01	4.729E+00	4.121E−01	4.602E−01	5.770E−01	7.869E−01
165.5	1.699E−02	9.566E−02	7.561E−02	1.016E−01	3.546E−01	9.878E−01	7.121E+00	4.151E−01	4.843E−01	6.134E−01	8.380E−01
166.0	2.116E−02	1.081E−01	7.544E−02	9.991E−02	3.428E−01	9.813E−01	4.171E+00	4.250E−01	5.110E−01	6.493E−01	8.902E−01
166.5	2.703E−02	1.247E−01	7.700E−02	9.910E−02	3.313E−01	9.788E−01	3.409E+00	4.418E−01	5.366E−01	6.932E−01	9.493E−01
167.0	3.492E−02	1.431E−01	7.918E−02	9.895E−02	3.221E−01	9.761E−01	3.027E+00	4.590E−01	5.671E−01	7.378E−01	1.017E+00
167.5	4.596E−02	1.627E−01	8.098E−02	1.001E−01	3.137E−01	9.754E−01	2.797E+00	4.780E−01	6.009E−01	7.757E−01	1.091E+00
168.0	6.027E−02	1.868E−01	8.287E−02	1.006E−01	2.999E−01	9.760E−01	2.648E+00	5.011E−01	6.387E−01	8.252E−01	1.176E+00
168.5	8.264E−02	2.171E−01	8.607E−02	9.964E−02	2.857E−01	9.779E−01	2.544E+00	5.218E−01	6.800E−01	8.846E−01	1.270E+00
169.0	1.284E−01	2.560E−01	8.972E−02	1.000E−01	2.769E−01	9.777E−01	2.475E+00	5.555E−01	7.556E−01	9.484E−01	1.370E+00
169.5	1.855E−01	3.045E−01	9.432E−02	1.012E−01	2.686E−01	9.824E−01	2.434E+00	5.863E−01	8.306E−01	1.021E+00	1.485E+00
170.0	1.120E−01	3.700E−01	1.031E−01	9.925E−02	2.573E−01	9.937E−01	2.416E+00	6.213E−01	8.678E−01	1.112E+00	1.620E+00
170.5	6.733E−03	4.718E−03	1.242E−01	9.915E−02	2.475E−01	9.983E−01	2.424E+00	6.596E−01	9.256E−01	1.212E+00	1.777E+00
171.0	6.454E−03	6.755E−01	1.082E−01	1.002E−01	2.381E−01	1.006E+00	2.438E+00	1.026E+01	1.002E+00	1.321E+00	1.948E+00
171.5	6.591E−03	1.014E+00	7.581E−02	1.001E−01	2.268E−01	1.023E+00	2.462E+00	1.547E+01	1.084E+00	1.444E+00	2.142E+00
172.0	6.696E−03	6.611E−01	7.684E−02	1.020E−01	2.187E−01	1.048E+00	2.504E+00	9.275E+00	1.192E+00	1.592E+00	2.374E+00
172.5	7.131E−03	5.861E−02	7.526E−02	1.042E−01	2.123E−01	1.074E+00	2.567E+00	7.766E+00	1.319E+00	1.766E+00	2.657E+00
173.0	7.216E−03	4.613E−02	7.400E−02	1.057E−01	2.065E−01	1.100E+00	2.652E+00	7.071E+00	1.462E+00	1.972E+00	2.994E+00
173.5	6.911E−03	4.121E−02	7.425E−02	1.132E−01	2.007E−01	1.133E+00	2.749E+00	6.728E+00	1.645E+00	2.221E+00	3.387E+00
174.0	7.375E−03	3.823E−02	7.388E−02	1.106E−01	1.923E−01	1.177E+00	2.867E+00	6.550E+00	1.895E+00	2.514E+00	3.859E+00
174.5	7.483E−03	3.570E−02	7.374E−02	1.003E−01	1.872E−01	1.238E+00	3.028E+00	6.509E+00	2.418E+00	2.879E+00	4.440E+00
175.0	7.876E−03	3.209E−02	7.376E−02	1.012E−01	1.847E−01	1.307E+00	3.223E+00	6.595E+00	5.801E+00	3.334E+00	5.181E+00
175.7	8.908E−03	3.103E−02	7.373E−02	9.885E−02	1.806E−01	1.384E+00	3.457E+00	6.792E+00	2.693E+01	3.921E+00	6.133E+00
176.0	8.514E−03	2.946E−02	7.467E−02	9.706E−02	1.770E−01	1.490E+00	3.767E+00	7.118E+00	3.687E+01	4.662E+00	7.397E+00
176.5	8.550E−03	2.908E−02	7.403E−02	9.884E−02	1.740E−01	1.631E+00	4.166E+00	7.611E+00	2.413E+01	5.632E+00	9.117E+00

(continued)

Table A5.2 (*cont.*)

	1.1	1.2	1.3	1.333	1.4	1.5	1.6	1.7	1.8	1.9	2.0
177.0	9.383E−03	2.903E−02	7.471E−02	1.006E−01	1.736E−01	1.824E+00	4.699E+00	8.333E+00	2.229E+01	7.036E+00	1.153E+01
177.5	8.902E−03	2.861E−02	7.554E−02	1.014E−01	1.745E−01	2.102E+00	5.461E+00	9.415E+00	2.262E+01	9.107E+00	1.518E+01
178.0	9.916E−03	3.001E−02	7.430E−02	1.006E−01	1.756E−01	2.510E+00	6.614E+00	1.108E+01	2.471E+01	1.235E+01	2.122E+01
178.5	1.278E−02	3.122E−02	7.490E−02	9.989E−02	1.780E−01	3.208E+00	8.527E+00	1.397E+01	2.937E+01	9.397E+01	3.246E+01
179.0	4.973E−02	3.096E−02	7.501E−02	9.844E−02	.1787E−01	4.605E+00	1.235E+01	1.984E+01	3.969E+01	1.850E+02	5.877E+01
180.0	1.368E−01	3.148E−02	7.756E−02	1.012E−01	1.840E−01	8.762E+00	2.385E+00	3.751E+01	7.219E+01	2.010E+02	1.837E+02

Table A5.3. Phase functions of large transparent oblate (4 : 4 : 3) spheroids for different refractive indices, calculated with ray tracing Monte Carlo code (Kokhanovsky and Nakajima, 1998).

	1.1	1.2	1.3	1.333	1.4	1.5	1.6	1.7	1.8	1.9	2.0
0.0	1.231E+02	3.749E+01	1.990E+01	1.761E+01	1.318E+01	9.917E+00	7.935E+00	6.691E+00	5.769E+00	5.182E+00	4.709E+00
1.0	1.215E+02	3.736E+01	1.989E+01	1.746E+01	1.315E+01	9.886E+00	7.896E+00	6.678E+00	5.731E+00	5.142E+00	4.690E+00
2.0	1.151E+02	3.672E+01	1.969E+01	1.720E+01	1.307E+01	9.797E+00	7.850E+00	6.590E+00	5.704E+00	5.095E+00	4.655E+00
3.0	1.057E+02	3.576E+01	1.940E+01	1.692E+01	1.292E+01	9.682E+00	7.756E+00	6.531E+00	5.641E+00	5.052E+00	4.580E+00
4.0	9.452E+01	3.456E+01	1.905E+01	1.663E+01	1.273E+01	9.554E+00	7.666E+00	6.455E+00	5.573E+00	5.000E+00	4.537E+00
5.0	8.272E+01	3.313E+01	1.864E+01	1.627E+01	1.254E+01	9.447E+00	7.573E+00	6.380E+00	5.515E+00	4.944E+00	4.456E+00
10.0	3.561E+01	2.418E+01	1.575E+01	1.404E+01	1.122E+01	8.651E+00	7.037E+00	5.946E+00	5.136E+00	4.587E+00	4.152E+00
20.0	6.191E+00	9.767E+00	9.149E+00	8.651E+00	7.746E+00	6.542E+00	5.616E+00	4.903E+00	4.396E+00	4.032E+00	3.846E+00
30.0	1.078E+00	3.496E+00	4.542E+00	4.603E+00	4.623E+00	4.376E+00	4.032E+00	3.705E+00	3.466E+00	3.192E+00	2.974E+00
40.0	7.429E−02	9.947E−01	1.924E+00	2.140E+00	2.422E+00	2.611E+00	2.612E+00	2.539E+00	2.420E+00	2.293E+00	2.154E+00
50.0	2.704E−02	1.471E−01	5.552E−01	7.178E−01	1.004E+00	1.279E+00	1.426E+00	1.499E+00	1.512E+00	1.486E+00	1.448E+00
60.0	1.336E−02	4.928E−02	1.177E−01	1.811E−01	3.140E−01	4.528E−01	6.057E−01	7.283E−01	8.126E−01	8.308E−01	8.538E−01
70.0	9.146E−03	3.048E−02	5.596E−02	6.831E−02	8.490E−02	1.870E−01	2.386E−01	3.254E−01	3.803E−01	4.496E−01	5.098E−01
80.0	1.203E−01	2.008E−02	5.015E−02	4.688E−02	4.692E−02	8.560E−02	1.293E−01	1.621E−01	2.344E−01	3.067E−01	3.737E−01
90.0	4.146E−02	1.432E−02	2.699E−02	3.257E−02	3.872E−02	7.152E−02	1.288E−01	1.535E−01	2.379E−01	3.370E−01	4.282E−01
100.0	1.595E−02	2.510E−01	2.337E−02	2.888E−02	3.728E−02	8.500E−02	1.571E−01	2.717E−01	3.668E−01	4.160E−01	5.198E−01
110.0	8.775E−03	1.402E−01	4.064E−01	4.588E−01	5.361E−02	1.262E−01	2.377E−01	3.517E−01	4.906E−01	6.054E−01	5.702E−01
120.0	9.956E−02	7.373E−02	2.889E−01	3.839E−01	5.679E−01	7.343E−01	2.721E−01	2.178E−01	1.085E−01	1.622E−01	1.869E−01
130.0	3.537E−02	3.942E−02	1.989E−01	2.961E−01	5.089E−01	6.208E−01	7.355E−01	7.501E−01	2.261E−01	1.212E−01	1.799E−01
140.0	1.853E−02	3.313E−02	1.742E−01	2.406E−01	3.089E−01	5.340E−01	7.155E−01	8.615E−01	9.413E−01	9.728E−01	9.639E−01
150.0	8.887E−03	3.755E−02	1.065E−01	1.300E−01	2.871E−01	4.108E−01	6.373E−01	8.323E−01	1.000E+00	1.122E+00	1.207E+00
160.0	9.610E−03	1.641E−01	1.551E−01	1.167E−01	1.945E−01	4.844E−01	5.307E−01	7.265E−01	9.969E−01	1.225E+00	1.401E+00
170.0	1.214E−01	3.867E−01	1.247E−01	1.567E−01	2.869E−01	4.445E−01	7.646E−01	1.368E+00	1.194E+00	1.261E+00	1.508E+00
180.0	1.077E−01	2.379E−01	1.496E−01	2.407E−01	1.369E+00	1.120E+00	1.336E+00	1.842E+00	2.523E+00	5.193E+00	4.067E+00

Table A5.4. Phase functions of large transparent prolate (5 : 4 : 4) spheroids for different refractive indices, calculated with ray tracing Monte Carlo code (Kokhanovsky and Nakajima, 1998).

	1.1	1.2	1.3	1.333	1.4	1.5	1.6	1.7	1.8	1.9	2.0
0.0	1.304E+02	3.870E+01	2.007E+01	1.720E+01	1.296E+01	9.505E+00	7.437E+00	6.076E+00	5.165E+00	4.579E+00	4.080E+00
1.0	1.285E+02	3.846E+01	2.000E+01	1.718E+01	1.298E+01	9.445E+00	7.410E+00	6.058E+00	5.158E+00	4.558E+00	4.066E+00
2.0	1.218E+02	3.793E+01	1.983E+01	1.706E+01	1.284E+01	9.398E+00	7.359E+00	6.048E+00	5.121E+00	4.484E+00	4.040E+00
3.0	1.116E+02	3.700E+01	1.956E+01	1.682E+01	1.276E+01	9.290E+00	7.280E+00	5.977E+00	5.067E+00	4.465E+00	3.970E+00
4.0	9.928E+01	3.576E+01	1.923E+01	1.655E+01	1.260E+01	9.200E+00	7.218E+00	5.914E+00	5.014E+00	4.387E+00	3.909E+00
5.0	8.626E+01	3.419E+01	1.881E+01	1.625E+01	1.239E+01	9.091E+00	7.121E+00	5.846E+00	4.959E+00	4.340E+00	3.857E+00
10.0	3.556E+01	2.460E+01	1.581E+01	1.395E+01	1.106E+01	8.349E+00	6.658E+00	5.483E+00	4.663E+00	4.065E+00	3.626E+00
20.0	5.809E+00	9.546E+00	8.988E+00	8.514E+00	7.542E+00	6.280E+00	5.301E+00	4.521E+00	3.939E+00	3.485E+00	3.142E+00
30.0	1.066E+00	3.376E+00	4.396E+00	4.496E+00	4.469E+00	4.214E+00	3.831E+00	3.467E+00	3.135E+00	2.862E+00	2.625E+00
40.0	9.551E−02	1.077E+00	1.982E+00	2.180E+00	2.430E+00	2.611E+00	2.590E+00	2.481E+00	2.351E+00	2.243E+00	2.089E+00
50.0	2.619E−02	2.321E−01	7.455E−01	9.088E−01	1.178E+00	1.482E+00	1.622E+00	1.653E+00	1.673E+00	1.627E+00	1.573E+00
60.0	1.332E−02	4.604E−02	1.702E−01	2.548E−01	4.587E−01	6.669E−01	8.458E−01	1.035E+00	1.059E+00	1.088E+00	1.105E+00
70.0	8.776E−03	2.324E−02	5.136E−02	7.908E−02	1.420E−01	2.302E−01	3.730E−01	4.951E−01	6.138E−01	6.565E−01	7.206E−01
80.0	9.820E−02	1.714E−02	4.417E−02	3.666E−02	5.419E−02	9.393E−02	1.577E−01	2.270E−01	3.315E−01	4.286E−01	5.279E−01
90.0	5.562E−02	1.416E−02	3.081E−02	3.162E−02	5.056E−02	1.102E−01	1.135E−01	1.617E−01	2.351E−01	3.386E−01	4.043E−01
100.0	1.433E−02	1.952E−01	2.548E−02	2.877E−02	4.710E−02	6.709E−02	1.054E−01	1.588E−01	1.915E−01	2.360E−01	3.009E−01
110.0	9.596E−03	1.549E−01	2.271E−01	3.061E−02	1.003E−01	8.428E−02	8.785E−02	9.031E−02	1.152E−01	1.292E−01	1.581E−01
120.0	1.649E−02	9.025E−02	3.189E−01	3.778E−01	3.576E−01	5.047E−02	6.293E−02	7.869E−02	1.054E−01	1.315E−01	1.521E−01
130.0	4.696E−02	4.037E−02	2.806E−01	4.748E−01	5.342E−01	5.940E−01	6.036E−02	7.757E−02	9.491E−02	1.277E−01	1.669E−01
140.0	2.227E−02	3.305E−02	2.910E−01	3.419E−01	4.690E−01	7.954E−01	1.025E+00	9.477E−01	9.917E−02	1.568E−01	2.496E−01
150.0	9.101E−03	3.353E−02	2.061E−01	1.485E−01	3.789E−01	7.283E−01	1.199E+00	1.747E+00	2.307E+00	2.640E+00	2.417E+00
160.0	1.397E−01	4.656E−01	8.488E−02	1.160E−01	2.442E−01	6.277E−01	9.072E−01	1.198E+00	1.469E+00	1.804E+00	2.175E+00
170.0	1.319E−01	2.486E−01	8.021E−02	1.214E−01	2.199E−01	4.838E−01	9.850E−01	1.457E+00	1.588E+00	1.851E+00	2,148E+00
180.0	4.260E−02	8.906E−02	7.591E−02	1.120E−01	3.684E−01	2.360E+00	2.472E+00	3.215E+00	4.722E+00	7.774E+00	6.969E+00

References

Aas, E. (1984) *Some Aspects of Light Scattering by Marine Particles*, Oslo: Institute of Geophysics.

Ackerman, S. A., and G. L. Stephens (1987) The absorption of solar radiation by cloud droplets: an application of anomalous diffraction theory, *J. Atmos. Sci.*, **44**, 1574–1588.

Aden, A. L., and M. Kerker (1951) Scattering of electromagnetic waves from two concentric spheres, *J. Appl. Phys.*, **22**, 1242–1246.

Ailawadi, N. K. (1973) Generalization of the Ashcroft–Lekner hard-sphere model for the structure factor, *Phys. Rev.*, **A7**, 2200–2203.

Al-Chalabi, S. A. M., and A. R. Jones (1994) Development of a mathematical model for light scattering by statistically irregular particles, *Particle & Particle Syst. Charact.*, **11**, 200–206.

Al-Chalabi, S. A. M., and A. R. Jones (1995) Light scattering by irregular particles in the Rayleigh–Gans–Debye approximation, *J. Phys. D*, **28**, 1304–1308.

Allen, T. (1990) *Particle Size Measurement*, London: Chapman & Hall.

Alexandrov, M. D., V. S. Remizovich, and D. B. Rogozkin (1993) Multiple light scattering in a two-dimensional medium with large scatterers, *J. Opt. Soc. Am.*, **A10**, 2602–2610.

Al-Nimr, M. A., and V. S. Arpaci (1994) Optical properties of interacting particles, *Appl. Opt.*, **33**, 8412–8416.

Ambarzumian, V. A. (1943) On diffuse reflection of light from turbid media, *Doklady AN SSSR*, **8**, 257–265.

Ambarzumian, V. A. (1961) *Scientific Papers*, Vol. 1, Erevan: Armenian Academy of Sciences.

Angelsky, O. V., and Maksimyak, P. P. (1993) Optical correlation method for studying disperse media, *Appl. Opt.*, **32**, 6137–6141.

Aoki, T., T. Aoki, M. Fukabori, Y. Tachibana, F. Nishio, and T. Oishi (1998) Spectral albedo observation of the snow field at Barrow, *Polar Meteorol. Glaciol.*, **12**, 1–9.

Aoki, T., T. Aoki, M. Fukabori, A. Hachikubo, Y. Tachibana, and F. Nishio (2000) Effects of snow physical parameters on spectral albedo and bi-directional reflectance of snow surface, *J. Geophys. Res.*, **105**, 10219–10236.

Arking, A., and J. D. Childs (1985) Retrieval of cloud cover parameters from multispectral satellite images, *J. Appl. Meteorol.*, **24**, 323–333.

Arnott, W. P., Ya. Y. Dong, and J. Halett (1995) Extinction efficiency in the infrared (2–18 mm) of laboratory ice clouds: observations of scattering minima in the Christiansen bands of ice, *Appl. Opt.*, **34,** 541–551.

Aronson, A. (1995) Boundary conditions for diffusion of light, *J. Opt. Soc. Am.*, **A12**, 2532–2539.

Apresyan, L. A., and Y. A. Kravzov (1983) *Radiative Transfer Theory*, Moscow: Nauka.

Asano, S., and G. Yamamoto (1975) Light scattering by a spheroidal particle, *Appl. Opt.*, **14**, 29–49.

Ashcroft, N. W., and J. Lekner (1966) Structure and resistivity of liquid metals, *Phys. Res.*, **145**, 89–90.

Auer, A. H., and D. L. Veal (1970) The dimension of ice crystals in natural clouds, *J. Atmos. Sci.*, **27**, 919–926.

Babenko, *et al.* (2003) *Electromagnetic Scattering in Dispersed Media: Inhomogeneous and Anisotropic Particles*, Chichester, UK.: Springer–Praxis.

Balescu, R. (1975) *Equilibrium and Nonequilibrium Statistical Mechanics*, New York: John Wiley.

Barabanenkov, Y. N. (1982) On relative increasing of the extinction length due to correlation of the weak scatterers, *Izv. AN SSSR, Fiz. Atmos. Okeana*, **18**, 720–726.

Baran, A. J., and S. Havemann (1999) Rapid computation of the optical properties of hexagonal columns using complex angular momentum theory, *J. Quant. Spectr. Rad. Transfer*, **63**, 499–519.

Barber, P. W., and C. Yeh (1975) Scattering of electromagnetic waves by arbitrary shaped dielectric bodies, *Appl. Opt.*, **14**, 2864–2872.

Barber, P. W., and S. C. Hill (1990) *Light Scattering by Particles: Computational Methods*, Singapore: World Scientific.

Barkey, B., M. Bailey, K.-N. Liou, and J. Hallett, (2002) Light-scattering properties of plate and column ice crystals generated in a laboratory cold chamber, *Appl. Opt.*, **41**, 5792–5796.

Barun, V. V., and A. B. Gavrilovich (1987) Spectral characteristics of light scattering by soil aerosol, *J. Appl. Spektr.*, **47**, 453–460.

Barun, V. V. (1995) Visual perception of retroreflective objects through light scattering media, *Proc. SPIE*, **2410**, 470–479.

Barun, V. V. (2000) Influence of cloud aerosol microstructure on the backscattering signal from the object shadow area, *Izvestiya Rus. Acad. Nauk, Atmos. and Oceanic Physics*, **36**, 258–265.

Baumgarten, G., K. H. Fricke, and G. von Cossart (2002) Investigation of the shape of noctilucent cloud particles by polarization lidar technique, *Geophys. Res. Let.,* **29**, doi: 10.1029/2001GL013877.

Bayvel, L. P. and A. R. Jones (1981) *Electromagnetic Scattering and its Applications*, London: Applied Science.

Beddow, J. K., and T. Meloy (1980) *Testing and Characterization of Powder and Fine Particles*, London: Heyden.

Belov, V. F., A. G. Borovoi, N. I. Wagin, and S. N. Volkov (1984) On small-angle method under single and multiple light scattering, *Izv. Acad. Nauk SSSR, Fizika Atmos. and Okeana*, **20**, 323–327.

Berry, M. V., and I. C. Percival (1986) Optics of fractal clusters such as smoke, *Opt. Acta*, **33**, 577–591.

Bohren, C. F., and B. R. Barkstrom (1974) Theory of optical properties of snow, *J. Geophys. Res.*, **79**, 4527–4535.
Bohren, C. F. (1974) Light scattering by an optically active sphere, *Chem. Phys. Lett.*, **29**, 458–462.
Bohren, C. F., and B. R. Barkstrom, (1974) Theory of the optical properties of snow, *J. Geophys. Res.*, **79**, 4527–4535.
Bohren, C. F. (1975) Scattering of electromagnetic waves by an optically active spherical shell, *J. Chem. Phys.*, **62**, 1566–1571.
Bohren, C. F. (1978) Scattering of electromagnetic waves by an optically active cylinder, *J. Colloid Interface Sci.*, **66**, 105–109.
Bohren, C. F., and R. L. Beschta, (1979) Snowpack albedo and snow density, *Cold Regions Sci. Tech.*, **1**, 47–50.
Bohren, C. F., and T. Nevitt (1983) Absorption by a sphere: a simple approximation, *Appl. Opt.*, **22**, 774–775.
Bohren, C. F. (1983) Colors of snow, frozen waterfalls and icebergs, *J. Opt. Soc. Am.*, **73**, 1646–1652.
Bohren, C. F., and D. R. Huffman (1983) *Absorption and Scattering of Light by Small Particles*, New York: John Wiley.
Bohren, C. F. and G. Koh (1985) Forward-scattering corrected extinction by nonspherical particles. *Appl. Opt.*, **29**, 1023–1029.
Bohren, C. F. (1987) Multiple scattering of light and some of its observable consequences, *Am. J. Phys.*, **55**, 524–533.
Bohren, C. F., and S. B. Singham (1991) Backscattering by nonspherical particles: review of methods and suggested new approaches, *J. Geophys. Res.*, **96**, 5269–5277.
Born, M., and E. Wolf (1965) *Principles of Optics*, 3rd edn, Oxford: Pergamon.
Borovoi, A. G. (1982) Light propagation in precipitation, *Izv. Vyssh. Uchebn. Zaved. SSSR, Radiofizika*, **25**, 391–400.
Borovoi, A. G., N. I. Wagin, and V. V. Veretenniker (1986) Method of spatial correlations of intensity in diagnostics of scattering media, *Opt. Spektorsk.*, **61**, 1326–1330.
Borovoi, A. G. (1995) Retrieval of particle size distribution and number density profile in clouds and fogs from multiply scattered radiation, *Proc. SPIE*, **2471**, 375–386.
Borovoi, A. G. (1998) Optical diagnostics of coarse particulate media by back scattered laser radiation, *Proceedings of the 7th European Symposium on Particle Characterization* (R. Weichert, ed.), Preprints II, 551–560, Nurnberg: Nurnberg Messe.
Borovoi, A. G., E. I. Naats, and U. G. Oppel (1998) Characterization of shape parameters of particles from light scattering, *Proceedings of the 7th European Symposium on Particle Characterization* (R. Weichert, ed), Preprints I, 201–210, Nurnberg: Nurnberg Messe.
Bourelly, C. P., Chiappetta, R. Deleuil, and B. Torrésani (1989) Approximations for electromagnetic scattering by homogeneous arbitrarily shaped bodies. *Proc. IEEE*, **77**, 741–749.
Bovensmann, H., Burrows, J. P., Buchwitz, M. Frerick, J., Noël, J., and V. V. Rozanov (1999) SCIAMACHY: mission objectives and measurement modes, *J. Atmos. Sci.*, **56**, 127–150.
Bremmer, H. (1964) Random volume scattering, *Radio Science*, **68D**, 967–981.
Bricaud, A., and D. Stramski (1990) Spectral absorption coefficients of living phytoplankton and non-algal biogenous matter: a comparison between the Peru upwelling area and the Sargasso Sea, *Limnol. Oceanogr.*, **35**, 562–582.
Bricaud, A., A. Morel, M. Babin, K. Allali, and H. Claustre, (1998) Variations of light absorption by suspended particles with chlorophyll A concentration in oceanic (case 1) waters: Analysis and implications for bio-optical models, *J. Geophys. Res.*, **103**, 31033–31044.

Bryant, F. D., and P. Latimer (1969) Optical efficiencies of large particles of arbitrary shape and orientation, *J. Colloid Interface Sci.*, **30**, 291–304.

Bucholtz, A. (1995) Rayleigh-scattering calculations for the terrestrial atmosphere, *Appl. Opt.*, **34**, 2765–2773.

Busygin, V. P., N. A. Yevstratov, and Y. M. Feigel son (1973) Optical properties of cumulus clouds, and radiant fluxes for cumulus cloud cover, *Izv. Acad. Sci. USSR, Atmos. Oceanic Phys.*, **9**, 1142–1151.

Cahalan, R. F., W. Ridgway, W. J. Wiscombe, and T. L. Bell (1994) The albedo of fractal stratocumulus clouds, *J. Atmos. Sci.*, **51**, 2434–2455.

Caorsi, S., A. Massa, and M. Pastorino (1996) Rytov approximation: application to scattering by two-dimensional weakly nonlinear dielectrics, *J. Opt. Soc. Am.*, **A13**, 509–516.

Card, J. B. A., and Jones, A. R. (1999) An Investigation of the potential of polarised light scattering for the characterisation of irregular particles, *J. Phys. D*, **32**, 2467–2474 .

Case, K. M., and P. F. Zweifel (1967) *Linear Transport Theory*, New York: Addison-Wesley.

Chandrasekhar, S. (1950) *Radiative Transfer*, Oxford: Oxford University Press.

Chang, T., and H. Yu (1990) Light scattering and electric birefringence studies of biomembrane vesicles, *Comments on Mol. Cell. Biophys.*, **7**, 25–27.

Chen, T. W. (1987) Scattering of light by a stratified sphere in high energy approximation, *Appl. Opt.*, **26**, 4155–4158.

Chen, T. W., and W. S. Smith (1992) Large-angle light scattering at large size parameters, *Appl. Opt.*, **31**, 6558–6560.

Chen, T. W. (1993) Simple formula for light scattering by a large spherical dielectric, *Appl. Opt.*, **32**, 7568–7571.

Chou, M.-D. (1998) Parametrizations for cloud overlapping and shortwave single scattering properties for use in general circulation and cloud ensemble models, *J. Climate*, **11**, 202–214.

Chylek, P., and J. D. Klett (1991a) Extinction cross sections of nonspherical particles in the anomalous diffraction approximation, *J. Opt. Soc. Am.*, **A8**, 274–281.

Chylek, P., and J. D. Klett (1991b) Absorption and scattering of electromagnetic radiation by prismatic columns: anomalous diffraction approximation, *J. Opt. Soc. Am.*, **A8**, 1713–1720.

Chylek, P., P. Damiano, and E. P. Shettle (1992) Infrared emittance of water clouds, *J. Atmos. Sci.*, **49**, 1459–1472.

Chylek, P., J. Zhan, and R.G. Pinnich (1993) Absorption and scattering of microwaves by falling snow, *Int. J. Infrared Millim. Waves*, **14**, 2295–2310.

Chylek, P., N. Kalyaniwalla, and E. P. Shettle (1995) Radiative properties of water clouds: simple approximations, *Atmos. Res.*, **35**, 139–156.

Cohen, A., and E. Tirosh (1990) Absorption by a large sphere with an arbitrary complex refractive index, *J. Opt. Soc. Am.*, **A7**, 323–325.

Conference on Light Scattering by Nonspherical Particles: Theory, Measurements, and Applications (1998), Boston: American Meteorological Society.

Cooray, M. F. R., and I. R. Ciric (1993) Wave scattering by a chiral spheroid, *J. Opt. Soc. Am.*, **A10**, 1197–1203.

Cross, D. A., and P. Latimer (1970) General solutions for the extinction and absorption efficiences of arbitrarily oriented cylinders by anomalous diffraction methods, *J. Opt. Soc. Am.*, **60**, 904–907.

d'Almeida, G. A., P. Koepke, E. P. Shettle (1991) *Atmospheric Aerosols: Global Climotology and Radiative Characteristics*, New York: A. Deepak.

Danielson, R. E. *et al.* (1969) The transfer of visible radiation through clouds, *J. Atmos. Sci.*, **26**, 1078–1087.

Dau-Sing, W., and P. W. Barber (1979) Scattering by inhomogeneous nonspherical objects, *Appl. Opt.*, **18**, 1190–1197.

Davies , A. B. *et al.* (1999) Off-beam lidar: an emerging technique in cloud remote sensing based on radiative Green-function theory in the diffusion domain, *Phys. Chem. Earth B*, **24**, 177–185.

De Haan, J. F. (1987) Effects of aerosols on the brightness and polartization of cloudless planetary atmospheres, PhD thesis, Free University of Amsterdam.

De Rooij, W. A. (1985) Reflection and transmission of polarized light by planetary atmospheres, PhD thesis, Free University of Amsterdam.

De Wolf, D. A. (1971) Electromagnetic reflection from an extended turbulent medium: cumulative forward-scatter single back-scatter approximation, *IEEE Trans., Antennas & Propag.*, **19**, 254–262.

Deirmendjian, A. (1969) *Electromagnetic Scattering on Spherical Polydispersions*, Amsterdam: Elsevier.

Deschamps, P.-Y., F.-M. Breon, M. Leroy, A. Podaire, A. Bricaud, J.-C. Buriez, and G. Sèze (1994) The POLDER mission: instrument characteristics and scientific objectives, *IEEE Trans.*, **GE32**, 598–614.

Dolginov, A. Z., Yu. N. Gnedin, and N. A. Silant'ev (1995) *Propagation and Polarization of Radiation in Cosmic Media*, Amsterdam: Gordon & Breach.

Dolin, L. S. (1964) Light beam scattering in a turbid medium mayer, *Izv. Vyssh. Uchebn. Zaved. Radiofiz.*, **7**, 471–478.

Dolin, L. S. and I. M. Levin (1991) *Theory of Underwater Vision (Reference Book)*, Leningrad: Gidrometeoizdat.

Domingue, D., D. B. Hartmon, and A. Verbiscer (1997) The scattering properties of natural terrestrial snow versus icy satellite surfaces, *Icarus*, **128**, 28–48.

Domke, H. (1978a) Linear Fredholm integral equations for radiative transfer problems in finite plane-parallel media. I. Imbedding in an infinite medium, *Astron. Nachr.*, **299**, 87–93.

Domke, H. (1978b) Linear Fredholm integral equations for radiative transfer problems in finite plane-parallel media. II. Imbedding in a semi-infinite medium, *Astron. Nachr.*, **299**, 95–102.

Draine, B. T. (1988) The discrete-dipole approximation and its application to interstellar graphite grains, *Astrophys. J.*, **333**, 848–872.

Draine, B. T., and P. J. Flatau (1994) Discrete-dipole approximation for scattering calculations, *J. Opt. Soc. Am.*, **A11**, 1491–1499.

Draine, B. T., and J. Goodman (1993) Beyond Clausius–Mossotti: wave propagation on a polarizable point lattice and the discrete dipole approximation, *Astrophys. J.*, **405**, 685–697.

Draine, B. T., and J. C. Weingartner (1996) Radiative torques on interstellar grains. Allignment with the magnetic field, *Astrophys. J.*, **480**, 633–646.

Drofa, A. S., and A. L. Usachev (1980) About vision in a cloudy medium, *Izv. Akad. Nauk SSSR, Fiz. Atmos. Okeana*, **16**, 933–938.

Dubova, G. S., A. Ya. Khairullina, and S. F. Shumilina (1977) Determination of the absorption spectrum of a hemoglobin by light scattering methods, *J. Appl. Spektr.*, **27**, 871–878.

Dubova, G. S., A. Ya. Khairullina, and S. F. Shumilina (1981) Retrieval of the spectrum of the imaginary part of the refractive index of pigments of soft closely packed particles, *J. Appl. Spektr.*, **34**, 1058–1064.

Evans, B. T. N., and G. R. Fournier (1990) A simple approximation to extinction efficiency valid over all size parameters, *Appl. Opt.*, **29**, 4666–4670.

Evans, B. T. N., and G. R. Fournier (1994) Analytic approximation to randomly oriented spheroid extinction, *Appl. Opt.*, **33**, 5796–5804.

Evans, B. T. N., and Fournier, G. R. (1996) Approximations of polydispersed extinction, *Appl. Opt.*, **35**, 3281–3285.

Farafonov, V. G. (1983) The scattering of a plane electromagnetic wave by a dielectric spheroid, *Differential Equations*, **19**, 1765–1777.

Farafonov, V. G. (1990) Optical properties of strongly prolate and oblate spheroidal particles, *Opt. Spektr.*, **69**, 866–872.

Farafonov, V. G., N. V. Voshchinnikov, and V. N. Somsikov (1994) Light scattering by a core-mantle spheroidal particle, *Appl. Opt.*, **35**, 5412–5426.

Farafonov, V. G. (2000) Light scattering bimultiplier ellipsoids in the Rayleigh approximation, *Opt. and Spectroscopy*, **88**, 441–443.

Fily, M., B. Bourdlles, J. P. Dedieu, and C. Sergent (1997) Comparison of *in situ* and Landsat Thematic Mapper derived snow grain characteristics in the Alps, *Remote Sens. Environ.*, **59**, 452–460.

Flatau, P. J. (1992) Scattering by irregular particles in anomalous diffraction and discrete dipole approximations, PhD thesis, Colorado: Colorado State University.

Fock, M. V. (1944) On some integral equations of mathematical physics, *Matem. Sbornik*, **14**, 3–50.

Foldy, L. L. (1945) The multiple scattering of waves, *Phys. Rev.*, **67**, 107–119.

Fournier, G. R., and B. T. N. Evans (1991) Approximation to extinction efficiency for randomly oriented spheroids, *Appl. Opt.*, **30**, 2042–2048.

Fournier, G. R., and B. T. N. Evans (1993) Bridging the gap between the Rayleigh and Thomson limits for spheres and spheroids, *Appl. Opt.*, **32**, 6159–6166.

Fournier, G. R., and B. T. N. Evans (1996) Approximations to extinction from randomly oriented circular and elliptical cylinders, *Appl. Opt.*, **35**, 4271–4282.

Frouin, R. M. Schwinling, and P.-Y. Deschamps (1996) Spectral reflectance of sea foam in the visible and near-infrared: In situ measurement and remote sensing applications, *J. Geophys. Res.*, **102**, 14361–14371.

Frouin, R., S. F. Iacobellis, and P. -Y. Deschamps (2001) Influence of oceanic whitecaps on the global radiation budget, *Geophys. Res. Let.*, **28**, 1523–1526.

Fu, Q. (1996) An accurate parameterization of the solar radiative properties of cirrus clouds for climate models, *J. Climate*, **9**, 2058–2082.

Furutsu, K. (1980a) Diffusion equation derived from space-time transport equation, *J. J. Opt. Soc. Am.*, **70**, 360–366.

Furutsu, K. (1980b) Diffusion equation derived from space-time transport equation in anisotropic media, *J. Math. Phys.*, 765–777.

Furutsu, K. (1997) Pulse wave scattering by an absorber and integrated attenuation in the diffusion approximation, *J. Opt. Soc. Am.*, **A14**, 267–274.

Gadsen, M. and W. Schröder (1989) *Noctilucent Clouds*, Berlin: Springer-Verlag.

Garcia, R. D. M., and C. E. Siewert (1985) Benchmark results in radiative transfer, *Transp. Theory Stat. Phys.*, **14**, 437–483.

Gershun, A. A. (1937) On the problem of diffuse light transmission, *GOI Proc.*, **4**, 12–40.

Gilbert, G. D. and J. C. Pernicka (1996) Improvement of underwater visibility by reduction of backscatter with a circular polarization technique, in *Underwater Photo-Optics, Seminar Proceedings*, Santa Barbara: SPIE, pp. A-111–1 to A-111–11.

Glautshing, W. J., and S.-H. Chen (1981) Light scattering from water droplets in the geometrical optics approximation, *Appl. Opt.*, **20**, 2499–2509.
Gobel, G., J. Kuhn, and J. Fricke (1995) Dependent scattering effects in latex-sphere suspensions and scattering powders, *Waves in Random Media*, **5**, 413–426.
Goloub, P., D. Tanre, J. L. Deuze, M. Marchand, F. M. Breon (1999) Validation of the first algorithm applied for the aerosol properties over the oceans using POLDER/ADEOS measurements, *IEEE Transact. Geoscience and Remote Sensing*, **37**, 1586–1595.
Goloub, P., M. Herman, H. Chepfer, J. Riedi, G. Brogniez, P. Couvert, and G. Sẽze (2000) Cloud thermodynamical phase classification from the POLDER spaceborne instrument, *J. Geophys. Res. D*, **105**, 14747–14759.
Golubitsky, B. M., I. M. Levin, and M. B. Tantashev (1974) Brightness coefficient of semi-infinite layer of seawater, *Izv. Acad. Sci. USSR, Atmos. Oceanic Physics*, **10**, 1235–1238.
Goodman, J. W. (1968) *Introduction to Fourier Optics*, New York: McGraw-Hill.
Gordon, D. S. (1972) Mie scattering by optically active particles, *Biochemistry*, **11**, 413–420.
Gordon, H. G. (1973) Simple calculation of the diffuse reflectance of the ocean, *Appl. Opt.*, **12**, 2803–2804.
Gordon, H. G., and M. Wang (1994) Influence of oceanic whitecaps on atmospheric correction of ocean-colour sensors, *Appl. Opt.*, **33**, 7754–7763.
Gordon, H. R. (1997) Atmospheric correction of ocean color imagery in the Earth Observing System era, *J. Geoph. Res.*, **D102**, 17081–17106.
Granovskii, Ya. I., and M. Ston (1994) Attenuation of light scattered by transparent particles, *J. Exp. Theor. Phys.*, **78**, 645–649.
Grenfell, T. C., D. K. Perovich, and J. A. Ogren (1981) Spectral albedos of an alpine snowpack, *Cold Regions Sci. Technol.*, **4**, 121–127.
Grenfell, T. C. *et al.* (1994) Reflection of solar radiation by the Antarctic snow surface at ultraviolet, visible, and near-infrared wavelengths, *J. Geophys. Res.*, **D99**, 18669–18684.
Hage, J. I. (1990) The optics of porous particles and the nature of comets, PhD thesis, Free University of Amsterdam.
Hage, J. I., J. M. Greenberg, and R. T. Wang (1991) Scattering from arbitrarily shaped particles: theory and experiment, *Appl. Opt.*, **30**, 1141–1152.
Hale, G. M., and M. R. Querry (1973) Optical contrasts of water in the 200-nm to 200-mm wavelength region, *Appl. Opt.*, **12**, 555–563.
Han, Q., W. B. Rossow, and A. A. Lecis (1994) Near global survey of effective droplet radii in liquid water clouds using ISCCP data, *J. Climate*, **7**, 465–497.
Hansen, J. E., and J. Hovenier (1974) Interpretation of the polarization of Venus, *J. Atmos. Sci.*, **31**, 1137–1160.
Hansen, J. E., and L. D. Travis (1974) Light scattering in planetary atmospheres, *Space Sci. Rev.*, **16**, 527–610.
Hansen, J. E., and L. Nazarenko (2004) Soot climate forcing via snow and ice albedo, *Proc. of National Acad. of Sciences of USA*, **101**, 423–428.
Hapke, B., (1981) Bidirectional reflectance spectroscopy. 1. Theory, *J. Geophys. Res.*, **86**, 3039–3054.
Hapke, B. (1993) *Theory of Reflectance and Emittance Spectroscopy*, Cambridge: Cambridge University Press.
Hart, R. W. and E. W. Montroll (1951) On the scattering of plane waves by soft particles. Spherical obstacles, *J. Appl. Phys.*, **22**, 376–386.
Heffels, C., D. Heitzmann, E. Dan Hirleman, and B. Scarlett (1995) Forward light scattering for arbitrarily sharp-edged crystals in Fraunhofer and anomalous diffraction approximations, *Appl. Opt.*, **34**, 6552–6560.

Hess, M., and M. Wiegner (1994): COP: a data library of optical properties of hexagonal ice crystals, *Appl. Opt.*, **33**, 7740–7746.
Hirleman, E. D. (1988) Modelling of multiple scattering effects in Fraunhofer diffraction particle size analysis, *Part. and Part. Syst. Charact.*, **5**, 57–65.
Hirleman, E. D. (1991) General solution to the inverse near-forward-scattering particle sizing problem in multiple-scattering environments: theory, *Appl. Opt.*, **30**, 4832–4838.
Hirleman, E. D., and C. F. Bohren (1991) Optical particle sizing: an introduction by the feature editors, *Appl. Opt.*, **30**, 4685–4687.
Hobbs, P. V. (1974) *Ice Physics*. London: Oxford University Press.
Hodkinson, J. R. (1963) Light scattering and extinction by irregular particles, *Electromagnetic Scattering* (M. Kerker, ed.), New York: Academic Press, pp. 87–100.
Holoubek, J. (1991) Small-angle light scattering from an anisotropic sphere in the Rayleigh–Gans–Debye approximation: the Mueller matrix formalism, *Appl. Opt.*, **30**, 4987–4992.
Hovenac, E. A. (1991) Calculation of far-field scattering from nonspherical particles using a geometrical optics approach, *Appl. Opt.*, **30**, 4739–4746.
Hovenier, J. W. (1969) Symmetry relationships for scattering of polarized light in a slab of randomly oriented particles, *J. Atmos. Sci.*, **26**, 488–499.
Hovenier, J. W. (1971) Multiple scattering of polarized light in planetary atmospheres, *Astron. & Astrophys.*, **13**, 7–29.
Hubley, R. C. (1955) Measurement of diurnal variations in snow albedo on Lemon Greek Glacier, Alaska, *J. Glaciol.*, **2**, 560–563.
Husar, R. B., J. M. Prospero, and L. L. Stowe (1997) Characterization of tropospheric aerosols over the oceans with the NOAA AVHRR optical thickness product, *J. Geoph. Res.*, **102**, 16889–16909.
Hutt, D. L., L. R. Bissonnette, D. St. Germain, and J. Oman (1992) Extinction of visible and infrared beams by falling snow, *Appl. Opt.*, **31**, 5121–5132.
Irvine, W. M. (1963) The asymmetry parameter of the scattering diagram of a spherical particle, *Bull. Astron. Inst. Netherlands*, **3**, 176–184.
Ishimaru, A. (1978) *Wave Propagation and Scattering in Random Media*, New York: Academic Press.
Ishimaru, A., and C. W. Yeh (1984) Matrix representations of the vector radiative-transfer theory for randomly distributed nonspherical particles, *J. Opt. Soc. Am.*, **A1**, 359–364.
Ito, S. (1993) Optical wave propagation in discrete random media with large particles: a treatment of the phase function, *Appl. Opt.*, **32**, 1652–1656.
Ivanov, A. P. (1969) *Optics of Scattering Media*, Minsk: Nauka i Tekhnika.
Ivanov, A. P. (1975) *Physical Principles of Hydrooptics*, Minsk: Nauka i Tekhnika.
Ivanov, A. P., S. A. Makarevich, and A. Ya. Khairullina (1987) On specific features of radiation propagation in tissues and bioliquids with close packed particles, *J. Appl. Spektr.*, **47**, 662–668.
Ivanov, A. P., V.A. Loiko, and V. P. Dick (1988) *Light Propagation in Close Packed Media*, Minsk: Nauka i Tekhnika.
Ivlev, L. S. (1986) *Optical Properties of Atmospheric Aerosols*, Leningrad: Leningrad State University.
Jaffe, J. S. (1995) Monte Carlo modeling of underwater image formation: validity of the linear and small-angle approximations, *Appl. Opt.*, **34**, 5413–5421.
John, S., G. Pang, and Y. Yang (1996) Optical coherence propagation and imaging in a multiply scattering medium, *J. Biomed. Opt.*, **1**, 180–191.
Johnson, B. R. (1996) Light scattering by a multilayer sphere, *Appl. Opt.*, **35**, 3286–3296.

Jones, A. R. (1977) Error contour charts relevant to particle sizing by forward scattered lobe methods, *J. Phys. D*, **10**, L163–L165.
Jones, A. R. (1987) Fraunhofer diffraction by random irregular particles, *Part. and Part. Syst. Charact.*, **4**, 123–127.
Jones, A. R. (1988) Fraunhofer diffraction by random irregular particles, in: *Optical Particle Sizing: Theory and Practice* (G. Gouesbet and G. Grehan, eds), pp. 301–319, New York: Plenum Press.
Jones, A. R., J. Koh, and A. Nasaruddin (1996) Error contour charts for the two-wave WKB approximation, *J. Phys. D*, **29**, 39–42.
Jones, A. R. (1999) Light scattering for particle characterization, *Progress in Energy and Combustion Sci.*, **25**, 1–53.
Jones, D. S. (1957) High-frequency scattering of electromagnetic waves, *Proc. Royal Soc. London*, **A240**, 206–213.
Joseph, J. H., W. J. Wiscombe, and J. A. Weinmann (1976) The delta-Eddington approximation for radiative flux transfer, *J. Atmos. Sci.*, **33**, 2452–2459.
Junge, C. E. (1963) *Air Chemistry and Radiochemistry*, New York: Academic Press.
Kadyshevich, Ye. A., Yu. S. Lubortsera, and G. V. Rozenberg (1976) Light scattering matrices of Pacific and Atlantic waters, *Izv. Acad. Sci. USSR, Atmos. Oceanic Phys.*, **12**, 106–111.
Karam, M. A. (1998) Polarimetric optical theorem, *J. Opt. Soc. Am.*, **A15**, 196–201.
Katsev, I. L., E. P. Zege, A. S. Prikhach, and I. N. Polensky (1997) Efficient technique to determine backscattered light power for various atmospheric and oceanic sounding and imaging systems, *J. Opt. Soc. Am.*, **A14**, 1338–1346.
Kaufman, Y. J. and C. Sendra (1988) Algorithm for automatic atmospheric corrections to visible and near-IR satellite imagery, *Int. J. Remote Sensing*, **9**, 1357–1381.
Kaufman, Y. J., D. Taure, L. A. Remer, E. F. Vermote, A. Chu, and B. N. Holben (1997) Operational remote sensing of tropospheric aerosol over land from EOS moderate resolution imaging spectrometer, *J. Geophys. Res,* **D102**, 17051–17067.
Kerker, M. (1969) *The Scattering of Light and Other Electromagnetic Radiation*, New York: Academic Press.
Kerker, M., ed. (1988) *Selected Papers on Light Scattering*, Bellingham: SPIE – The International Society for Optical Engineering.
Khlebtsov, N. G. (1984) Role of multiple scattering in turbidimetric investigations of disperse systems, *J. Appl. Spectrosk.*, **40**, 243–247.
Khlebtsov, N. G. (1999) On light scattering by Gaussian spheres and ellipsoids [comments on the papers of V. M. Rysakov and M. Ston], *Optics and Spectr.*, **87**, 909–913.
Khlebtsov, N. G. (1993) Optics of fractal clusters in the anomalous diffraction approximation, *J. Mod. Opt.*, **40**, 2221–2235.
Khlebtsov, N. G. (1996) Spectroturbidimetry of fractal clusters: test of density correlation function cutoff, *Appl. Opt.*, **21**, 4261–4270.
Kim, C., N. Lior, and K. Okuyama (1996) Simple mathematical expression for spectral extinction and scattering properties of small size-parameter particles, including examples for soot and TiO_2, *J. Quant. Spectr. & Rad. Transfer,* **55**, 391–411.
King, M. D. (1981) A method for determining the single scattering albedo of clouds through observation of the internal scattered radiation field, *J. Atmos. Sci.*, **38**, 2031–2044.
King, M., and Harshvardhan (1986) Comparative accuracy of selected multiple scattering approximations, *J. Atmos. Sci.*, **43**, 784–801.
King, M. D. (1987) Determination of the scaled optical thickness of clouds from reflected solar radiation measurements, *J. Atmos. Sci.*, **44**, 1734–1751.

King, M. D., Y. Kaufman, Menzel, P., and D. Taure (1992) Remote sensing of cloud, aerosol, and water vapour properties from the moderate resolution imaging spectrometer (MODIS), *IEEE Trans.*, **GE30**, 2–27.

Klett, J. D., and R. A. Sutherland (1992) Approximate methods for modeling the scattering properties of nonspherical particles: evaluation of the Wentzel–Kramers–Brillouin method, *Appl. Opt.*, **31**, 373–386.

Kluskens, M. S., and E. H. Newman (1991) Scattering by a multilayer chiral cylinder, *IEEE Trans. Antennas Propag.*, **39**, 91–96.

Koenderink, J. J., and W. A. Richards (1992) Why is snow so bright, *J. Opt. Soc. Am.*, **A9**, 643–648.

Koepke, P., and H. Quenzel (1981) Turbidity of the atmosphere determined from satellite: Calculation of of the optimum wavelength, *J. Geophys. Res.*, **86**, 9801–9805.

Koepke, P. (1984) Effective reflectance of oceanic whitecaps, *Appl. Opt.*, **23**, 1816–1824.

Kokhanovsky, A. A. (1988) Integral characteristics of light scattering by big spherical particles with refractive index bellow unity, *Optics & Spectroscopy*, **67**, 165–169.

Kokhanovsky, A. A. (1989) Geometrical optics approximation for absorption cross-section of a layered sphere, *Atmospheric Optics*, **2**, 908–912.

Kokhanovsky, A. A. (1990) Light scattering by large two-layered particles: geometrical optics approximation, *Izvestiya RAN, Fizika Atmosfer i Okeana*, **17**, 949–957.

Kokhanovsky, A. A., and E. P. Zege (1995) Local optical parameters of spherical polydispersions: simple approximations, *Appl. Opt.*, **34**, 5513–5519.

Kokhanovsky, A. A., and E. P. Zege (1996) The determination of the effective radius of drops and liquid water path of water clouds from satellite measurements, *Earth Research from Space*, **2**, 33–44.

Kokhanovsky, A. A. (1997) Small-angle approximations of the radiative transfer theory, *J. Phys. D. Appl. Phys.*, **30**, 2837–2840.

Kokhanovsky, A. A., and E. P. Zege (1997a) Optical properties of aerosol particles: a review of approximate analytical solutions, *J. Aerosol Sci.*, **28**, 1–21.

Kokhanovsky, A. A., and E. P. Zege (1997b) Physical parametrization of local optical characteristics of cloudy media. *Izvestiya RAN, Fizika Atmosfer i Okeana*, **33**, 209–218.

Kokhanovsky. A. A., and A. Macke (1997) Integral light scattering and absorption characteristics of large nonspherical particles, *Appl. Opt.*, **36**, 8785–8790.

Kokhanovsky, A. A. (1998) On variability of phase function of atmospheric aerosol at large scattering angles, *J. Atmos. Sci.*, **55**, 314–320.

Kokhanovsky, A. A., T. Nakajima, and E. P. Zege (1998) Physically-based parametrizations of the shortwave radiative characteristics of weakly absorbing optically thick media: application to liquid water clouds, *Appl. Opt.*, **37**, 9750–9757.

Kokhanovsky, A. A., and T. Y. Nakajima (1998) The dependence of phase functions of large transparent particles on their refractive index and shape, *J. Phys. D.*, **31**, 1329–1335.

Kokhanovsky, A. A. (2001) The reflection and transmission matrices of weakly absorbing media, *J. Opt. Soc. Am.*, **18**, 883–887.

Kokhanovsky, A. A., R. Weichert, M. Heuer, and W. Witt (2001) The angular spectrum of light transmitted through optically dense media, *Appl. Opt.*, **40**, 2595–2600.

Kokhanovsky, A. A. (2002a) Reflection and polarization of light by semi-infinite turbid media: simple approximations, *J. Col. Interf. Sci.*, **251**, 429–431.

Kokhanovsky, A. A. (2002b) The accuracy of selected approximations for the reflection function of a semi-infinite turbid medium, *J. Appl. Phys.*, **D35**, 1057–1062.

Kokhanovsky, A. A. (2003a) Optical properties of irregularly shaped particles, *J. Appl. Phys.*, **D36**, 915–923.

Kokhanovsky, A. A. (2003b) *Polarization Optics of Random Media*, Chichester, U.K.: Springer-Praxis.

Kokhanovsky, A. A., and V. V. Rozanov (2003) The reflection function of optically thick weakly absorbing turbid layers: A single approximation, *J. Quart. Spectr. and Rad. Transfer*, **77**, 165–175.

Kokhanovsky, A. A., V. V. Rozanov, E. P. Zege, H. Bovensmann, and J. P. Burrows (2003) A semi-analytical cloud retrieval algorithm using backscattering radiation in 04–2.4 m spectral range, *J. Geophys. Res.*, **D108**, 10.1029/2001JD001543.

Kokhanovsky, A. A. (2004) Reflection of light from nonabsorbing semi-infinite cloudy media: A simple approximation, *J. Quant. Spectr. and Rad. Transfer*, **85**, 25–33.

Kokhanovsky, A. A. and V. V. Rozanov (2004) The physical parameterization of the top-of-atmosphere reflectrion function for a cloudy atmosphere-underlying surface system: The oxygen A-band case study, *J. Quant. Spectr. Rad. Transfer*, **85**, 35–55.

Kokhanovsky, A. A., W. von Hoyningen–Huene, H. Bovensmann, and V. P. Burrows (2004) The determination of the atmospheric optical thickness over Western Europe using SeaWiFS imagery, *IEEE Trans. Geosc. Rem. Sens.*, **42**, 824–832.

Kondratyev, K. Ya., and V. I. Binenko (1984) *Impact of Cloudiness on Radiation and Climate*, Leningrad: Gidrometeoizdat.

Konovalov, N. V. (1975) On range of applicability of asymptotical formulae for calculations of monochromatic radiation in a nonuniform optically thick plane-parallel layer, *Izv. AN SSSR, FAO*, **11**, 1263–1271.

Kravtsov, Y. A. (1993) New effects in wave propagation and scattering in random media (a mini review), *Appl. Opt.*, **32**, 2681–2691.

Kruglyakov, P. M., and D. R. Ekserora (1990) *Foams and Foamy Films*, Moscow: Nauka.

Kuik, F., J. F. de Haan and J. W. Hovenier (1992) Benchmark results for single scattering by spheroids, *J. Quant. Spectr. Radiative Transfer*, **47**, 477–489.

Kuscher, I. and M. Ribaric (1959) Matrix formalism in the theory of diffusion of light, *Optica Acta*, **6**, 42–51.

Kvien, K. (1995) Validity of weak-scattering models in forward two-dimensional optical scattering, *Appl. Opt.*, **34**, 8447–8459.

Laczik, Z. (1996) Discrete-dipole-approximation-based light scattering calculations for particles with a real refractive index smaller than unity, *Appl. Opt.*, **19**, 3736–3745.

Lagendijk, Ad., and B. A. van Tiggelen (1996) Resonant multiple scattering of light, *Physics Reports*, **270**, 143–215.

Landolt-Börnstein (1988) Numerical data and functional relationships in science and technology (Group V: Geophysics and Space Research. V.5: Meteorology. Subvolume b), *Physical and Chemical Properties of the Air*, 570 pp. (G. Fischer, ed.), Berlin: Springer-Verlag.

Latimer, P. (1980) Predicted scattering by spheroids. Comparison of approximate and exact methods, *Appl. Opt.*, **19**, 3039–3041.

Lax, M. (1952) Multiple scattering of waves. II. The effective field in dense systems, *Phys. Rev.*, **85**, 261–269.

Lee, R. L. (1990) Green icebergs and remote sensing, *J. Opt. Soc. Am.*, **A7**, 1862–1874.

Lenoble, J., ed. (1985) *Radiative Transfer in Scattering and Absorbing Atmospheres: Standard Computational Procedures*, Hampton: A. Deepak.

Lenoble, J. (1993) *Atmospheric Radiative Transfer*, Hampton: A. Deepak.

Levoni, C., A. Cervino, R. Gutti, and F. Torricella (1997) Atmospheric aerosol optical properties: a database of radiative characteristics for different components and classes, *Appl. Opt.*, **36**, 8031–8041.

Li, W., K. Stamnes, and B. Chen (2001) Snow grain size retrieved from near-infrared radiances at multiple wavelengths, *Geophys. Res. Let.*, **28**, 1699–1702.

Lilienfeld, P. (1991) Gustav Mie: the person, *Appl. Opt.*, **30**, 4696–4698.

Lind, A. C., and J. M. Greenberg (1966) Electromagnetic scattering by obliquely oriented spheroids, *J. Appl. Phys.*, **37**, 3195–3203.

Liou, K.-N. *et al.* (1983) Scattering phase matrix comparison for randomly oriented hexagonal cylinders and spheroids, *Appl. Opt.*, **22**, 1684–1687.

Liou, K. N. (1992) *Radiation and Cloud Processes in the Atmosphere*, Oxford: Oxford University Press.

Liou, K. N. (2002) *Introduction to Atmospheric Radiation*, N.Y.: Academic Press.

Lock, J. A., and C. L. Chiu (1994) Correlated light scattering by a dense distribution of condensation droplets on a window pane, *Appl. Opt.*, **33**, 4663–4671.

Lock, J. A. (1996a) Ray scattering by an arbitrarily oriented spheroid. I. Diffraction and specular reflection, *Appl. Opt.*, **35**, 500–514.

Lock, J. A. (1996b) Ray scattering by an arbitrarily oriented spheroid. II. Transmission and cross-polarization effects, *Appl. Opt.*, **35**, 515–531.

Loeb, N. G., and R. Davies (1996) Observational evidence of plane parallel model biases: Apparent dependence of cloud optical depth on solar zenith angle, *J. Geophys. Res.*, **101**, 1621–1634.

Loeb, N. G., and J. A. Coakley, Jr (1998) Inference of marine stratus cloud optical depths from satellite measurements: Does 1D theory apply? *J. Atmos. Sci.*, **11**, 215–233.

Logan, N. A. (1965) Survey of some early studies of the scattering of plane waves by a sphere, *Proc. IEEE*, **53**, 773–785.

Loiko, V. A. and A. V. Konkolovich (2000) Interference effect of coherent transmittance quenching: theoretical study of optical modulation by surface ferroelectric liquid crystal droplets, *J. Phys. D: Appl. Phys.*, **33**, 2201–2210.

Loiko, V. A. and G. I. Ruban (2000) Absorption and scattering of light by a photolayer with close-packed particles, *Optics and Spectr.*, **88**, 834–839.

Lopatin, V. N., and F. Ya. Sid'ko (1988) *Introduction to Optics of Cell Suspensions*, Moscow: Nauka.

Lui, C. W., M. Clarkson, and R. W. Nicholls (1996) An approximation for spectral extinction of atmospheric aerosols, *J. Quant. Spectr. & Rad. Transfer*, **55**, 519–531.

Lumme, K., J. Rahola, and J. W. Hovenier (1997) Light scattering by dense clusters of spheres, *Icarus*, **126**, 455–469.

Lutomirski, R. F., A. P. Ciervo, and J. H. Gainford (1995) Moments of multiple scattering, *Appl. Opt.*, **34**, 7125–7136.

Macke, A. and F. Tzschihholz (1992) Scattering of light by fractal particles: aqualitative estimate exemplary for two-dimensional triadic Koch island, *Physica A*, **191**, 159–170.

Macke, A. (1993) Scattering of light by polyhedral ice crystals, *Appl. Opt.*, **32**, 2780–2788.

Macke, A. (1994) Modellierung der optischen Eigenschaften von Cirruswolken, PhD thesis, University of Hamburg.

Macke, A., M. I. Mishchenko, K. Muinonen, and B. E. Carlson (1995) Scattering of light by large nonspherical particles: ray-tracing approximation versus T-matrix method, *Opt. Lett.*, **20**, 1934–1936.

Macke, A., and M. Mishchenko (1996) Applicability of regular particle shapes in light scattering calculations for atmospheric ice particles, *Appl. Optics*, **35**, 4291–4296.

Macke, A., J. Mueller, and E. Raschke (1996) Scattering properties of atmospheric ice crystals. *J. Atmos. Sci.*, **53**, 2813–2825.

Macke, A. (2000) Monte Carlo calculations of light scattering by large particles with multiple internal inclusions, in *Light Scattering by Nonspherical Particles* (ed. by M. I. Mishchenko, J. W. Hoverier, and L. D. Travis, New York: Academic Press, pp. 309–322.

Magono, C., and C. V. Lee (1966) Meteorological classification of natural snow crystals, *J. Fac. Sci. Hokkaido Univ.*, **7**, 321–362.

Mackowski, D. W. (1995) Electrostatics analyses of radiative absorption by sphere clusters in the Rayleigh limit: application to soot particles, *Appl. Opt.*, **34**, 3535–3553.

Markel, V. A., and E. Y. Polyakov (1997) Radiative relaxation of quasinormal modes in small dielectric particles, *Phil. Mag.*, **B76**, 895–909.

Markel, V. A., and T. George, eds (2001) *Optics of Nanostructured Materials*, New York: John Wiley.

Marshall, S. F., D. S. Covert, and R. J. Charlson (1995) Relationship between asymmetry parameter and hemispheric backscatter ratio: implications for climate forcing by aerosols, *Appl. Opt.*, **34**, 6306–6311.

Massom, R. A., H. Eicken, C. Haas, M. O. Jeffris, M. R. Drinkwater, M. Sturm, A. P. Worby, X. Wu, V. I. Lytle, and S. Ushio (2001) Snow on Antarctic ice, *Rev. Geophys.*, **39**, 413–445.

McCartney, E. J. (1977) *Optics of the Atmosphere*. New York: John Wiley.

McGraw, R., S. Nemesure, and S. E. Schwartz (1998) Properties and evolution of aerosols with size distributions having identical moments, *J. Aerosol Sci.*, **29**, 761–772.

Meeten, G. H. (1980) The birefringence of colloidal dispersions in the Rayleigh and anomalous diffraction approximations, *J. Colloid Interface Sci.*, **73**, 38–44.

Michel, B. (1995) Statistical method to calculate extinction by small irregularly shaped particles, *J. Opt. Soc. Am.*, **A12**, 2471–2481.

Michel, B. and A. Lakhtakia (1995) Strong property fluctuation theory for homogenizing chiral particulate composites, *Phys. Rev.*, **E51**, 5101–5107.

Middleton, W. E. K., and A. G. Mungall (1952) The luminous directional reflectance of snow, *J. Opt. Soc. America*, **42**, 572–579.

Mie, G. (1908) Beiträge zur optik truber Medien speziell kolloidaler Metallösungen, *Ann. Phys.*, **25**, 377–445.

Minin, I. N. (1988) *Radiative Transfer Theory in Planetary Atmospheres*, Moscow: Nauka.

Mishchenko, M. I. (1990) Multiple scattering of polarized light in anisotropic plane-parallel media, *Transport Theory and Statistical Physics*, **19**, 293–316.

Mishchenko, M. I. (1991) Light scattering by nonspherical ice grains: An application to noctilucent cloud particles, *Earth, Moon, and Planets*, **57**, 203–211.

Mishchenko, M. I. (1994a) Transfer of polarized infrared radiation in optically anisotropic media: application to horizontally oriented ice crystals: comment, *J. Opt. Soc. Am.*, **A11**, 1376–1377.

Mishchenko, M. I. (1994b) Asymmetry parameters of the phase function for densely packed scattering grains. *J. Quant. Spectr. and Radiative Transfer*, **52**, 95–110.

Mishchenko, M. I., and L. D. Travis (1994a): Light scattering by polydispersions of randomly oriented spheroids with sizes comparable to wavelength of observation, *Appl. Opt.*, **33**, 7206–7225.

Mishchenko, M. I., and L. D. Travis (1994b) T-matrix computations of light scattering by large spheroidal particles, *Opt. Commun.*, **109**, 16–21.

Mishchenko, M. I., D. W. Mackowski, and L. D. Travis (1995a) Scattering of light by bispheres with touching and separated components, *Appl. Opt.*, **34**, 4589–4599.

Mishchenko, M. I. *et al.* (1995b) Effect of particle nonsphericity on bidirectional reflectance of cirrus clouds, in: *Proceedings of the 1995 ARM Science Meeting, San Diego, CA*, 19–23 March.

Mishchenko, M. I., L. D. Travis, and A. Macke (1996a) Scattering of light by polydisperse, randomly oriented, finite circular cylinders, *Appl. Opt.*, **35**, 4927–4940.

Mishchenko, M. I., W. B. Rossow, A. Macke, and A. A. Lacis (1996b) Sensitivity of cirrus cloud albedo, bidirectional reflectance and optical thickness retrieval accuracy to ice particle shape, *J. Geophys. Res. D*, **101**, 16973–16985.

Mishchenko, M. I., and A. Macke (1997) Asymmetry parameters of the phase function for isolated and densely packed spherical particles with multiple internal packed spherical particles with multipole internal inclusions in the geometrical optics limit, *J. Quant. Spectr. and Radiative Transfer*, **57**, 767–794.

Mishchenko, M. I., and A. Macke (1998) Incorporation of physical optics effects and computation of the Legendre expansion for ray-tracing phase functions involving δ-function transmission, *J. Geophys. Res. D*, **103**, 1799–1805.

Mishchenko, M. I., and L. D. Travis (1999) Polarization and depolarization of light by small particles, in: *Lecture Notes in Physics: Scattering by Microstructures*, Berlin: Springer-Verlag.

Mishchenko, M. I., J. M. Dlugach, E. G. Yanovitskij, and N. T. Zakharova, (1999) Bidirectional reflectance of flat, optically thick particulate layers: An efficient radiative transfer solution and applications to snow and soil surfaces, *J. Quant. Spectrosc. Radiat. Transfer*, **63**, 409–432.

Mishchenko, M. I., J. W. Hovenier, and L. D. Travis, eds. (2000) *Light Scattering by Nonspherical Particles: Theory, Measurements, and Applications*, New York: Academic Press.

Mishchenko, M. I., *et al.* (2002) *Absorption, Scattering, and Emission of Light by Small Particles*, Cambridge, U.K.: Cambridge University Press.

Mishchenko, M. I. (2002) Vector radiative transfer equation for arbitrarily shaped and arbitrarily oriented particles: A microphysical deviation from statistical electromagnetics, *Appl. Optics*, **33**, 7114–7134.

Mitchell, D. L., and W. P. Arnott (1994) A model predicting the evolution of ice particle spectra and radiative properties of cirrus clouds. II. Dependence of absorption and extinction on ice crystal morphology, *J. Atmos. Sci.*, **51**, 817–832.

Mitchell, D. L. (2000) Parametrization of the Mie extinction and absorption coefficients for water clouds, *J. Atmos. Sci.*, **57**, 1311–1326.

Mobley, C. D. (1994) *Light and Water: Radiative Transfer in Natural Waters*, San Diego: Academic Press.

Momota, M., H. Miike, and H. Hashimoto (1994) Measuring particle size distribution by digital image processing with inverse Fourier–Bessel transformation, *Jpn. J. Appl. Phys.*, **33,** 1189–1194.

Moore, K. D., K. J. Voss, and K. G. Gordon (1997) Whitecaps: spectral reflectance in the open ocean and their contribution to water-living radiance, *Proc. SPIE Int. Soc. Opt. Eng.*, **2693**, 246–251.

Moore, K. D., A. K. V. Voss, and H. R. Gordon (1998) Spectral reflectance of whitecaps: instrumentation, calibration and performance in coastal waters, *Atmos. Oceanic Technol.*, **15**, 496–509.

Moore, K. D., A. K. V. Voss, and H. R. Gordon (2000) Spectral reflectance of whitecaps: Their contribution to water-leaving radiance, *J. Geophys. Res.*, **C105**, 6493–6499.

Monahan, E. C., and Macniocaill (eds) (1986) *Oceanic Whitecaps: Their Role in Air–Sea Exchange Processes*, Dordrecht: D. Reidel.

Monahan, E. C., and I. G. O'Muircheartaigh (1986) Whitecaps in the passive remote sensing of the ocean surface, *Int. J. Remote Sensing*, **7**, 627–692.

Morel, A., and B. Gentili (1991) Diffuse reflectance of ocean waters: its dependence on sun angle as influenced by the molecular scattering contribution, *Appl. Opt.*, **30**, 4427–4438.

Morel, A. and B. Gentili (1993) Diffuse reflectance of oceanic waters. II. Bidirectional aspects, *Appl. Opt.*, **32**, 6864–6879.

Morel, A. and S. Maritorena (2001) Bio-optical properties of oceanic waters: A reppraisal, *J. Geophys. Res.*, **C106**, 7163–7180.

Muinonen, K. (1996) Light scattering by Gaussian random particles: Rayleigh and Rayleigh–Gans approximations, *J. Quant. Spectr. and Radiative Transfer*, **55**, 603–613.

Muinonen, K., T. Nousiainen, P. Fast, K. Lumme, and J. E. Peltoniemi (1996) Light scattering by Gaussian random particles: ray optics approximation, *J. Quant. Spectr. and Radiative Transfer*, **55**, 577–601.

Muinonen, K., L. Lamberg, P. Fast, and K. Lumme (1997) Ray optics regime for Gaussian random spheres, *J. Quant. Spectr. and Radiative Transfer*, **57**, 197–205.

Muhlenweg, H., and E. D. Hirleman (1998) The influence of particle shape on diffraction spectroscopy, *Proceedings 7th European Symposium on Particle Characterization* (R. Weichert, ed.), Nurnberg: Nurnberg Messe, 181–190.

Nakajima, T., and M. Tanaka (1988) Algorithms for radiative intensity calculations in moderately thick atmospheres using a truncation approximation. *J. Quant. Spectr. and Radiative Transfer*, **40**, 51–69.

Nakajima, T., and M. D. King (1990) Determination of the optical thickness and effective particle radius of clouds from reflected solar radiation measurements. Part 1. Theory, *J. Atmos. Sci.*, **47**, 1878–1893.

Nakajima, T., and M. D. King (1992) Asymptotic theory for optically thick layers: application to the discrete ordinates method, *Appl. Opt.*, **31**, 7669–7683.

Nakajima, T., G. Tonna, R. Rao, P. Boi, J. Kaufman, and B. Holben (1996) Use of sky brightness measurements from ground for remote sensing of particulate polydispersions, *Appl. Opt.*, **35**, 2672–2686.

Natsyama, H. H., Ueno, S., and A. P. Wang (1998) *Terrestrial Radiative Transfer: Modeling, Computation, and Data Analysis*, Tokyo: Springer-Verlag.

Nevitt, T. J. and C. Bohren (1984) Infrared backscattering by irregularly shaped particles: a statistical approach, *J. Climate and Appl. Meteor.*, **23**, 1342–1349.

Newton, R. (1982) *Scattering Theory of Waves and Particles*, New York: Springer-Verlag.

Nicolas, J. -M., P. -Y. Deschamps, and R. Frouin (2001) Spectral reflectance of oceanic whitecaps in the visible and near-infrared: Aircraft measurements over open ocean, *Geophys. Res. Let.*, **28**, 4445–4448.

Nolin A. W., and J. Dozier (2000) A hyperspectral method for remotely sensing the grain size of snow, *Remote Sens. Environ.*, **74**, 207–216.

Nolin, A. W., and S. Liang (2000) Progress in bi-directional reflectance modeling and applications for surface particulate media: Snow and soils, *Remote Sens. Rev.*, **18**, 307–342.

Nousiainen, T., and K. Muinonen (1999) Light scattering by Gaussian, randomly oscillating raindrops, *J. Quant. Spectr. Rad. Transfer*, **63**, 643–666.

Nussenzweig, H. M., and W. J. Wiscombe (1980) Efficiency factors in Mie scattering, *Phys. Rev. Lett.*, **45**, 1490–1494.

Nussenzveig, H. M. (1992) *Diffracrtion Effects in Semiclassical Scattering*, London: Cambridge University Press.

Onaka, T. (1980) Light scattering by spheroidal grains, *Ann. Tokyo Astron. Obs.*, **18**, 1–54.

Osborn, J. A. (1945) Demagnetizing factors for the general ellipsoid, *Phys. Rev.*, **67**, 351–357.

Oshchepkov, S., and H. Isaka (1997) Inverse scattering problem for mixed-phase and ice clouds. I. Numerical simulation of particle sizing from phase function measurements, *Appl. Opt.*, **36**, 8765–8774.

Painter, T. H., J. Dozier, D. A. Roberts, R. E. Davis, and R. O. Green (2003) Retrieval of subpixel snow-covered area and grain size from imaging spectrometer data, *Remote Sens. Environ.*, **85**, 64–77.

Papadakis, S. N., N. K. Uzunoglu, and C. N. Capsalis (1990) Scattering of a plane wave by a general anisotropic dielectric ellipsoid, *J. Opt. Soc. Am.*, **A7**, 991–997.

Papanicolaou, G. C. and R. Burridge (1975) Transport equations for the Stokes parameters from Maxwell's equations in a random media, *J. Math. Phys.*, **16**, 2074–2082.

Paramonov, L. E., V. N. Lopatin, and F. Ya. Sid'ko (1986) Light scattering by soft spheroidal particles, *Opt. Spektrosk.*, **61**, 570–576.

Paramonov, L. E. (1994) On optical equivalence of randomly oriented ellipsoidal and polydisperse spherical particles. The extinction, scattering and absorption cross sections, *Opt. Spektrosk.*, **77**, 660–663.

Paramonov, L. E. (1995) T-matrix approach and the angular momentum theory in light scattering problems by ensembles of arbitrarily shaped particles, *J. Opt. Soc. Amer.*, **A12**, 2698–2707.

Patterson, M. S., Chance, B., and C. Wilson (1989) Time resolved reflectance and transmittance for the non-invasive measurement of tissue optical properties, *Appl. Opt.*, **28**, 2331–2336.

Peltoniemi, J., K., Lumme, V. Muinonen, and W. M. Irvine (1989) Scattering of light by stochastically rough particles, *Appl. Opt.*, **28**, 4088–4095.

Peltoniemi, J. I. (1993) Light scattering in planetary regolits and cloudy atmospheres, PhD thesis, University of Helsinki.

Pendorf, R. (1960) Scattering coefficients for absorbing and nonabsorbing aerosols. *Technical Report RAD-TR-60-27*, Air Force Cambridge Research Laboratory, Bedford, Massachusetts.

Percus, J. K., and Yevick, G. J. (1958) Analysis of classical statistical mechanics by means of collective coordinates, *Phys. Rev.*, **110**, 1–13.

Perelman, A. Ya. (1986) Extinction efficiency factor of particles of sea suspensions. *Izv. AN SSSR, Fiz. Atmos. Okeana*, **22**, 242–250.

Perelman, A. Ya. (1991) Extinction and scattering by soft spheres, *Appl. Opt.*, **30**, 475–484.

Perelman, A. Ya. (1994) Improvement of the convergence of series for absorption cross section of a soft sphere, *Opt. and Spektrosk.*, **77**, 643–647.

Perelman, A. Ya. (1996) Scattering by particles with radially variable refractive indices, *Appl. Opt.*, **35**, 5452–5460.

Perrin, F. (1942) Polarization of light scattered by isotropic opalescent media, *J. Chem. Phys.*, **10**, 415–427.

Petzold, T. J. (1972) *Volume Scattering Functions for Selected Ocean Waters*, SIO Ref. 72–78, La Jolla: Scripps Institute of Oceanography, 79 pp.

POAN Research Group (ed.) (1998) *New Aspects of Electromagnetic and Acoustic Diffusion*, Springer Tractates in Modern Physics, Vol. 144, Berlin: Springer.

Pope, R. M., and E. S. Fry (1997) Absorption spectrum (380–700 nm) of pure water. II: Integrating cavity measurements, *Appl. Opt.*, **36**, 8710–8723.

Popov, A. A., and O. V. Shefer (1995) Theoretical and numerical investigations of the polarization properties of a lidar signal scattered by a set of oriented ice plates, *Appl. Opt.*, **34**, 1488–1492.

Prieur, L., and S. Sathyendranath (1981) An optical classification of coastal and oceanic waters based on the specific absorption curves of phytoplankton pigments, dissolved organic matter, and other particulate materials, *Limnol. Oceanogr.*, **26**, 671–689.

Prishivalko, A. P., V. A. Babenko, and V. N. Kuzmin (1984) *Scattering and Absorption of Light by Inhomogeneous and Anisotropic Spherical Particles*, Minsk: Nauka i Tekhnika.

Purcell, E. M., and C. R. Pennypacker (1973) Scattering and absorption of light by nonspherical dielectric grains, *Astrophys. J.*, **186**, 705–714.

Rayleigh, Lord (1871) On the light from the sky, its polarization and colour, *Philos. Mag.*, **41**, 107–120, 274–279.

Rayleigh, Lord (1881) On the electromagnetic theory of light, *Phil. Mag.*, **12**, 81–101.

Reiss, H. (1988) *Radiative Transfer in Nontransparent, Dispersed Media*, Berlin: Springer-Verlag.

Rinneberg, H. (1995) Scattering of laser light in turbid media: optical tomography for medical diagnostics? In: *The Inverse Problem Symposium in Memoriam of Hermann von Helmboltz* (H. Lubbigied.), Berlin: Academie Verlag, p. 107–141.

Rogovtsov, N. N. (1991) Radiative transfer in scattering media of different configurations, in: *Scattering and Absorption of Light in Natural and Artificial Dispersed Media* (A. P. Ivanov, ed.), pp. 58–81, Minsk: Nauka i Tekhnika.

Rogovtsov, N. N. (1999) *Properties and Principles of Invariance*, Minsk: Belarussian Polytechnical Academy.

Rolland, P., K. N. Liou, M. D. King, S. C. Tsay, and G. M. McFarguher (2000) Remote sensing of optical and microphysical properties of cirrus clouds using Moderate-Resolution Imaging Spectroradiometer channels: Methodology and sensitivity to physical assumptions, *J. Geophys. Res. D*, **105**, 11721–11738.

Rosen, D. L., and J. D. Pendleton (1995) Detection of biological particles by the use of circular dichroism measurements improved by scattering theory, *Appl. Opt.*, **34**, 5875–5884.

Rozenberg, G. V. (1962) Optical characteristics of thick weakly absorbing scattering layers, *Doklady AN SSSR*, **145**, 775–777.

Rozenberg, G. V. (1967) Physical foundations of light scattering media spectroscopy, *Soviet Physics Uspekhi*, **91**, 569–608.

Rozenberg, G. V. (1977) Light ray, *Soviet Physics Uspekhi*, **121**, 97–138.

Rossow, W. B. (1989) Measuring cloud properties from space: a review, *J. Climate*, **2**, 419–458.

Rossow, W. B., L. C. Garder, and A. A. Lacis (1989) Global, seasonal cloud variations from satellite radiance measurements. Part I: Sensitivity of analysis. *J. Climate*, **2**, 419–458.

Rusanov, A. I., V. V. Krotov, and A. G. Nekrasov (1998) New methods for studying foams: foaminess and foam stability, *A. Col. Interface Sci.*, **206**, 392–396.

Rusin, N. P. (1961) *Meteorological and Radiative Regime of Antarctica*, Leningrad: Gidrometeoizdat.

Rytov, S. M., Yu. I., Kravzov, and V. I. Tatarskii (1978) *Introduction to Statistical Radiophysics*, Vol. 2, Moscow: Nauka.

Sasano, Y. (1996) Tropospheric aerosol extinction coefficient profiles derived from scanning lidar measurements over Tsukuba, Japan, from 1990 to 1993, *Appl. Opt.*, **35**, 4941–4952.

Sathyendranath, S., L. Prieur, and A. Morel (1989) A three-component model of ocean colour and its application to remote sensing of phytoplankton pigments in coastal waters, *Int. J. Remote Sensing*, **10**, 1373–1394.

Saulnier, P. M., M. P. Zinkin, and G. H. Watson (1990) Scatterer correlation effects on photon transport in dense random media, *Phys. Rev.*, **B42**, 2621–2623.

Saxon, D. S. (1955) *Lectures on the Scattering of Light*, Los Angeles: UCLA Department of Meteorology Scientific Report No. 9.

Schnablegger, H., and O. Glatter (1995) Sizing of colloidal particles with light scattering: corrections to beginning multiple scattering, *Appl. Opt.*, **34**, 3489–3501.

Schotland, J. C. (1977) Continuous wave diffusion imaging, *J. Opt. Soc. Am.*, **A14**, 275–279.

Segelstein, D. (1981) The Complex Refractive Index of Water (Thesis M.S.), Kansas City: University of Missouri.

Sergent, C., C. Leroux, E. Pougatch, and F. Guirado (1998) Hemispherical–directional reflectance measurements of natural snow in the 0.9–1.45 mm range; Comparison with adding-doubling modelling, *Ann. Glaciol.*, **26**, 59–63.

Shah, G. A. (1972) Scattering of plane elctromagnetic waves by infinite concentric circular cylinders at oblique incidence, *Mon. Not. R. Astron. Soc.*, **148**, 93–102.

Sharma, S. K. and D. J. Somerford (1989) A comparison of extinction efficiencies in the eikonal and the anomalous diffraction approximations for spheres, *J. Mod. Opt.*, **36**, 1411–1413.

Sharma, S. K. and D. J. Somerford (1990) The eikonal approximation revised, *Nuovo Cimento*, **12D**, 711–748.

Sharma, S. K., and D. J. Somerford (1996) Relationship between the S approximation and the Hart–Montroll approximation, *J. Opt. Soc. Am.*, **13**, 1285–1286.

Sharma, S. K., and D. J. Somerford (1997) S-approximation for light scattering by an infinitely long cylinder, *Appl. Opt.*, **36**, 6109–6114.

Sharma, S. K., and D. J. Somerford (1999) Scattering of light in the eikonal approximation, in: *Progress in Optics*, (E. Wolf, ed.), New York: Elsevier.

Sharma, S. K. and A. R. Jones, (2000) On the validity of an approximate formula for absorption and scattering of light by a large sphere with highly absorbing spherical inclusions, *J. Phys. D*, **33**, 584–588.

Shifrin, K. S. (1951) *Scattering of Light in a Turbid Media*, Moscow: Gostekhteorizdat [English translation: NASA Tech. Trans. TT F-447, 1968, NASA, Washinghton, DC].

Shifrin, K. S. (1955) On calculation of radiative properties of water clouds, *Trudy Glavnoi Geophys. Observ.*, **46**, 5–33.

Shifrin, K. S., and I. A. Mikulinskii (1982) Light scattering by a system of particles in Rayleigh–Gans approximation, *Optics and Spectoscopy*, **52**, 359–366.

Shifrin, K. S., Ya. S. Shifrin, and I. A. Mikulinski (1984) Light scattering by an ensemble of large particles of arbitrary shape, *Doklady Akad. Nauk SSSR*, **277**, 582–585.

Shifrin, K. S. (1988) *Introduction to Ocean Optics*, Leningrad: Gidrometeoizdat.

Shifrin, K. S., and G. Tonna (1992) Simple formula for absorption coefficient of weakly refracting particles, *Opt. Spektr.*, **72**, 487–490.

Shifrin, K. S., and G. Tonna (1993) Inverse problems related to light scattering in the atmosphere and ocean, *Adv. Geophys.*, **34**, 175–252.

Shifrin, K. S., and I. G. Zolotov (1996) Information content of the spectral transmittance of the marine atmospheric boundary layer, *Appl. Opt.*, **35**, 4835–4842.

Shuerman, D., ed. (1983) *Light Scattering by Irregularly Shaped Particles*, New York: Plenum Press.

Slingo, A. (1989) A GCM parametrization for the shortwave radiative properties of water clouds, *J. Atmos. Sci.*, **46**, 1419–1427.

Slingo, A., and H. M. Schrecker (1982) On the shortwave radiative properties of stratiform water clouds, *Quart. J. Roy. Meteor. Soc.*, **108**, 407–426.

Smolyar, I. (2000) Biological atlas of the Barents and Kara seas, *Earth System Monitor*, **11**, 1–9.

Sobolev, V. V. (1956) *Radiative Transfer in Stellar and Planetary atmospheres*, Moscow: Gostekhteorizdat.

Sobolev, V. V. (1972) *Light Scattering in Planetary Atmospheres*, Moscow: Nauka.

Sobolev, V. V. (1984) Integral relations and asymptotic expressions in the theory of radiative transfer, *Astrofizika*, **20**, 123–132.

Sorensen, C. M., J. Cai, and N. Lu (1992) Light scattering measurements of monomer size, monomers per aggregate, and fractal dimension for soot aggregates in flames, *Appl. Opt.*, **31**, 6547–6557.

Spinhirne, J. D., and T. Nakajima (1994) Glory of clouds in the near infrared, *Appl. Opt.*, **33**, 4652–4662.

Stephens, G. L. (1984) Scattering of plane waves by soft obstacles: anomalous diffraction theory for circular cylinders, *Appl. Opt.*, **23**, 954–959.

Stephens, G. L. (1994) *Remote Sensing of the Lower Atmosphere*, Oxford: Oxford University Press.

Stoyan, D., and H. Stoyan (1994) *Fractals, Random Shapes and Point Fields*, New York: John Wiley.

Streekstra, G. J., A. G. Hoekstra, and R. M. Heethaar (1994) Anomalous diffraction by arbitrarily oriented ellipsoids: applications in ektacytometry, *Appl. Opt.*, **33**, 7288–7296.

Stuhlmann, R., P. Minnis, and G. L. Smith (1985) Cloud bidirectional reflectance functions: a comparison of experimental and theoretical results, *Appl. Opt.*, **24**, 396–401.

Sturm, B. (1981) Correction of remotely sensed data and the quantitative determination of suspended matter in marine water surface layers, *Remote Sensing in Meteorology, Oceanography and Hydrology* (A. P. Cracknell, ed.), pp. 163–197, New York: John Wiley.

Sun, W., and Q. Fu (1999) Anomalous diffraction theory for arbitrarily oriented hexagonal crystals, *J. Quant. Spectr. Rad. Transfer*, **63**, 727–737.

Takano, Y., and K.-N. Liou (1989) Solar radiative transfer in cirrus clouds, 1. Single scattering and optical properties of hexagonal ice crystals, *J. Atmos. Sci.*, **46**, 3–19.

Tanre, D., M. Herman, and Y. J. Kaufman (1996) Information on aerosol size distribution contained in solar reflected spectral radiances, *J. Geophys. Res. D*, **101**, 19043–19060.

Tanre, D., L. A. Remer, Y. J. Kaufman, S. Mattoo, P. V. Hobbs, J. M. Livingston, P. B. Russel, and A. Smirnov (1999) Retrieval of aerosol optical thickness and size distribution from the MODIS airborne simulator during TARFOX, *J. Geophys. Res.*, **D104**, 2261–2278.

Thomas, G. E., and K. Stamnes (1996) *Radiative Transfer in the Atmosphere and Ocean*, Cambridge: Cambridge University Press.

Tsang, L., and J. A. Kong (1983) Scattering of electromagnetic waves from a half space of densely distributed dielectric scatterers, *Rad. Sci.*, **18**, 1260–1272.

Tsang, L., J. A. Kong, and R. T. Shin (1985) *Theory of Microwave Remote Sensing*, New York: Wiley-Interscience.

Tsang, L., J. A. Kong, and K-H Ding (2000) *Scattering of Electromagnetic Waves*, New York: Wiley-Interscience.

Tuchin, V. V. (1997) Light scattering studies of tissues, *Uspekhi Fiz. Nauk*, **167**, 517–539.

Tuchin, V. V. (1998) *Lasers and Fiber Optics in Biomedical Studies*, Saratov: Saratov State University Publishing.

Twersky,V. (1962) On scattering of waves by random distributions. I. Free space scatterer formulation, *J. Math. Phys.*, **3**, 700–715.

Twomey, S. (1977) *Atmospheric Aerosols*, London: Elsevier.

Umow, N. (1905) Chromatische Depolarisation durch Lichtzerstreuung, *Phys. Z.*, **6**, 674–676.

Uno, K., J. Uozumi, and T. Asakura (1993) Statistical properties of the Frunhofer diffraction field produced by random fractals, *Appl. Opt.*, **32**, 2722–2733.
Vagin, N. I., and V. V. Veretennikov (1989) Optical diagnostics of disperse media under multiple scattering in small angle approximation, *Izv. Acad. Nauk SSSR, Fizika Atmos. and Okeana*, **25**, 723–731.
Van de Bosch, H. F. M., K. J. Ptasinski, and P. J. A. M. Kerkhof (1996) Edge contributions to forward scattering by spheres, *Appl. Opt.*, **35**, 2285–2291.
Van de Hulst, H. C. (1957) *Light Scattering by Small Particles*, New York: John Wiley.
Van de Hulst, H. C. (1980) *Multiple Light Scattering: Tables, Formulas and Applications*, New York: Academic Press.
Van Rossum, M. C. W., and T. M. Nieuwenhuizen (1999) Multiple scattering of classical waves: microscopy, mesoscopy and diffusion, *Rev. Mod. Phys.*, **71**, 313–371.
Vera, M. U., A. Saint–Jalmes, and D. D. Durian (2001) Scattering optics of foam, *Appl. Opt.*, **40**, 4210–4214.
Vitkin, I. A., and R. C. N. Studinski (2001) Polarization preservation in diffusive scattering from in vivo turbid media: effects of tissue optical absorption in the exact backscattering direction, *Optics Communications*, **190**, 37–43.
Volkovitsky, O. A., L. N. Pavlova, and A. G. Petrushin (1984) *Optical Properties of the Ice Clouds*, Leningrad: Gidrometeoizdat.
Volnistova, L. P., and A. S. Drofa (1986) Quality of image transfer through a light scattering media, *Opt. Spektrosk.*, **61**, 116–121.
von Hoyningen–Huene, W., M. Freitag, and J. P. Burrows (2003) Retrieval of aerosol optical thickness over land surfaces from top-of-atmosphere radiance, *J. Geophys. Res.*, **D108**, doi: 10.1029/2001JD002018.
Voshchinnikov, N. V., and V. G. Farafonov (1993) Optical properties of spheroidal particles, *Astrophys. Space Sci.*, **204**, 19–86.
Voss, K. J. and E. S. Fry (1989) Measurement of the Mueller matrix for ocean water, *Appl. Opt.*, **23**, 4427–4439.
Wait, J. R. (1955) Scattering of a plane wave from a circular dielectric cylinder at oblique incidence, *Canad. J. of Phys.*, **33**, 189–195.
Wang, M. C., and E. Guth (1951) On the theory of multiple scattering of particularly charged particles, *Phys. Rev.*, **84**, 1092–1111.
Wang, M., and H. R. Gordon (1993) Retrieval of the columnar aerosol phase function and single scattering albedo from sky radiance over the ocean: simulation, *Appl. Opt.*, **32**, 4598–4609.
Wang, M., and H. R. Gordon (1994) Estimating aerosol optical properties over the oceans with the multiangle imaging spectroradiometer: some preliminary studies, *Appl. Opt.*, **33**, 4042–4057.
Wang, P. C. (1997) Characterization of ice crystals in clouds by simple mathematical expressions based on successive modification of simple shapes, *J. Atmos. Sci.*, **54**, 2035–2041.
Warren, S. G., and W. J. Wiscombe (1980) A model of the spectral albedo of snow, II: Snow containing atmospheric aerosols, *J. Atmos. Sci.*, **37**, 2734–2745.
Warren, S. G. (1982) Optical properties of snow, *Rev. Geophys.*, **20**, 67–89.
Warren, S. G. (1984) Optical constants of ice from the ultraviolet to the microwave, *Appl. Opt.*, **23**, 1206–1225.
Waterman, P. C., and R. Truell (1961) Multiple scattering of waves, *J. Math. Phys.*, **2**, 512–537.
Waterman, P. C. (1971) Symmetry, unitarity, and geometry in electromagnetic scattering, *Phys. Rev. D.*, **3**, 825–839.

Waterman, T. H. (1954) Polarization patterns in submarine illumination, *Science*, **120**, 927–932.

Waterman, T. H. (1988) Polarization of marine light fields and animal orientation, *Ocean Optics*, ix, (M. A. Blizard, ed.), *Proc. SPIE*, **925**, 431–437.

Wauben, W. M .F. (1992) Multiple scattering of polarized radiation in planetary atmospheres, PhD thesis, Free University of Amsterdam.

Wax, A., and J. E. Thomas (1998) Measurement of smoothed Wigner phase-space distributions for small-angle scattering in a turbid medium, *J. Opt. Soc. Am.*, **A15**, 1896–1908.

WCP-112 (1986) *A Preliminary Cloudless Standard Atmosphere for Radiation Computation*, 53 pp., Geneva: World Meteorological Organization.

Weiss-Wrana, K. (1983) Optical properties of interplanetary dust: comparison with light scattering by larger meteoritic and terrestrial grains, *Astron. Astrophys.*, **126**, 240–250.

Wells, W. H. (1969) Loss of resolution in water as a result of multiple small-angle scattering, *J. Opt. Soc. Am.*, **59**, 686–691.

West, R., D. Gibbs, L. T. Sang, and A. K. Fung (1994) Comparison of optical scattering experiments and the quasi-crystalline approximation for dense media, *J. Opt. Soc. Am.*, **A11**, 1854–1858.

Whitby, K. T. (1978) The physical characteristics of sulfur aerosols, *Atmos. Environ.*, **12**, 135–159.

Whitlock, C. H., D. S. Barlett, and R. A. Guganus (1982) Sea foam rEflectance and influence of optimum wavelength for remote sensing of ocean aerosols, *Geophys. Res. Lett.*, **7**, 719–722.

Wickramasinghe, N. C. (1973) *Light Scattering Functions for Small Particles with Applications in Astronomy*. New York: John Wiley.

Wiscombe, W. J., and G. W. Grams (1976) The backscattered fraction in two-stream approximations, *J. Atmos. Sci.*, **33**, 2440–2451.

Wiscombe, W. J. (1977) The delta-M method: rapid yet accurate radiative flux calculations for strongly asymmetric phase function, *J. Atmos. Sci.*, **34**, 1408–1422.

Wiscombe, W. J., and S. G. Warren (1980) A model for the spectral albedo of snow. I. Pure snow, *J. Atmos. Sci.*, **37**, 2712–2733.

Wiscombe, W. J., and A. Mugnai (1986) *Single Scattering from Nonspherical Chebeshev Particles: A Copendium of Calculations*, NASA Ref. Pub. 1157, Greenhelt, MD: NASA Goddard Space Flight Center.

Witt, W., and S. Rothele (1996) Laser diffraction–unlimited, *Part. Part. Syst. Charact.*, **13**, 280–286.

Xu, M. (2003) Light extinction and absorption by arbitrarily oriented finite circular cylinders by use of geometrical path statistics or rays, *Applied Optics*, **42**, 6710–6723.

Xu, M., M. Lax, and R. R. Alfano (2003) Anomalous diffraction of light with geometrical path statistics of rays and a Gaussian ray approximation, *Opt. Let.*, **28**, 179–181.

Xu, M., Z. Chen, J. Gong, A. Taflove, and V. Backman (2004a) Analytical techniques for addressing forwad and inverse problems of light scattering by irregularly shaped particles, *Opt. Let.*, **29**, 1239–1241.

Xu, F., X. Cai, and K. Ren (2004b) Geometrical-optics approximation of forward scattering by coated particles, *Appl. Opt.*, **43**, 1870–1879.

Yang, P., and K. N. Liou (1995) Light scattering by hexagonal ice crystals: comparison of finite-difference time domain and geometric optics models, *J. Opt. Soc. Am.*, **12**, 162–176.

Yang, P., and K. N. Liou (1996a) Finite-difference time domain method for light scattering by small ice crystals in three-dimensional space, *J. Opt. Soc. Am.*, **A13**, 2072–2085.

Yang, P., and K. N. Liou (1996b) Geometric-optics-integral-equation method for light scattering by nonspherical ice crystals, *Appl. Opt.*, **35**, 6568–6584.
Yang, P., and K. N. Liou (1998) Single-scattering properties of complex ice crystals in terrestrial atmosphere, *Contr. Atmos. Phys.*, **71**, 223–248.
Yang, P., K. N. Liou, K. Wyser, and D. Mitchell (2000) Parametrization of the scattering and absorption parameters of individual ice crystals, *J. Geophys. Res.*, **105**, 4699–4718.
Yang, P., B.-C. Gao, B. A. Baum, Y. X. Hu, W. J. Wiscombe, M. I. Mishchenko, D. M. Winker, and S. L. Nasiri (2001) Asymptotic solutions for optical properties of large particles with strong absorption, *Appl. Opt.*, **40**, 1532–1547.
Yanovitskij, E. G. (1997) *Light Scattering in Inhomogeneous Atmospheres*, New York: Springer-Verlag.
Zege, E. P. (1969) Radiative transfer in a plane-papallel layer with account for the nonlinearity in absorption and scattering, *J. Appl. Spectr.*, **10**, 940–947.
Zege, E. P., and I. L. Katsev (1974) On the relationship of time-dependent light fields in absorbing and nonabsorbing media, *Astrofizika*, **10**, 219–225.
Zege, E. P., and I. L. Katsev (1978) *Time-dependent Asymptotic Solutions of the Radiative Transfer Equation and Their Applications*, Reprint, Minsk: Institute of Physics.
Zege, E. P., A. A. Kokhanovsky, and L. I. Chaikovskaya (1987) Light field in deep layers of a gyrotropic medium with optically soft particles, *Vesti AN BSSR*, **5**, 61–69.
Zege, E. P., and A. A. Kokhanovsky (1989) On anomalous diffraction approximation for two-layered spheres, *Izv. AN SSSR, Fiz. Atmos. & Okeana*, **27**, 949–954.
Zege, E. P., A. P. Ivanov, and I. L. Katsev (1991) *Image Transfer through a Scattering Medium*, New York: Springer-Verlag.
Zege, E. P., and A. A. Kokhanovsky (1992) On sizing of big particles under multiple light scattering in a medium, *Optics and Spectroscopy*, **72**, 220–226.
Zege, E. P., I. L. Katsev, and I. N. Polonsky (1993) Multicomponent approach to light propagation in clouds and mists, *Appl. Opt.*, **32**, 2803–2812.
Zege, E. P., and A. A. Kokhanovsky (1994) Analytical solution for optical transfer function of a light scattering medium with large particles, *Appl. Opt.*, **33**, 6547–6554.
Zege, E. P., I. L. Katsev, and I. N. Polonsky (1995) Analytical solution to LIDAR return signals from clouds with regard to multiple scattering, *J. Appl. Phys.*, **B60**, 345–353.
Zege, E. P. and A. A. Kokhanovsky (1995) Influence of parameters of coarse aerosols on properties of light fields and optical transfer functions: analytical solutions, *Izv. Acad. Nauk, Atmos. and Oceanic Physics*, **30**, 777–783.
Zege, E. P., A. A. Kokhanovsky, I. L. Katsev, I. N. Polonsky, and A. S. Prikhach (1998) The retrieval of the effective radius of snow grains and control of snow pollution with GLI data, *Proceedings of Conference on Light Scattering by Nonspherical Particles: Theory, Measurements, and Applications* (M. I. Mishchenko, L. D. Travis, and J. W. Hovenier, eds), Boston, MA, USA: Amer. Meteor. Soc., 288-290.
Zhang, J., and L. Xu (1995) Light scattering by absorbing hexagonal ice crystals in cirrus clouds, *Appl. Opt.*, **34**, 5867–5874.
Zhang, T., and H. R. Gordon (1997) Retrieval of elements of the columnar aerosol scattering phase matrix from polarized sky radiance over the ocean: simulations, *Appl. Opt.*, **36**, 7948–7959.
Zhang, X., M. Lewis, and B. Jonson (1998) Influence of bubbles on scattering of light in the ocean, *Appl. Opt.*, **37**, 6525–6536.
Zhao, F., Z. Gong, H. Hu, M. Tanaka, and T. Hayasaka (1997) Simultaneous determination of the aerosol complex index of refraction and size distribution from scattering measurements of polarized light, *Appl. Opt.*, **36**, 7992–8001.

Zolotova, Zh. K., and Shifrin, K. S. (1993) Scattering and absorption of radiation by large water droplets in spectral range 0.2–200 μm, *Izv. AN SSSR, Fizika Atmos. Okeana*, **29**, 532–537.

Zurk, L. M., L. Tsang, K. H. Ding, and D. P. Winebrenner (1995) Monte Carlo simulations of the extinction rate of densely packed spheres with clustered and nonclustered geometries, *J. Opt. Soc. Am.*, **12**, 1772–1781.

Zumer, S. (1988) Light scattering from nematic droplets: anomalous diffraction approach, *Phys. Rev.*, **A37**, 4006–4015.

Index

Printing: Mercedes-Druck, Berlin
Binding: Stein+Lehmann, Berlin